Maximizing Fiber Utilization for Sustainable, Efficient Ruminant Production

Maximizing Fiber Utilization for Sustainable, Efficient Ruminant Production

Guest Editors

Zhiyuan Ma
Ting Liu
Weiwei Wang

MDPI

Basel • Beijing • Wuhan • Barcelona • Belgrade • Novi Sad • Cluj • Manchester

Guest Editors

Zhiyuan Ma
College of Pastoral Agriculture Science and Technology
Lanzhou University
Lanzhou
China

Ting Liu
College of Animal Science and Technology
Gansu Agricultural University
Lanzhou
China

Weiwei Wang
College of Animal Science
Guizhou University
Guiyang
China

Editorial Office
MDPI AG
Grosspeteranlage 5
4052 Basel, Switzerland

This is a reprint of the Special Issue, published open access by the journal *Animals* (ISSN 2076-2615), freely accessible at: https://www.mdpi.com/journal/animals/special_issues/Q974VU04G9.

For citation purposes, cite each article independently as indicated on the article page online and as indicated below:

Lastname, A.A.; Lastname, B.B. Article Title. *Journal Name* **Year**, *Volume Number*, Page Range.

ISBN 978-3-7258-7875-8 (Hbk)
ISBN 978-3-7258-7876-5 (PDF)
https://doi.org/10.3390/books978-3-7258-7876-5

Cover image courtesy of Ting Liu

Contents

About the Editors

Zhiyuan Ma

Zhiyuan Ma is affiliated with the College of Pastoral Agriculture Science and Technology, Lanzhou University. Research interests cover ruminant nutritional regulation and rumen energy metabolism. Research achievements include clarification of key rate-limiting pathways in rumen methanogenesis, construction of theoretical models for directional regulation of rumen hydrogen metabolic flux, and illustration of molecular mechanisms underlying microbial functional reconstruction. Research outputs also involve the development of low-carbon green additive systems for mutton sheep production and the formulation of provincial livestock carbon emission reduction strategies. Compilation of multiple technical manuals on low-carbon animal husbandry has been completed. A total of 38 academic publications have been completed, among which 15 SCI papers have been published as first or corresponding author in journals, including *Microbiome, Journal of Dairy Science, ISME Journal* and *Journal of Animal Science.* Research projects include two national and four provincial grant programs, together with two second prizes for the provincial science and technology progress award.

Ting Liu

Ting Liu is affiliated with the College of Animal Science and Technology, Gansu Agricultural University, and serves as a flexible introduced expert at the Feed Research Institute of Xinjiang Academy of Animal Sciences. Research interests focus on nutrition regulation of young ruminants and ruminant nutritional physiology. Research projects include granted programs from the National Natural Science Foundation of China, provincial science and technology authorities and open laboratory funds. A total of sixteen academic papers have been published, including eight SCI papers, with ongoing supervision of five master's students. Academic honors cover provincial science and technology awards and multiple teaching and instructional commendations at university and national levels.

Weiwei Wang

Weiwei Wang holds a Doctor of Animal Science and is affiliated with Guizhou University. Research interests concentrate on rumen microecology, rumen fermentation metabolism and ruminant methane emission reduction. Research findings have been published in a series of high-profile journals, including *Journal of Integrative Agriculture, Frontiers in Microbiology, Animal Feed Science and Technology, Current Biology, Nature Communications* and *Animal Microbiome.* Research undertakings include presiding over multiple national, provincial and university-level research projects, as well as participation in national key R&D programs and international cooperative projects. More than twenty academic papers have been published in relevant research fields, with sustained research efforts on the structure and function of gastrointestinal and rumen microorganisms.

Preface

To address the urgent need for sustainable ruminant production amid global food security and climate change challenges, we are pleased to present this Special Issue Reprint, entitled "Maximizing Fiber Utilization for Sustainable, Efficient Ruminant Production". Curated by Guest Editors Zhiyuan Ma, Ting Liu, and Weiwei Wang, this collection brings together twelve peer-reviewed research articles and reviews that advance scientific understanding and practical solutions for unlocking the potential of feedstuffs in ruminant agriculture.

Ruminants' unique ability to convert human-inedible lignocellulosic biomass—such as crop residues and forages—into high-quality meat and milk is a cornerstone of sustainable food systems. However, inefficient fiber digestibility, suboptimal rumen fermentation, and associated greenhouse gas emissions remain critical bottlenecks. This Reprint addresses these gaps by focusing on three core objectives: enhancing fiber degradation through microbiome manipulation and feed processing, optimizing dietary formulations for rumen health and productivity, and mitigating environmental impacts via integrated nutritional strategies.

The scope of this collection is deliberately multidisciplinary, bridging microbial ecology, animal nutrition, feed technology, and on-farm management. It features innovative research, such as the use of mixed ensiling to improve roughage degradability, yeast culture supplementation to stabilize rumen pH and enrich fibrolytic microbes, and tailored dietary balances to prevent acidosis. Studies on adapted species like Tibetan sheep and yaks provide insights into host-microbiome adaptation, while investigations into nitrate additives and fermentation modifiers offer actionable methane mitigation strategies that do not compromise production efficiency.

Motivated by the need to translate basic science into practical solutions, this Reprint serves as a valuable resource for animal nutritionists, ruminant production researchers, livestock farmers, and agricultural policymakers. It synthesizes cutting-edge findings to support evidence-based decision-making, from optimizing feed processing techniques to designing low-carbon feeding systems. By emphasizing nutritional balance and integrated management, the collection aims to drive sustainable intensification—reducing feed waste, lowering environmental footprints, and enhancing the economic resilience of ruminant production globally.

This Reprint provides a comprehensive overview of current advancements and future directions, making it an indispensable reference for anyone committed to advancing efficient, sustainable, and climate-smart ruminant agriculture.

Zhiyuan Ma, Ting Liu, and Weiwei Wang
Guest Editors

animals

Editorial

Maximizing Fiber Utilization for Sustainable, Efficient Ruminant Production

Zhiyuan Ma [1,*], Ting Liu [2] and Weiwei Wang [3]

[1] College of Pastoral Agriculture Science and Technology, Lanzhou University, Lanzhou 730020, China
[2] College of Animal Science and Technology, Gansu Agricultural University, Lanzhou 730070, China; liuting@gsau.edu.cn
[3] College of Animal Science, Guizhou University, Guiyang 550025, China; wwwang3@gzu.edu.cn
* Correspondence: mzy@lzu.edu.cn:

1. Introduction

The dual pressures of a growing global population and the urgent need to mitigate climate change necessitate a fundamental shift in livestock production. For ruminants—a critical source of meat and milk—this shift hinges on their unique ability to convert fibrous, inedible plant materials into human food. However, the inherent inefficiency in fiber digestibility within the ruminant digestive system represents a major bottleneck [1]. Therefore, further research on ruminant fiber utilization represents not merely an incremental improvement but a cornerstone for sustainable intensification. Maximizing the ruminant's natural foregut fermentation process directly enhances feed efficiency [2], reduces the environmental footprint [3], and improves the economic resilience of production systems by transforming waste streams into nutrients.

We are pleased to present this Special Issue, which comprises twelve insightful research articles and reviews addressing this core challenge. This collection reflects the vibrant, multidisciplinary efforts that are being made to unravel the complex interactions between feed processing, rumen function, microbial ecology, and animal performance.

2. An Overview of Published Papers

2.1. *Enhancing Fiber Degradation Through Rumen Microbiome Manipulation and Feed Processing*

This theme focuses on directly improving the accessibility and fermentability of fibrous substrates. Research demonstrates that the physical and chemical structure of fiber is a primary determinant of its utilization. Mixing ensiled rape straw with whole crop corn can effectively disrupt its lignocellulose matrices, significantly enhancing their ruminal degradability [Pu, et al., Contribution 8]. Beyond physical processing, targeted nutritional modulation of the rumen ecosystem is key. Another contribution [Xu et al., Contribution 4] reveals that yeast culture supplementation in pelleted diets promotes a more favorable rumen environment for fibrolytic microbes, aiding rumen development and metabolic function. These results are consistent with reports from other researchers that yeast cultures help stabilize ruminal pH [1], since excessively low pH is a major factor that impairs fibrolysis in the rumen [4]. Furthermore, foundational work [Lv et al., Contribution 7] systematically reviews the rumen as a reservoir of specialized lignocellulolytic microorganisms, providing a crucial resource for future bioaugmentation strategies. Importantly, research on adapted species like Tibetan sheep [Zheng et al., Contribution 11] highlights how the rumen microbiome dynamically responds to dietary shifts, offering insights into host–microbiome co-evolution. A study on dairy cows [Xia et al., Contribution 9] establishes a direct link between specific rumen microbiota features and host feed efficiency, underscoring the microbiome's functional role in fiber energy harvest. Interestingly, nutritional strategies traditionally believed to inhibit the proliferation of fibrolytic bacteria in the rumen, such as increasing concentrate supplementation [5], can also promote ruminal fiber degradation when added at appropriate levels. Concentrate supplementation can increase the abundance of fiber-degrading bacteria and enhance the fiber-degrading capacity in the rumen of lactating Tibetan sheep during the cold season [Yang et al., Contribution 10]. Fats, especially those rich in polyunsaturated fatty acids, are generally regarded as a major nutrient that inhibits fibrolytic bacteria [6]. However, supplementing hempseed oil (up to 200 g/d) in the diet of lactating buffaloes slightly reduced their dry matter intake, but maintained the stability of rumen fermentation and specifically altered the abundance of *Acetobacter* associated with acetate production [Gu et al., Contribution 2]. This indicates that the hempseed oil exerted no negative effects on the rumen environment required for fiber degradation. Collectively, these results suggest that nutritional balance in the rumen is crucial for maintaining fibrolytic function.

2.2. Optimizing Dietary Formulation for Rumen Function, Health, and Productivity

The papers with this theme aim to determine how to balance dietary components to sustain optimal rumen conditions for fiber digestion while supporting animal performance. A central tenet is maintaining rumen health and stability. Research on sheep [Zhang et al., Contribution 1] precisely defines the optimal balance between forage neutral detergent fiber and rumen-degradable starch, providing a dietary formulation guideline to prevent acidosis and support continuous fermentation. The inclusion of high-quality forages remains vital, as shown in [Xia et al., Contribution 9], where partial replacement of alfalfa hay with silage improves dairy cow performance, demonstrating the value of forage quality and preservation. Concurrently, addressing anti-nutritional factors in alternative protein sources is essential for overall diet quality; a comprehensive review [Yan et al., Contribution 5] details strategies to mitigate these factors in plant protein feeds, ensuring a balanced nutrient supply. This holistic approach to diet formulation directly supports sustainable intensification by maximizing nutrient extraction from the complete diet.

2.3. Mitigating Environmental Impact Through Integrated Nutritional Strategies

A core objective of modern ruminant production is to reduce its environmental footprint, particularly methane emissions, without compromising productivity. The research presented offers several promising pathways for achieving this. This bibliometric analysis by Zheng et al. [Contribution 3] identifies that research on rumen microbiota and methane mitigation has evolved from traditional animal nutrition studies focused on phenotypic indicators like digestibility to a deeper focus on microbial mechanisms (e.g., protozoa and methanogens), with enteric methane recognized as being primarily produced via the anaerobic fermentation of ingested carbohydrates, including fiber, in the rumen. One study [Li et al., Contribution 12] investigates nitrate supplementation as a hydrogen sink, directly competing with methanogenesis and demonstrating significant methane reduction potential. Similarly, work on yeast culture [Yi et al., Contribution 6] confirms its role in modifying fermentation patterns, leading to lower methane production. Crucially, these mitigation strategies are examined in the context of overall system efficiency. The aforementioned study on dairy cow feed efficiency [Xia et al., Contribution 9] explicitly connects dietary fiber–starch ratios with both methane yield and milk output, illustrating the possibility of synergistic gains. This integrated perspective is essential for developing practical, adoption-ready solutions that align environmental stewardship with economic viability for producers.

https://doi.org/10.3390/ani1609134C

3. Conclusions

Bringing together twelve insightful research articles and reviews, this collection represents a multidisciplinary effort to unravel the complex interactions between feed processing, rumen function, microbial ecology, and animal performance. This work aims to enhance the efficiency of converting fibrous, inedible plant materials into human food. The published papers are organized around three key themes, collectively offering valuable insights for optimizing ruminant production systems. The first theme focuses on enhancing fiber degradation through rumen microbiome manipulation and feed processing, demonstrating that physical processing (e.g., mixed ensiling of rape straw and whole crop corn) can disrupt lignocellulose matrices, while targeted nutritional modulation (e.g., yeast culture supplementation, appropriate concentrate addition, and hempseed oil supplementation) can create a favorable rumen environment for fibrolytic microbes without compromising fiber degradation. Research on adapted species like Tibetan sheep and dairy cows further highlights the dynamic response of the rumen microbiome to dietary shifts and its functional role in fiber energy harvest. The second theme is the optimization of dietary formulation, emphasizing the importance of balancing forage-neutral detergent fiber and rumen-degradable starch to maintain rumen health, as well as the value of high-quality forages and strategies to mitigate anti-nutritional factors in plant protein feeds, ensuring balanced nutrient supply and supporting animal performance. The third theme focuses on mitigating the environmental impact through integrated nutritional strategies, identifying the shift in methane mitigation research from phenotypic indicators to microbial mechanisms and presenting promising solutions such as nitrate supplementation and yeast culture to reduce enteric methane emissions while maintaining system efficiency. Collectively, the findings in this Special Issue emphasize that nutritional balance in the rumen is crucial for maintaining fibrolytic function. Integrated approaches combining feed processing, microbiome manipulation, and dietary optimization are key to achieving sustainable intensification. These strategies enhance feed efficiency, reduce environmental footprint, and improve the economic resilience of ruminant production systems.

Acknowledgments: We sincerely acknowledge the Science and Technology Planning Project of Gansu Province [26ZDNG002] and Key Research Project of Guizhou Province [ZD (2026) 066].

Conflicts of Interest: The authors declare no conflicts of interest.

List of Contributions:

1. Zhang, Z.; Li, F.; Li, F.; Wang, Z.; Guo, L.; Weng, X.; Sun, X.; He, Z.; Meng, X.; Liang, Z.; Li, X. Influence of Dietary Forage Neutral Detergent Fiber on Ruminal Fermentation, Chewing Activity, Nutrient Digestion, and Ruminal Microbiota of Hu Sheep. *Animals* **2025**, *15*, 314; https://doi.org/10.3390/ani15030314.
2. Gu, Q.; Lin, B.; Wan, D.; Kong, Z.; Tang, Q.; Yan, Q.; Cai, X.; Ding, H.; Qin, G.; Zou, C. Effect of Hemp Seed Oil on Milk Performance, Blood Parameters, Milk Fatty Acid Profile, and Rumen Microbial Population in Milk-Producing Buffalo: Preliminary Study. *Animals 2025, 15*, 514; https://doi.org/10.3390/ani15040514.
3. Zheng, X.; Tang, L.; Wang, R.; Zhang, X.; Wang, M.; Wu, D. Hotspot Analysis of Rumen Microbiota and Methane Mitigation in Ruminants: A Bibliometric Analysis from 1998 to 2023. *Animals* **2025**, *15*, 681; https://doi.org/10.3390/ani15050681.
4. Xu, J.; Li, X.; Fan, Q.; Zhao, S.; Jiao, T. Effects of Yeast Culture on Lamb Growth Performance, Rumen Microbiota, and Metabolites. *Animals* **2025**, *15*, 738; https://doi.org/10.3390/ani15050738.
5. Yan, Z.; Liu, Z.; Zhou, C.; Tan, Z. Anti-Nutritional Factors of Plant Protein Feeds for Ruminants and Methods for Their Elimination. *Animals* **2025**, *15*, 1107; https://doi.org/10.3390/ani15081107.
6. Yi, S.; Tian, X.; Qin, X.; Zhang, Y.; Guan, S.; Chen, Z.; Cai, D.; Wu, D.; Wang, R.; Ma, Z.; Wang, M.; Zhang, X. Effects of Yeast Cultures on Growth Performance, Fiber Digestibility, Ruminal

Dissolved Gases, Antioxidant Capacity and Immune Activity of Beef Cattle. *Animals* **2025**, *15*, 1452; https://doi.org/10.3390/ani15101452.

7. Lv, X.; Qiao, Z.; Chen, C.; Hua, J.; Zhou, C. Exploration of Multi-Source Lignocellulose-Degrading Microbial Resources and Bioaugmentation Strategies: Implications for Rumen Efficiency. *Animals* **2025**, *15*, 1920; https://doi.org/10.3390/ani15131920.
8. Pu, X.; Zhang, M.; Zhang, J.; Zhang, X.; Zhang, S.; Lin, B.; Wang, T.; Tan, Z.; Wang, M. Mixed Ensiling Increases Degradation Without Altering Attached Microbiota Through In Situ Ruminal Incubation Technique. *Animals* **2025**, *15*, 2131; https://doi.org/10.3390/ani15142131.
9. Xia, T.; Liu, Z.; Yang, Z.; Jiang, A.; Zhou, C.; Tan, Z. Effects of Partial Replacement of Alfalfa Hay with Alfalfa Silage in Dairy Cows: Impacts on Production Performance and Rumen Microbiota. *Animals* **2025**, *15*, 2748; https://doi.org/10.3390/ani15182748.
10. Yang, C.; Ma, Q.; Wang, J.; Wang, Z.; Hou, S. Multi-Omics Analysis Reveals Concentrate Supplementation Alleviates Body Weight Loss by Regulating Rumen Function in Lactating Tibetan Sheep During the Cold Season. *Animals* **2025**, *15*, 2791; https://doi.org/10.3390/ani15192791.
11. Zheng, S.; Ru, Y.; Xu, M.; Ma, Y.; Ma, Y.; Guo, N. The Seasonal Dietary Shift and Niche Resilience of Yaks on the Qinghai–Tibetan Plateau. *Animals* **2026**, *16*, 613; https://doi.org/10.3390/ani16040613.
12. Li, M.; Liu, T.; Zheng, C.; Nan, X.; Wang, J.; Chen, B.; Zeng, H. Effect of Dietary Calcium Nitrate Addition on Methane Emission, Nitrogen Excretion, and Ruminal Fermentation Parameters and Microbiota in Liuyang Black Goats. *Animals* **2026**, *16*, 1150; https://doi.org/10.3390/ani16081150.

References

1. Wang, X.; Li, F.; Zhang, N.; Ungerfeld, E.; Guo, L.; Zhang, X.; Wang, M.; Ma, Z. Effects of supplementing a yeast culture in a pelleted total mixed ration on fiber degradation, fermentation parameters, and the bacterial community in the rumen of sheep. *Anim. Feed. Sci. Technol.* **2023**, *296*, 115565. [CrossRef]
2. Liu, H.; Li, F.; Ma, Z.; Ma, M.; Ungerfeld, E.; Zhang, Z.; Weng, X.; Liu, B.; Deng, X.; Guo, L. Effects of active dry yeast supplementation in in vitro and in vivo nutrient digestibility, rumen fermentation, and bacterial community. *Animals* **2024**, *14*, 2916. [CrossRef]
3. Zhang, X.M.; Wang, M.; Yu, Q.; Ma, Z.Y.; Beauchemin, K.A.; Wang, R.; Wen, J.N.; Lukuyu, B.A.; Tan, Z.L. Liquid hot water treatment of rice straw enhances anaerobic degradation and inhibits methane production during in vitro ruminal fermentation. *J. Dairy. Sci.* **2020**, *103*, 4252–4261. [CrossRef] [PubMed]
4. Wang, R.; Wang, M.; Lin, B.; Ungerfeld, E.M.; Ma, Z.Y.; Wu, T.T.; Wen, J.N.; Zhang, X.M.; Deng, J.P.; Tan, Z.L. Associations of ruminal hydrogen and ph with fiber digestibility and microbiota composition induced by increasing starch intake in beef cattle. *Anim. Feed. Sci. Technol.* **2021**, *278*, 114980. [CrossRef]
5. Zhang, Z.; Wang, L.; Li, Q.; Li, F.; Ma, Z.; Li, F.; Wang, Z.; Chen, L.; Yang, X.; Wang, X.; et al. Effects of dietary forage neutral detergent fiber and rumen degradable starch ratios on chewing activity, ruminal fermentation, ruminal microbes and nutrient digestibility ofhu sheep fed a pelleted total mixed ration. *J. Anim. Sci.* **2024**, *102*, skae100. [CrossRef] [PubMed]
6. Liu, H.; Li, F.; Li, F.; Ma, Z.; Wang, T.; Li, Q.; Wang, X.; Li, K. Rumen-protected fat and rumen-protected choline co-supplementation: Impacts on performance and meat quality of growing lambs. *Vet. Sci.* **2025**, *12*, 525. [CrossRef] [PubMed]

https://doi.org/10.3390/ani16091340

animals

Article

Effect of Dietary Calcium Nitrate Addition on Methane Emission, Nitrogen Excretion, and Ruminal Fermentation Parameters and Microbiota in Liuyang Black Goats

Mingming Li [1], Ting Liu [1], Chen Zheng [1], Xuan Nan [2], Jun Wang [3], Baicong Chen [3] and Hanfang Zeng [1,*]

1 College of Animal Science and Technology, Gansu Agricultural University, Lanzhou 730070, China; 15293945390@163.com (M.L.); liuting@gsau.edu.cn (T.L.); zhengc@gsau.edu.cn (C.Z.)
2 Lanzhou Animal Disease Prevention and Control Center, Lanzhou 730046, China; nanxuan6150@163.com
3 Animal Husbandry Station of Beijing, Beijing 100107, China; wj552510_78@163.com (J.W.); allpass0907@sina.com (B.C.)
* Correspondence: 14236@gsau.edu.cn; Tel.: +86-931-7631225

Simple Summary

This study explored the effects of incorporating calcium nitrate into the diets of Liuyang black goats on methane emissions and their digestive processes. Results revealed that the addition of calcium nitrate significantly reduced both methane and carbon dioxide emissions from these goats. Furthermore, three hours after feeding, an increase in rumen pH was observed, accompanied by a decrease in ammonia concentration and a reduction in the acetate-to-propionate ratio. Additionally, calcium nitrate facilitated a decline in the population of specific microorganisms within the rumen. In summary, research indicates that calcium nitrate is highly effective in reducing methane emissions in goats and optimizing their gut microbial environment. These findings provide valuable theoretical perspectives on the role of calcium nitrate in the feeding regimens of Liuyang black goats.

Abstract

This research explored how dietary supplementation of calcium nitrate influences methane emissions, nitrogen excretion, ruminal fermentation parameters, and microbiota in Liuyang black goats. A total of twelve male goats from this breed were divided into two groups: one serving as a control group (CON), while the other received a treatment of 3% calcium nitrate (CAL). The research was conducted over a period of 40 days and comprised two separate trial phases. A 10-day adaptation period and a 5-day sampling period (days 11–15) for each stage. Results showed that incorporating calcium nitrate significantly reduced the emissions of methane (CH_4) ($p < 0.05$) and carbon dioxide (CO_2) ($p < 0.05$). Moreover, the use of calcium nitrate modified the trends in ruminal fermentation, resulting in an increase in pH ($p < 0.05$). Moreover, the ratio of acetate to propionate (A:P) was notably reduced in the CAL group ($p < 0.05$), indicating a shift toward enhanced production of propionate. At the microbial level, an increased presence of *Bacteroidota* and Prevotella was observed in the CAL group ($p < 0.05$). In contrast, the CON group exhibited elevated levels of *Firmicutes* and *Methanobrevibacter* ($p < 0.05$). This finding suggests that calcium nitrate plays a significant role in reducing methane emissions and also affects the fermentation processes in the rumen along with the microbiota of Liuyang black goats. Further research is needed to examine the long-term implications of calcium nitrate supplementation on the health and productivity of these goats.

Keywords: calcium nitrate; greenhouse gas; nutrient digestion; 16S rRNA; rumen fermentation

1. Introduction

Global warming stands as one of the most urgent challenges facing the world in the 21st century, primarily driven by greenhouse gases (GHGs) [1]. Notably, methane (CH_4) is of significant relevance, given that its ability to contribute to global warming is 28 times more potent than that of carbon dioxide (CO_2) when assessed over a period of one hundred years [2]. Ruminants, including cattle, sheep and goats, are significant sources of CH_4 emissions due to their unique digestive process involving enteric fermentation in the rumen [3]. Enteric CH_4 emissions, which primarily result from enteric methanogenesis, a microbial process that takes place in the digestive tract, are receiving increasing attention. This is due to their significantly higher global warming potential compared to CO_2, as well as their significant contribution to total agricultural emissions, accounting for approximately 39% [4]. Rumen fermentation in ruminants creates optimal growth conditions and a nutrient-rich environment for microorganisms. It also enables efficient utilization of fibrous feedstuffs that are typically difficult to digest [5]. While rumen fermentation promotes fiber degradation, 2% to 15% of ingested dietary energy is concurrently lost as CH_4. Therefore, practical feeding strategies are needed to reduce CH_4 emissions from ruminants. Earlier research indicates that nitrate demonstrates a potential to mitigate CH_4 levels in goats [6]. In the rumen, nitrates are reduced to nitrites and ammonia, which compete with rumen methanogenesis for reducing equivalents and reduce CH_4 emissions [7]. Although nitrate supplementation holds potential advantages, its use in ruminant diets has been constrained by worries regarding toxicity and the significant inconsistencies in CH_4 reduction reported in various studies. Nitrate reduction in the rumen requires an adaptation period to increase the population of nitrate-nitrite reducing bacteria, thereby preventing poisoning from nitrite accumulation [8]. Reasonable and appropriate supplementation does not adversely affect the production health of goats [9,10]. Calcium nitrate, a soluble form of nitrate, has emerged as a potential alternative that can be rapidly absorbed in the rumen and utilized by rumen microbes to produce less CH_4-intensive fermentation products, such as propionate, thereby mitigating the problems associated with other forms of nitrate while still providing the desired reduction in CH_4 emissions [11]. Additionally, calcium nitrate may improve nitrogen utilization by providing a readily available nitrogen source for microbial protein synthesis, thereby reducing nitrogen excretion and enhancing overall feed efficiency [6]. However, the effects of calcium nitrate on CH_4 emission, nitrogen metabolism, and rumen fermentation dynamics in specific ruminant species, such as Liuyang black goats, remain understudied. As an important local breed in China, Liuyang black goats are known for their high meat quality and adaptability to local environments. Understanding the impact of calcium nitrate on these goats is crucial for developing sustainable feeding strategies that can reduce environmental impacts while maintaining or enhancing animal productivity.

The objective of this research was to investigate how dietary supplementation with calcium nitrate influences CH_4 emissions, nitrogen excretion patterns, ruminal fermentation parameters, and the microbial community composition in Liuyang black goats. We hypothesized that adding calcium nitrate will not adversely affect the production health of goats, can reduce methane production, and optimize rumen fermentation. The findings will contribute to optimizing dietary strategies for environmentally sustainable goat production while maintaining animal health.

https://doi.org/10.3390/ani1608115C

2. Materials and Methods

2.1. Animal, Diets and Treatments

We selected 12 male Liuyang Black goats sourced from the Subtropical Agricultural Research Institute of the Chinese Academy of Sciences in Changsha. The goats were randomly assigned to two groups: a control group (CON) and a treatment group (CAL) that was fed a diet containing 3% calcium nitrate (dry weight basis, grams per kilogram). Each group consisted of 6 goats with an average weight of 28 ± 0.2 kg; each goat served as an independent replicate. The experiment consisted of two 20-day periods, totaling 40 days. Feeding conditions remained consistent throughout the entire experiment. During the first 20 days, samples were collected from 3 goats in each group; during the second 20 days, samples were collected from the remaining 3 goats in each group. The calcium nitrate used in the treatment group was a commercial compound obtained from Rhus Chinensis Mill, with a purity of 99.9%, supplied by the National Pharmaceutical Chemical Reagent Company in Shanghai. Each phase included a 10-day pre-experimental period. On the 15th day of each phase, respiratory metabolism tests were performed using a single-chamber system, measuring the dynamic emissions of CH_4 and CO_2 over two consecutive days. Each goat was housed in a spacious, well-ventilated pen measuring 2.5 m in length and 1.5 m in width. They received feed two times a day, specifically at 8:00 AM and 5:00 PM, and they always had free access to clean drinking water.

The formulation of the experimental diet and the calculation of its nutritional components were based on the Agricultural Industry Standard of the People's Republic of China (NY/T816-2021) [12], which outlines the nutritional requirements for male goat weighing approximately 30 ± 0.2 kg. The experimental diet consisted of fully mixed pellet feed with a diameter of approximately 3.5 mm and a length of 1 to 2 cm, provided by Gansu Runmu Biological Engineering Co., Ltd., Jinchang, China. Detailed information related to the composition and chemical analysis of the experimental diet can be found in Table 1.

Table 1. Composition and nutrition level of the experiment diets (DM basis).

Items	Treatments [1]	
	CON	CAL
Ingredients, %		
Corn straw	17.76	17.76
Soy husk	10.34	10.34
Sunflower hull	2.00	2.00
Corn	48.60	49.65
Soybean meal	11.35	7.55
Sesame cake	3.00	3.00
Molasses	5.00	5.00
Salt	0.70	0.70
Expanded urea	0.80	0.00
Calcium carbonate	0.25	0
Premix [2]	1.00	1.00
Calcium nitrate	0.00	3.00
Total	100.00	100.00
Nutrient level [3]		
DE (MJ/kg)	12.31	11.92
CP (%)	14.34	14.70
EE (%)	2.48	2.95
NDF (%)	39.22	38.09

https://doi.org/10.3390/ani16081150

Table 1. *Cont.*

Items	Treatments [1]	
	CON	CAL
ADF (%)	17.20	16.89
Ca (%)	0.31	0.81
P (%)	0.21	0.22

[1] Treatments; control group (CON); calcium nitrate group (CAL). [2] Premix ingredients: Fe = 25 mg, Mn = 40 mg, Zn = 40 mg, Cu =8 mg, I = 0.3 mg, Se = 0.3 mg, Co = 0.1 mg, Vit A = 940 IU, Vit D = 111 IU, Vit E = 20 IU. [3] Digestible energy (DE), the digestive energy in the nutritional level was calculated, and the rest were measured; crude protein (CP), ether extract (EE), neutral detergent fiber (NDF), acid detergent fiber (ADF), calcium (Ca), phosphorus (P).

2.2. Sample Collection

The experimental design comprised an initial 10-day adaptation period, succeeded by a 5-day sampling phase (days 11–15) in each phase. Sampling methods were adapted from Wang et al. [13]. During the sampling period, daily measurements had been conducted to gather essential data: (1) Daily records had been maintained for the quantity of feed provided and the residual amount to calculate dry matter intake (DMI). (2) Fecal samples were collected using plastic buckets placed behind the test sheep. (3) Urine output had been captured via a collection system featuring a PVC tube fitted onto the penis, draining into a storage bucket. All collected samples had undergone standardized processing procedures. Fecal samples had been homogenized, with 10% (w/w) of the aliquot reserved for further analysis. Urine output had been quantified by volume, and a 10% (v/v) subsample had been taken for additional examination.

Throughout the formal trial period, body weight (BW) measurements had been taken every morning prior to feeding. The goats' average daily weight gain (ADG) was calculated by evaluating the differences between their starting and ending BW.

On the thirteenth day of each experimental phase, rumen fluid samples were collected using a rumen tube and vacuum pump prior to feeding, and again three hours after feeding. To avoid contamination from saliva, approximately 20 mL of the initial sample was discarded. The leftover fluid was subsequently filtered through four layers of gauze and examined right away for pH levels with the help of a portable pH meter (pH-HJ90, Beijing Aerospace Computer Co, Beijing, China). In order to facilitate additional analysis of ammonia nitrogen in conjunction with volatile fatty acids (VFAs), two samples were preserved at a temperature of −20 °C. Additionally, another two aliquots had been preserved at −80 °C for 16S rRNA pyrosequencing.

On the 15th day of each experimental phase following the goats' adaptation to the respiration chamber, methane CH_4 and carbon dioxide CO_2 production was measured using a mobile open-circuit respiratory calorimeter system. The specific methodology followed the experimental protocol established by Wang et al. [14]. The measurement process had been designed to span two days and was conducted within an open respiration chamber, which maintains an average airflow rate of 40 m^3/h as regulated by a gas flow meter (C100L-CRWE-DD, SIYA, Shanghai, China). During this period, a sophisticated greenhouse gas analyzer (MIU-374-8, Los Gatos Research, San Jose, CA, USA) had assessed gas concentrations at both the chamber outlet and in the surrounding environment at hourly intervals. Each measurement consisted of a 9 min detection phase followed by a 9 min control phase. Additionally, a gas flow meter (C100L-CRWE-DD, SIYA, Shanghai, China) had continuously monitored the gas production rate of the Liuyang black goats throughout the entire day. The data collected had been transmitted to a computer system for processing and calibration with pure CH_4 and CO_2 at a flow rate of 100 mL/min.

https://doi.org/10.3390/ani16081150

2.3. Chemical Analysis

Fecal samples were subjected to drying in an oven set at 60 °C for a duration of 48 h, after which they were ground with a vertical Wiley mill produced by Arthur H. Thomas Company located in Philadelphia, PA, USA, until they could pass through a sieve of 0.45 mm. The assessment of the chemical composition involved quantifying the DM content (utilizing method 930.15; AOAC, 2012), assessing CP levels (applying method 984.13; AOAC, 2012), and measuring the concentrations of calcium and phosphorus (methods 942.05 and 965.17; AOAC, 2012) [15]. The Ankom A200i fiber analyzer, developed by ANKOM Technology Co. (New York, NY, USA), was employed to evaluate acid detergent fiber (ADF) and neutral detergent fiber (NDF), following the protocols set forth by Van Soest et al. [16]. The calculation of digestible energy (DE) was conducted in line with the methods suggested by the National Research Council (NRC) and associated tables for goats [17].

To investigate the VFAs found in ruminal fluid, a high-performance gas chromatograph (HPGC; model GC-2014; produced by Shimadzu Corporation, Kyoto, Japan) was employed. This apparatus features a flame ionization detector (FID) that operates using hydrogen and includes a capillary column supplied by Agilent Technologies, Inc. (Wilmington, DE, USA). The column has a length of 30 m, an inner diameter of 0.32 mm, and a film thickness of 0.50 μm. The temperatures for both the column and FID were set at 120 °C and 200 °C, respectively, with high-purity nitrogen gas (99.99%) serving as the carrier gas at a flow rate of 1.2 mL/min. Hydrogen and air were delivered at rates of 30 mL/min and 300 mL/min, respectively. For the preparation of samples, 1 mL of ruminal fluid was combined with 0.25 mL of metaphosphoric acid (25%, *w*/*v*) and then centrifuged at 4 °C and 3000× *g* for 15 min to extract the supernatant for analysis. The ammonia nitrogen (NH_3-N) concentration in the ruminal fluid was quantified using a UV/Visible spectrophotometer (UV-1801), manufactured by Beijing Beifen-Ruili Analytical Instrument Co, Ltd. (Beijing, China), in accordance with the procedure outlined by Broderick and Kang [18], utilizing the phenol-chloramine colorimetric method.

2.4. Sequencing of the 16S rRNA Gene and Bioinformatics Analysis

Samples of rumen content were obtained from six black goats in each experimental group. The experimental method referenced that of Yang et al. [19]. Genomic DNA was fully extracted with the QIAamp DNA Stool Mini Kit provided by QIAGEN (Hilden, Germany). The amplification of hypervariable regions V3 and V4 of the bacterial 16S rRNA gene was executed using barcode-labeled primers 515F (5′-GTGCCAGCMGCCGCGGTAA-3′) and 806R (5′-GGACTACHVGGGTWTCTTAAT-3′) [20,21]. The PCR reaction consisted of 15 μL of Phusion® High-Fidelity PCR Master Mix (New England Biolabs, Ipswich, MA, USA), along with 0.2 μM of each primer and 10 ng of template DNA. The thermal cycling protocol included an initial denaturation step at 98 °C for one minute, followed by 30 cycles of annealing and extension, concluding with a final extension at 72 °C for a duration of 5 min [22]. The size of the amplified products was assessed using 2% agarose gel electrophoresis and then purified with the Qiagen gel extraction kit (QIAGEN GmbH, Hilden, Germany). Finally, paired-end sequencing of 250 bp was conducted on the Illumina NovaSeq platform (Illumina, San Diego, CA, USA). The assembly of raw reads was carried out employing FLASH software (Version 1.2.7, http://ccb.jhu.edu/software/FLASH/ (accessed on 3 April 2025)) [23]. Subsequently, QIIME (Version 1.9.1, http://qiime.org/scripts/split_libraries_fastq.html (accessed on 3 April 2025)) [24] was employed for quality control to eliminate chimeric sequences, resulting in effective tags. The sequences were subsequently grouped into operational taxonomic units (OTUs) based on a similarity threshold of 97% [25]. The UPARSE tool (Version 7.0.1001,

https://doi.org/10.3390/ani16081150

http://www.drive5.com/uparse/ (accessed on 3 April 2025)) was utilized, with species classification based on the UCHIME algorithm [26]. Furthermore, the Chao1 and Shannon diversity indices were computed [27], while principal coordinates analysis (PCoA) and visualization were performed utilizing R software (version 3.5.2, R Foundation for Statistical Computing, Vienna, Austria) [28]. Lastly, taxonomic classification was conducted using Mothur software (version 1.41.1), referencing the Silva.nr.132 database [29]. We performed linear discriminant analysis effect size (LEfSe) analysis with the Micro Biome Process package (version 1.1.463, RStudio, Inc., Boston, MA, USA) in RStudio (version 3.5.2, R Foundation for Statistical Computing, Vienna, Austria) to evaluate all classification levels ranging from phylum to species, establishing a threshold for LDA scores ($\geq$3.5) to ascertain the significance of the identified biomarkers. Additionally, Spearman rank correlation analysis was applied to examine the relationship between rumen fermentation characteristics and the relative abundances of the top 10 dominant bacterial genera and phyla.

2.5. Calculations and Statistical Analysis

The CH_4 emission, CO_2 emission, growth performance and apparent nutrient digestibility were statistically analyzed with the aid of SPSS software (IBM Corp. Released 2019, IBM SPSS Statistics for Windows, Version 26.0, Armonk, NY, USA: IBM Corp.). The linear model applied was organized in the following manner:

$$Y_i = \mu + T_i + \beta_t \cdot t_j + \varepsilon_i$$

In this context, *Yi* represents the mean treatment observed during the 5-day collection period for the ith dietary group. The overall mean is indicated by μ. The fixed effects of the two treatments (CON and CAL, corresponding to $i = 1$ and 2) are denoted by T_i. The term β_t denotes the effect associated with the time stage, illustrating how time influences the outcome variable. The time covariate, t_j, is defined such that $t_j = 1$ corresponds to the initial time stage, while $t_j = 2$ refers to the subsequent time stage. Lastly, ε_i represents the random residual error.

SPSS software (IBM Corporation 2019 edition, IBM SPSS Statistics for Windows, Version 26.0, Armonk, NY, USA: IBM Corporation) was used to examine the normality assumption, homogeneity of variance, and outlier handling for data on average daily gain (ADG), apparent nutrient digestibility, rumen fermentation indicators, and relative abundance of rumen bacteria. Independent samples *t*-tests were performed. The models used are as follows:

$$t = \frac{\overline{x_1} - \overline{x_2}}{S_{\overline{x_1} - \overline{x_2}}}$$

In this context, $\overline{x_1}$ and $\overline{x_2}$ denote the average values of various treatment groups, whereas $S_{\overline{x_1} - \overline{x_2}}$ represents the standard error associated with the mean difference. A significance threshold of $p < 0.05$ was determined. Additionally, Spearman's correlation coefficients were utilized to assess the associations between microbial species and rumen VFAs, with the same statistical significance threshold of $p < 0.05$ applied.

3. Results

3.1. Effects of Dietary Calcium Nitrate Supplementation on Methane and Carbon Dioxide Emissions

The emissions of CH_4 and CO_2 were tracked and visually displayed (Figures 1 and 2). The findings indicated that the CAL group had consistently lower emissions of both CH_4 and CO_2 compared to the CON group across all time points measured ($p < 0.05$). Specifically, inclusion of calcium nitrate in the diet led to a significant decrease in the emission rates of CH_4 and CO_2 in goats ($p < 0.05$). Furthermore, the temporal patterns of CH_4 and CO_2

https://doi.org/10.3390/ani1608115

production in goats exhibited a distinct diurnal variation, with emissions peaking within 2 to 3 h after feeding and subsequently decreasing until the next feeding event.

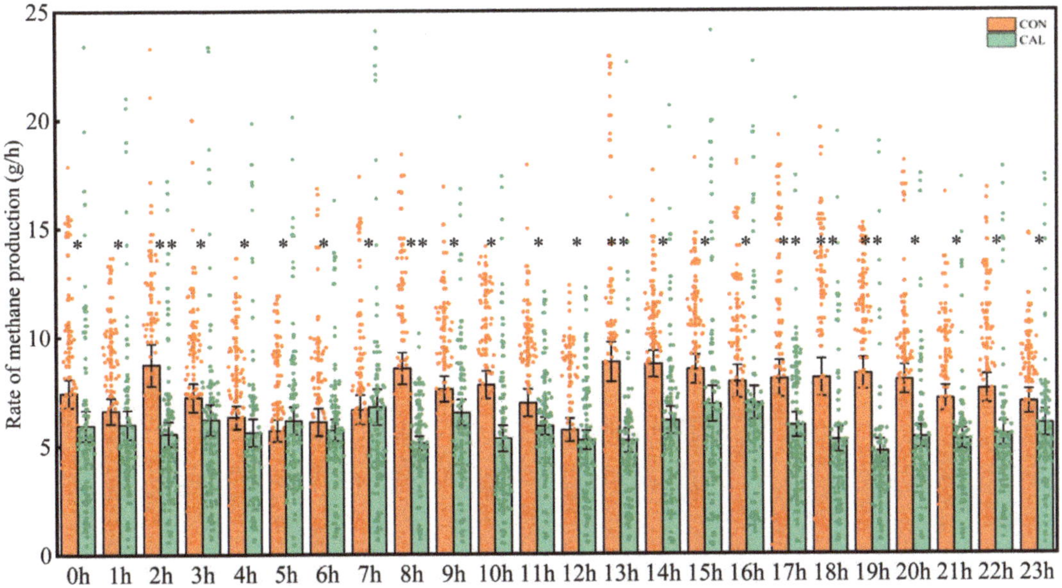

Figure 1. Diurnal pattern in CH_4 emission in respiration chambers (g/h). At the same time, * above the histogram indicated significant difference ($p < 0.05$), and ** indicated extremely significant difference ($p \leq 0.01$). Abbreviation: CON = control group; CAL = calcium nitrate group.

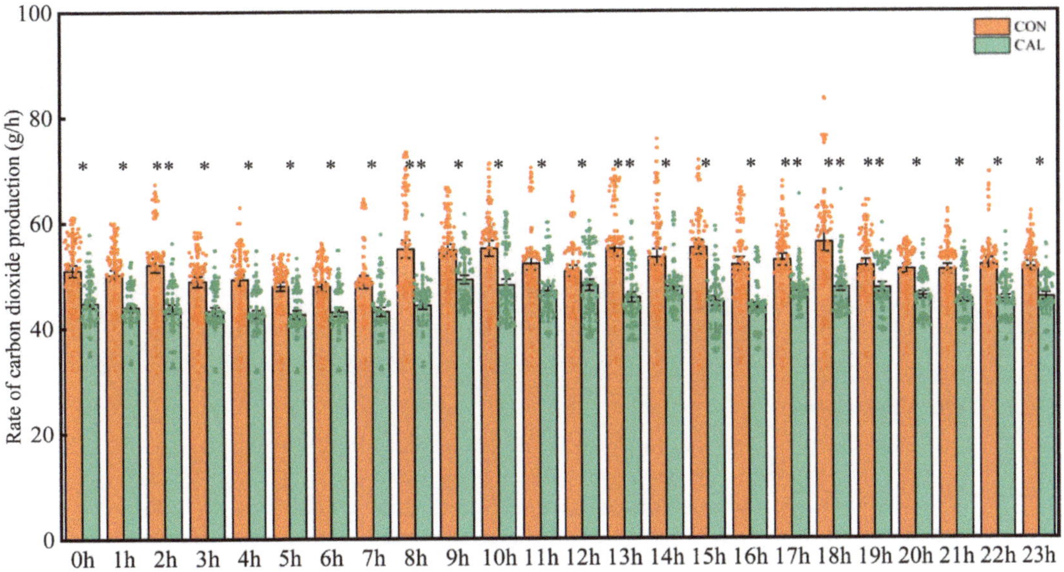

Figure 2. Diurnal pattern in CO_2 emission in respiration chambers (g/h). At the same time, * above the histogram indicated significant difference ($p < 0.05$), and ** indicated extremely significant difference ($p \leq 0.01$). Abbreviation: CON = control group; CAL = calcium nitrate group.

https://doi.org/10.3390/ani16081150

The diet of goats that included calcium nitrate led to a noteworthy reduction in daily CH_4 emissions, as shown in Table 2. Specifically, the production of CH_4 per unit of dry matter intake (DMI) was significantly lower in the CAL group when compared to the CON group (CH_4/DMI, $p < 0.05$). Conversely, no significant differences were observed in methane emissions per unit of average daily gain between the two groups (CH_4/ADG, $p > 0.05$). Similarly, methane emissions per unit of neutral detergent fiber intake did not differ significantly between the groups (CH_4/NDFI, $p > 0.05$). Additionally, no significant differences were found in methane emissions per unit of acidic detergent fiber intake (CH_4/ADFI, $p > 0.05$). Moreover, incorporating calcium nitrate into the diet resulted in significant decreases in both CH_4 energy loss and emissions per metabolic BW relative to the CON group (CH_4-E, $CH_4/BW^{0.75}$, $p < 0.05$). Similarly, the CAL group showed substantially lower daily CO_2 emissions compared to the CON group ($p < 0.05$). In addition, the CAL group exhibited significant reductions across all individual indicators of CO_2 emissions ($p < 0.05$).

Table 2. Effects of calcium nitrate on methane emission in goats.

Items [1]	Treatment [2]		SEM [3]	*p*-Value
	CON	CAL		
CH_4				
CH_4 (g/d)	13.738	6.272	1.73	0.041
CH_4 (g/kg DMI)	17.704	8.029	1.19	0.036
CH_4 (g/kg ADG)	75.251	28.936	1.31	0.055
CH_4 (g/kg NDFI)	33.997	19.077	1.88	0.072
CH4 (g/kg ADFI)	49.461	39.828	4.19	0.283
CH_4-E (MJ/d)	0.765	0.349	0.10	0.041
$CH_4/BW^{0.75}$	1.372	0.644	0.17	0.042
CO_2				
CO_2 (g/d)	546.606	275.253	50.90	0.007
CO_2 (g/kg DMI)	713.616	352.072	62.87	0.001
CO_2 (g/kg ADG)	3133.288	1267.421	407.05	0.013
CO_2 (g/kg NDFI)	1308.544	836.508	91.80	0.003
CO_2 (g/kg ADFI)	2934.239	1886.476	204.63	0.003
CO_2-E (MJ/d)	4.888	2.462	0.46	0.007
$CO_2/BW^{0.75}$	54.248	28.177	4.66	0.003

[1] Items; CH_4, daily methane emission; CH_4/DMI, methane emission from dry matter intake; CH_4/ADG, methane emission from average daily gain; CH_4/NDFI, methane emission per neutral detergent fiber intake; CH_4/ADFI, methane emission per acid detergent fiber intake; CH_4-E, methane energy; $BW^{0.75}$, metabolic body weight; CO_2, daily carbon dioxide emission; CO_2/DMI, carbon dioxide emission from dry matter intake; CO_2/ADG, carbon dioxide emission from average daily gain; CO_2/NDFI, carbon dioxide emission per neutral detergent fiber intake; CO_2/ADFI, carbon dioxide emission per acid detergent fiber intake; CO_2-E, carbon dioxide energy. [2] Treatments; control group (CON); calcium nitrate group (CAL). [3] SEM, standard error of means for treatments.

3.2. *Effects of Dietary Calcium Nitrate Supplementation on Apparent Nutrient Digestibility and Growth Performance*

As shown in Table 3, the addition of calcium nitrate to the diets of goats did not significantly impact their growth performance or nutrient digestibility ($p > 0.05$). Calcium nitrate supplementation, in particular, did not have a significant impact on ADG, feed conversion ratio (F/G), or DMI ($p > 0.05$). However, the apparent nutrient digestibility was reduced with calcium nitrate supplementation, even though the differences between the supplemented group and the control group were not statistically significant ($p > 0.05$).

https://doi.org/10.3390/ani16081150

Table 3. Effects of calcium nitrate addition on apparent digestibility of nutrients in goats.

Items	Treatment [2]		SEM [3]	p-Value
	CON	CAL		
Growth performance [1]				
DMI, g/d	764.750	784.108	29.62	0.761
ADG, g/d	193.770	217.725	49.46	0.423
F/G	4.674	3.998	0.30	0.297
Apparent nutrient digestibility, %				
DM	40.863	35.134	0.03	0.374
NDF	45.714	39.549	0.03	0.258
ADF	24.483	19.064	0.03	0.385
CP	67.502	62.487	0.02	0.188

[1] Growth performance; DMI, dry matter intake; ADG, average daily gain; F/G, feed to gain ratio DM, dry matter; NDF, neutral detergent fiber; ADF, acid detergent fiber; CP, crude protein. [2] Treatments; control group (CON); calcium nitrate group (CAL). [3] SEM, standard error of means for treatments.

3.3. Effects of Dietary Calcium Nitrate Supplementation on Nitrogen Metabolism

Table 4 presents the effects of calcium nitrate supplementation on nitrogen metabolism in goats. The data indicated that dietary calcium nitrate did not significantly influence nitrogen intake, faecal nitrogen, urine nitrogen, digestive nitrogen, nitrogen deposition, nitrogen digestibility, net protein utilization, or the biological value of proteins in goats ($p > 0.05$).

Table 4. Effects of calcium nitrate addition on nitrogen metabolism in goats.

Items	Treatment [1]		SEM [2]	p-Value
	CON	CAL		
Nitrogen intake, g/d	18.365	17.153	0.58	0.321
Faecal nitrogen, g/d	6.373	6.955	0.39	0.493
Urine nitrogen, g/d	6.954	6.255	0.55	0.553
Digestive nitrogen, g/d	11.993	10.200	0.56	0.113
Nitrogen deposition, g/d	5.039	3.945	0.31	0.074
Nitrogen digestibility, %	65.325	59.343	0.02	0.149
Net protein utilization, %	27.887	22.900	0.02	0.165
Biological value of protein, %	43.567	38.433	0.03	0.436

[1] Treatments; control group (CON); calcium nitrate group (CAL). [2] SEM, standard error of means for treatments.

3.4. Effects of Dietary Calcium Nitrate Supplementation on Rumen Fermentation Parameters

Table 5 presents the effects of calcium nitrate supplementation on ruminal pH, NH_3-N, and VFAs in goats. Before feeding, no significant differences were observed between the two groups in terms of pH, NH_3-N, total volatile fatty acids (TVFAs), the proportion of individual VFAs, and the acetate to propionate ratio (A:P) ($p > 0.05$). However, a significant difference was observed in valerate proportions ($p < 0.05$), suggesting a specific effect of calcium nitrate on this VFAs.

Three hours after feeding, significant differences in rumen fermentation patterns were observed in goats. The CAL group had a higher pH compared to the CON group ($p < 0.05$). Additionally, NH_3-N production increased, but no significant difference was observed ($p > 0.05$). The CAL group also exhibited a lower acetate proportion and higher propionate proportion, along with increased concentrations of valerate, isobutyrate, and isovalerate ($p < 0.05$). These findings suggest that calcium nitrate supplementation modulates ruminal fermentation patterns, favoring propionate production over acetate ($p < 0.05$).

https://doi.org/10.3390/ani16081150

Table 5. Effect of calcium nitrate on ruminal pH, NH_3-N, VFAs in goats.

Items	Treatment [1]		SEM [2]	*p*-Value
	CON	CAL		
Before feeding				
pH	6.563	6.683	0.15	0.489
NH_3-N, mg/100 mL	10.809	12.044	0.68	0.378
TVFAs [3], mmol	58.360	67.318	4.45	0.325
VFA proportion, mol/100 mol				
Acetate	60.266	62.099	0.78	0.249
Propionate	19.647	17.814	0.95	0.350
Butyrate	15.686	15.742	0.44	0.950
Valerate	0.755	1.055	0.73	0.045
Isobutyrate	1.374	1.241	0.07	0.361
Isovalerate	2.273	2.049	0.11	0.316
Acetate/Propionate	3.354	3.567	0.17	0.551
3 h after feeding				
pH	5.838	6.290	0.15	0.022
NH_3-N, mg/100 mL	13.461	15.434	0.72	0.184
TVFAs [3], mmol	111.159	101.901	4.60	0.325
VFA proportion, mol/100 mol				
Acetate	62.753	53.014	1.51	0.001
Propionate	21.718	32.354	1.52	0.001
Butyrate	13.120	13.112	0.40	0.917
Valerate	0.696	1.022	0.07	0.014
Isobutyrate	0.295	0.560	0.04	0.001
Isovalerate	0.441	0.835	0.06	0.001
Acetate/Propionate	3.033	1.747	0.20	0.001

[1] Treatments; control group (CON); calcium nitrate group (CAL). [2] SEM, standard error of means for treatments. [3] TVFAs, Total volatile fatty acids.

3.5. Effects of Dietary Calcium Nitrate Supplementation on Diversity of Rumen Bacterial Community

The incorporation of calcium nitrate did not significantly influence the alpha diversity metrics, such as ACE, Chao1, Shannon, and Simpson, related to the bacterial microflora found in the rumen fluid of Liuyang black goats ($p > 0.05$, Figure 3). Meanwhile, the beta diversity assessment utilizing weighted UniFrac distances showed a more centralized spatial arrangement of samples within the CAL group, suggesting enhanced internal consistency. In contrast, the CON group displayed a more dispersed pattern. This suggests that the CAL group exhibited a significantly higher degree of sample aggregation compared to the CON group (Figure 4).

Among bacteria, the study's findings indicated that the microbial community composition and relative abundance varied between the two treatment groups, with *Bacteroidota* and *Firmicutes* being the dominant phyla. Notably, *Bacteroidota* was most abundant in the CAL group and *Firmicutes* in the CON group ($p < 0.05$, Figure 5a and Table 6). At the genus level, *Prevotella* and *Methanobrevibacter* were particularly prominent, with *Prevotella* showing a higher relative abundance in the CAL group and *Methanobrevibacter* in the CON group ($p < 0.05$, Figure 5b and Table 7), and some genera such as *Prevotellaceae_UCG-001* appeared to be more abundant in the CAL group compared to the CON group ($p < 0.05$).

https://doi.org/10.3390/ani1608115(

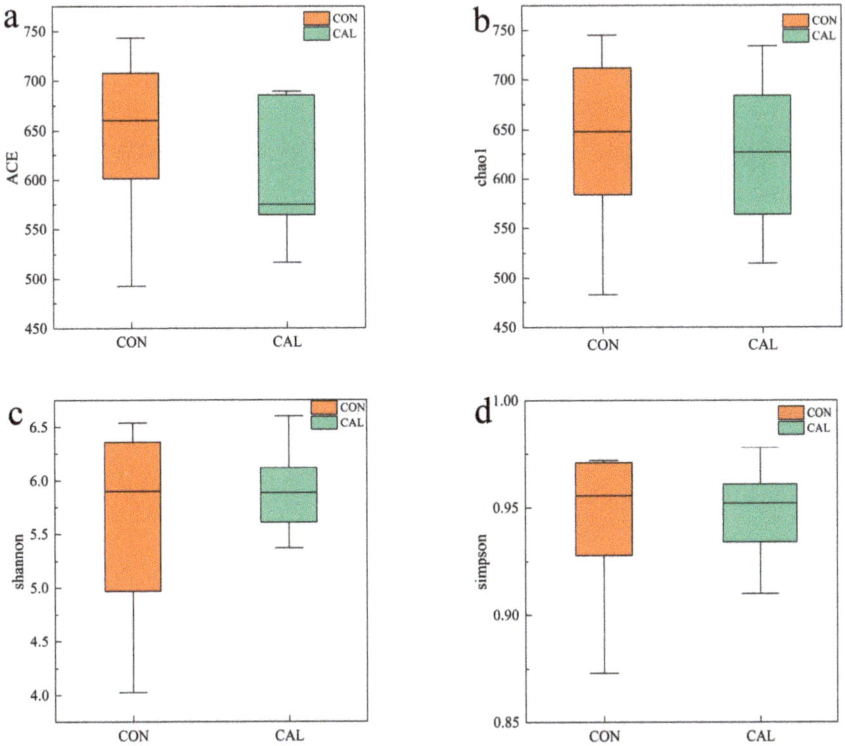

Figure 3. The microbial community's alpha diversity in the rumen is influenced by the species observed. The indexes used to measure alpha diversity comprise (**a**) the ACE index, (**b**) the chao1 index, (**c**) the Shannon index, and (**d**) the Simpson index. Abbreviations used include: CON = control group; CAL = calcium nitrate group.

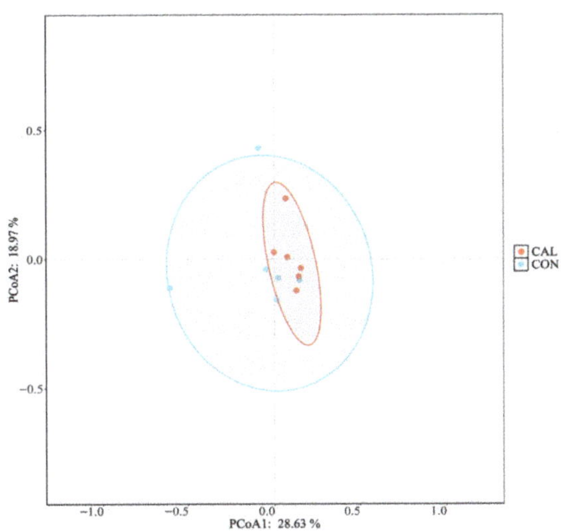

Figure 4. A principal coordinate analysis (PCoA) was performed to illustrate the makeup of the rumen bacterial community subjected to different treatments, Abbreviations: CON = control group; CAL = calcium nitrate group.

https://doi.org/10.3390/ani16081150

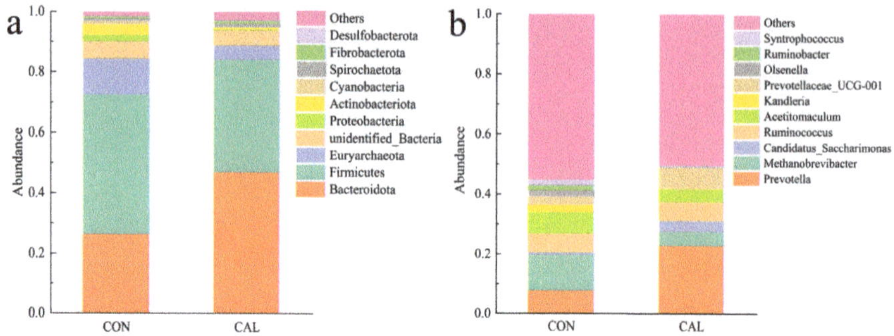

Figure 5. The structure of the rumen microbiome at both the phylum and genus levels: (**a**) the composition of rumen microbiome at phylum level, (**b**) the composition of major rumen genus level; Abbreviations: CON = control group; CAL = calcium nitrate group.

Table 6. Relative abundance of rumen microorganisms (phylum level, %).

Items	Treatments [1]		SEM [2]	*p*-Value
	CON	CAL		
Firmicutes	46.107	37.168	0.03	0.180
Bacteroidota	26.483	47.036	0.04	0.007
Euryarchaeota	11.911	4.664	0.02	0.094
Unidentified_Bacteria	5.677	5.057	0.02	0.871
Actinobacteriota	3.410	0.849	0.01	0.173
Proteobacteria	2.123	0.141	0.01	0.366
Spirochaetota	0.841	1.227	0.01	0.436
Fibrobacterota	0.568	0.814	0.01	0.473
Cyanobacteria	1.199	0.062	0.01	0.377
Desulfobacterota	0.130	0.092	0.01	0.272

[1] Treatments; control group (CON); calcium nitrate group (CAL). [2] SEM, standard error of means for treatments.

Table 7. Relative abundance of rumen microorganisms (genus level, %).

Items	Treatments [1]		SEM [2]	*p*-Value
	CON	CAL		
Prevotella	7.970	22.952	0.03	0.001
Methanobrevibacter	11.860	4.644	0.02	0.005
Ruminococcus	6.330	6.295	0.01	0.991
Acetitomaculum	7.146	4.276	0.01	0.254
Prevotellaceae_UCG-001	2.840	7.115	0.01	0.044
Candidatus_Saccharimonas	0.764	3.579	0.02	0.434
Kandleria	2.358	0.162	0.01	0.367
Syntrophococcus	1.899	0.445	0.01	0.335
Olsenella	2.117	0.150	0.01	0.327
Ruminobacter	1.633	0.018	0.01	0.363

[1] Treatments; control group (CON); calcium nitrate group (CAL). [2] SEM, standard error of means for treatments.

The analysis using Linear discriminant analysis effect size (LEfSe) demonstrated clear differences in the characteristics of the microbial communities between the various treatment groups. In the rumen, the CAL group was significantly enriched with a diverse array of taxa, including *Sphingomonadaceae*, Unidentified_chloroplasts, *Lachnospirales*, *Lachnospiraceae*, and *Sphingomonadales*. In contrast, the CON group showed enrichment with

https://doi.org/10.3390/ani16081150

three primary taxa, including *Succinivibrionaceae*, *Gammaproteobacteria*, and *Aeromonadales* (Figure 6).

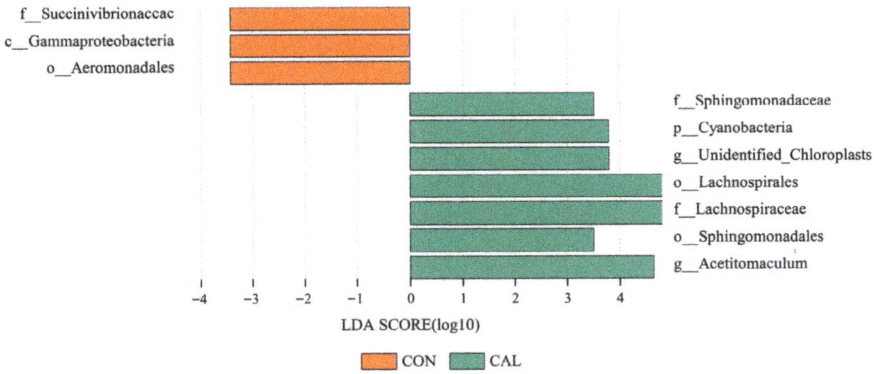

Figure 6. The comparison of the microbial community composition in goat's rumen using the LEfSe analysis (LDA = 3.5); Abbreviations: CON = control group; CAL = calcium nitrate group; LDA = linear discriminant analysis.

Spearman correlation analysis revealed meaningful relationships between parameters of rumen fermentation and the relative abundance of various microbial taxa (Figure 7). The pH level in the rumen was positively correlated with the abundance of *Bacteroidota* ($p < 0.05$). The concentration of acetate was positively associated with *Ruminobacter* ($p < 0.05$), *Actinobacteriota* ($p < 0.05$), and *Desulfobacterota* ($p < 0.01$), whereas it showed negative correlations with *Bacteroidota* ($p < 0.05$) and *Prevotella* ($p < 0.01$). Propionate concentration was positively associated with *Prevotella* abundance ($p < 0.05$). Butyrate and valerate concentrations both correlated positively with *Spirochaetota* ($p < 0.05$). The TVFA showed a significant positive relationship with *Acetitomaculum* abundance ($p < 0.05$).

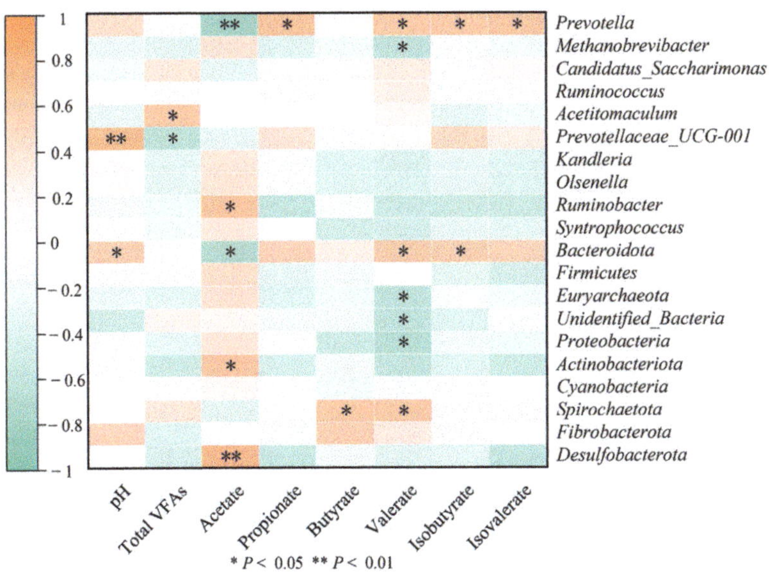

Figure 7. Analysis of the Spearman correlation between microbiota at the genus and phylum levels and parameters of rumen fermentation. * $p < 0.05$. ** $p < 0.01$.

https://doi.org/10.3390/ani16081150

4. Discussion

CH_4 is responsible for about 16% of worldwide human-made greenhouse gas emissions and has a significantly greater warming potential compared to CO_2 [30]. The production of ruminal CH_4 is a multifaceted trait affected by various factors, such as DMI, the composition of feed, the microbiota present in the rumen, the ratio of fermentation byproducts, genetic traits of the host, and environmental influences [31]. Upon entering the rumen of ruminants, cellulose undergoes fermentation, yielding VFAs and reduced H_2. Within the rumen ecosystem, methanogenic archaea efficiently utilize the reduced H_2 produced by other microorganisms through an interspecies H_2 transfer mechanism [32]. Research indicates that CH_4 production exhibits a gradual decline trend as nitrate supplementation in ruminant diets increases [33]. Research has shown that incorporating NO_3^- into ruminant diets can effectively reduce CH_4 emissions in vivo, with observed reduction potentials ranging from 6.8% to 12.5% for each 1% of NO_3^- added on a dry DM basis per day [34]. Asanuma et al. [35] reported a sharp decrease in the number of methanogens in the rumen of goats following the addition of nitrate, suggesting that nitrate may be toxic to methanogens, thereby reducing CH_4 emissions by inhibiting their activity. However, findings on the effects of nitrate on CH_4 emissions are not consistent across studies. Meller et al. [36] and Rebelo et al. [37] observed reduced CH_4 emissions with dietary nitrate supplementation but noted no significant differences in CH_4 production. This difference is mainly linked to reduced DMI resulting from the nitrate effect, as there is a positive correlation between feed intake and intestinal CH_4 emissions. In this study, the reduction in methane emissions was accompanied by no significant differences in DMI among treatment groups. Specifically, the dietary addition of 3% calcium nitrate significantly reduced CH_4 production, which is in agreement with the findings of the aforementioned researchers. There was no significant difference in methane emissions per unit of acid-detergent fiber intake, which may be attributed to the influence of individual variation. Furthermore, the lack of data on methemoglobin in this study represents a limitation, and there may be differences in the safety thresholds across different species, highlighting the need for further studies on the effects of nitrate in the rumen.

In the present study, DMI and ADG were slightly higher in the CAL group compared to the CON group, although no significant differences were observed between the two groups. Meller et al. [36] and Rebelo et al. [37] noted that nitrate addition decreased DMI by approximately 8.0% at similar levels of inclusion, which was initially attributed to nitrate toxicity, characterized by elevated blood methemoglobin levels exceeding 20% of total hemoglobin. When animals were gradually acclimated to nitrate or protective nitrates were used, the decrease in DMI was primarily due to the bitter taste of nitrates, rather than toxicity. This suggests that gradual acclimatization is crucial for maintaining DMI levels without compromising animal performance or health. For instance, Halmemies et al. [9] fed growing goats (initial BW of 10 kg) diets supplemented with 5.0% calcium nitrate or 2.6% urea for 12 weeks and observed that growth rates were comparable when calcium nitrate was used as the main nitrogen source. Similarly, Van Zijderveld et al. [38] found no detrimental effects on DMI and ADG when crossbred Texel lambs (initial BW 43 kg) were fed 26 g/kg DM calcium nitrate for 4 weeks. Seyyedsalehi et al. [39] concluded that DMI reductions and signs of nitrate toxicity only occurred when diets contained more than 30 g/kg nitrate for sheep and more than 10 g/kg nitrate for beef cattle. Li et al. [10] also demonstrated that, compared to urea, nitrate (21 g/kg DM) did not significantly affect DMI but resulted in a 19% decrease in ADG, highlighting the need for further studies on the effects of nitrate in the rumen.

Enhancing the efficiency of microbial protein production in the rumen markedly reduces the urinary excretion of nitrogen-containing metabolites. The mechanism underlying

https://doi.org/10.3390/ani1608115(

this process is the efficient utilization properties of microbial proteins during digestion, which enable the nitrogen in their metabolites to be fully absorbed by the host, thereby minimizing urinary nitrogen losses [40]. In this investigation, significant differences were not found between the CAL and CON groups regarding nitrogen metabolism. These results align with those presented by Li et al. [10], who indicated that the inclusion of nitrate did not affect the digestibility of DM or nitrogen (N) when compared to that of control diets. The findings from the current study further suggest that nitrate may act as a nitrogen source for microbial development in the rumen. Thus, it is logical to propose that nitrate is efficiently utilized as a nitrogen source for microbial growth within the rumen. Additionally, Almeida et al. [41] noted an increase in rumen microbial nitrogen outflow from 9.3 to 11.8 g N daily as the dietary nitrate concentration was elevated from 0% to 4%. Similarly, some researchers have found that sheep fed nitrate-supplemented diets consumed more N compared to those fed urea-supplemented diets. However, this increased intake did not affect the apparent digestibility of N in sheep [42]. When considering energy efficiency, energy losses through CH_4 were reduced by 3.5% in sheep consuming nitrate-supplemented diets. Although the proportion of energy lost as CH_4 was lower for nitrate-added diets, no significant differences in metabolic energy intake were observed between treatments.

The rumen is a central component of the digestive system in ruminants, playing a vital role in the degradation of fibrous feeds and serving as the primary site for the absorption of key nutrients, such as proteins and trace elements [43]. Maintaining a healthy and balanced rumen environment is crucial for optimizing digestive efficiency and nutrient absorption in ruminants. The pH of rumen fluid is a key indicator of the fermentation environment within the rumen, with changes in pH directly reflecting the conditions of fermentation. The normal pH range for rumen fluid is typically 6.0 to 7.0 [44]. Following feeding, ruminal pH typically decreases and then gradually increases due to the absorption of VFAs, rumination, and salivation [45]. The present study also observed this pattern of pH fluctuation. In contrast to our results, Mahmoudi et al. [46] did not find a significant difference in rumen fluid pH when dietary nitrate was introduced for fattening rams. Nevertheless, other research has yielded contrasting outcomes. Ungerfeld et al. [47] detected a notable increase in rumen fluid pH with the supplementation of nitrate in an in vitro batch culture. Likewise, Hassan et al. [48] reported that incorporating nitrate into diets at concentrations of 6.84 g/kg DM and 6.80 g/kg DM led to higher pH levels. Additionally, an in vitro experiment by Zhou et al. [49] revealed that significant pH increases occurred when nitrate levels exceeded 24 μmol/L. These findings align with those of the current experiment.

NH_3-N is released by rumen microorganisms during the decomposition of nitrogenous substances in feed. It serves not only as a key nitrogen source during rumen microbial fermentation but also as a fundamental indicator for assessing the efficiency of converting feed nitrogen to microbial nitrogen [50]. In the current study, the concentrations of NH_3-N rose at 3 h after feeding when compared to pre-feeding levels, though no significant differences were found among the groups. The CAL group's NH_3-N concentration was greater than that of the CON group, mainly because calcium nitrate was included in the diet, resulting in nitrite production and subsequently higher rumen NH_3-N levels [51]. Almeida et al. [41] noted that incorporating nitrate into the diet enhanced NH_3-N production in the rumen, which aligns with the findings of this study. However, previous in vitro studies have shown that nitrate supplementation had no effect on NH_3-N concentration [52,53]. Conversely, other reports indicate that nitrate supplementation may reduce NH_3-N concentrations [54]. These conflicting results suggest that nitrate metabolism in the rumen is not exclusively converted to NH_3, and its impact on NH_3-N levels may vary depending on the experimental conditions and dietary formulations.

https://doi.org/10.3390/ani16081150

Ruminants derive 60% to 80% of their energy from VFAs produced through rumen digestion of nutrients [55]. Under standard conditions, the minimum hydrogen requirement for acetate synthesis by acetogenic bacteria is approximately 10 to 100 times higher than that for methanogenesis by methanogenic archaea [56]. If ruminal VFA production is shifted towards propionate (lower acetate-to-propionate ratio, A:P), the net balance of H_2 in the rumen is reduced, thereby decreasing CH_4 production [57]. Research indicates that the supplementation of nitrate enhances the growth of specific bacteria that degrade fiber in the rumen, as well as boosts the production of valeric acid within this environment [58]. The impact of nitrate supplementation on the composition of VFAs has shown variability among different studies. For instance, Feng et al. [57] observed a linear increase in the proportion of propionic acid with the administration of three calcium nitrate doses at 5.3, 13.6, and 21.1 g/kg DM, a finding that aligns with the conclusions of the present study. This variation may be attributed to differences in sampling time and the dose-dependent effect of nitrate supplementation on CH_4 mitigation [57]. Numerous investigations highlighted a notable impact of nitrate supplementation on the profiles of VFAs, with sampling often conducted shortly after feeding—especially in grazing cows given dietary nitrate at concentrations of 22.5, 21.5, and 18 g/kg DM [59]. Conversely, Li et al. [10] reported no influence of nitrate supplementation on rumen VFAs when samples were taken from goats over six hours post-feeding. In our current study, we noted alterations in acetate proportions due to treatment and a marked reduction in the A:P ratio, which somewhat diverges from certain earlier findings [60]. This reduction was linked to an increase in H_2 release and a boost in the abundance and functioning of microorganisms that produce acetate [51]. Moreover, a noteworthy decrease in the A:P ratio was noted in the CAL group in contrast to the CON group, which could be linked to changes in the fermentation process resulting from variations in the relative abundance of rumen microorganisms, including *Fibrobacterium* and *Ruminococcus* species.

In the current study, it was observed that the relative abundance of the methanogenic genus *Methanobrevibacter* diminished at the genus level. Bharanidharan et al. [61] similarly noted that the addition of NO_3^- resulted in a 54% decrease in the population of methanogenic bacteria in ruminants. Furthermore, Li et al. [10] indicated that nitrate could significantly lower both *Methanobrevibacter* levels and CH_4 emissions in sheep. Other research has also indicated that nitrate supplementation helps reduce the numbers of methanogenic bacteria, consequently leading to a decrease in CH_4 production during fermentation. This has important implications for controlling greenhouse gas emissions from ruminants and reducing the energy losses associated with CH_4 emissions from animal feeds, which can range from 2% to 12% [62]. Furthermore, these studies suggest that nitrates can exert long-term effects on rumen methanogenic bacteria [63,64]. In the rumen, CH_4 is primarily produced by methanogenic archaea, which use H_2 and CO_2 as substrates for CH_4 synthesis [65]. When nitrates enter the rumen, nitrate-reducing bacteria utilize H_2 as an electron donor and nitrate ions as an electron acceptor. Through the action of nitrate reductase, nitrates are reduced to nitrites [66]. During this reduction process, nitrates compete with methanogens for electrons from H_2 donors, thereby inhibiting CH_4 production, reducing greenhouse gas emissions, and minimizing feed energy waste [67]. Patra and Yu [68] observed that nitrate has a high electron affinity, which allows it to outcompete methanogens for electrons, thereby further inhibiting CH_4 production.

5. Conclusions

The research illustrated that the supplementation of dietary calcium nitrate in Liuyang black goats led to a significant decrease in CH_4 and CO_2 emissions, thus emphasizing its potential as a method for reducing greenhouse gas emissions from ruminant animals.

https://doi.org/10.3390/ani16081150

Additionally, the results indicated that calcium nitrate influenced ruminal fermentation dynamics and modified the makeup of the rumen microbiota. These findings highlight the possible advantages of using calcium nitrate as a feed additive to enhance the sustainability of goat production systems. Future investigations should aim to refine the dosage of calcium nitrate and examine its long-term implications for animal health, productivity, and overall environmental effects.

Author Contributions: Conceptualization, M.L. and C.Z.; Formal analysis, M.L. and C.Z.; Funding acquisition, C.Z. and H.Z.; Investigation, M.L., X.N., J.W. and B.C.; Methodology, T.L., C.Z. and H.Z.; Project administration, T.L., C.Z. and H.Z.; Supervision, T.L., C.Z. and H.Z.; Writing—original draft, M.L.; Writing—review and editing, C.Z. and H.Z. All authors have read and agreed to the published version of the manuscript.

Funding: This research received funding from several sources, including the National Natural Science Foundation of China (grant numbers 32472938 and 32160803), the Publicly Recruited Doctoral Research Startup Fund Project at Gansu Agricultural University (grant number GAU-KYQD-2024-07), the Key Talent Project of Gansu Province (grant number 2024RCXM60), the Longyuan Youth Talent Special Support Program of Gansu Province (Document No. 11 from the Provincial Committee Talent Work Group, 2024), and the Discipline Team Project at Gansu Agricultural University (GAU-XKTD-2022-20).

Institutional Review Board Statement: All experiments in this study were conducted following the approved guidelines of the Regulation Standing Committee of the Gansu People's Congress. The Ethics Committee of the Gansu Agriculture University approved all experimental protocols and sample collections under permit number GSAU-Eth-AST-2024-023.

Informed Consent Statement: Informed consent was obtained from all subjects involved in the study.

Data Availability Statement: The data will be made available from the corresponding author upon reasonable request.

Acknowledgments: We sincerely thank Ying Yun for her valuable input in probability design, formal analysis, survey research, and the development of the initial draft. We are also grateful to Yi Wang for his assistance with fundraising and the execution of the survey. Our heartfelt appreciation extends to all individuals who participated in this experiment.

Conflicts of Interest: The authors declare that there is no conflict of interest.

Abbreviations

In this manuscript, the abbreviations listed below are utilized:

CAL	Calcium Nitrate Group
CON	Control Group
CH_4	Methane
NH_3-N	Ammonia Nitrogen
A:P	Acetate-to-propionate Ratio
VFAs	Volatile Fatty Acids
CO_2	Carbon Dioxide
GHGs	Greenhouse Gases
ADG	Average Daily Weight Gain
BW	Body Weight

https://doi.org/10.3390/ani16081150

References

1. Wang, W.; Lund, P.; Larsen, M.; Weisbjerg, M.R. Effect of nitrate supplementation, dietary protein supply, and genetic yield inde on performance, methane emission, and nitrogen efficiency in dairy cows. *J. Dairy Sci.* **2023**, *106*, 5433–5451. [CrossRef] [PubMed
2. Dlamini, J.C.; Cárdenas, L.; Tesfamariam, E.H.; Dunn, R.; Hawkins, J.; Blackwell, M.; Evans, J.; Collins, A. Soil methane (CH4 fluxes in cropland with permanent pasture and riparian buffer strips with different vegetation. *J. Plant Nutr. Soil Sci.* **2021**, *18* 132–144. [CrossRef]
3. Chang, J.; Peng, S.; Ciais, P.; Saunois, M.; Dangal, S.R.S.; Herrero, M.; Havlík, P.; Tian, H.; Bousquet, P. Revisiting enteric methan emissions from domestic ruminants and their $\delta^{13}C_{CH4}$ source signature. *Nat. Commun.* **2019**, *10*, 3420. [CrossRef] [PubMed]
4. Cheng, M.; McCarl, B.; Fei, C. Climate change and livestock production: A literature review. *Atmosphere* **2022**, *13*, 140. [CrossRef
5. Jafari, S.; Meng, G.Y.; Rajion, M.A.; Ebrahimi, M. The use of plant by-products as non-conventional feedstuff for livestock feedin with reference to rumen methanogenesis. *Agrofor. Syst.* **2019**, *10*, 1491–1500. [CrossRef]
6. Adejoro, F.A.; Hassen, A.; Akanmu, A.M.; Morgavi, D.P. Replacing urea with nitrate as a non-protein nitrogen source increase lambs' growth and reduces methane production, whereas acacia tannin has no effect. *Anim. Feed. Sci. Technol.* **2020**, *259*, 11436 [CrossRef]
7. Honan, M.; Feng, X.; Tricarico, J.M.; Kebreab, E. Feed additives as a strategic approach to reduce enteric methane production i cattle: Modes of action, effectiveness and safety. *Anim. Prod. Sci.* **2021**, *62*, 1303–1317. [CrossRef]
8. Lee, C.; Beauchemin, K.A. A review of feeding supplementary nitrate to ruminant animals: Nitrate toxicity, methane emission and production performance. *Can. J. Anim. Sci.* **2014**, *94*, 557–570. [CrossRef]
9. Halmemies-Beauchet-Filleau, A.; Rinne, M.; Lamminen, M.; Mapato, C.; Ampapon, T.; Wanapat, M.; Vanhatalo, A. Review Alternative and novel feeds for ruminants: Nutritive value, product quality and environmental aspects. *Animal* **2018**, *12*, 295–30 [CrossRef]
10. Li, L.; Davis, J.; Nolan, J.; Hegarty, R. An initial investigation on rumen fermentation pattern and methane emission of shee offered diets containing urea or nitrate as the nitrogen source. *Anim. Prod. Sci.* **2012**, *52*, 653–658. [CrossRef]
11. Hristov, A.N.; Oh, J.; Giallongo, F.; Frederick, T.W.; Harper, M.T.; Weeks, H.L.; Branco, A.F.; Moate, P.J.; Deighton, M.H.; Williams S.R.O.; et al. An inhibitor persistently decreased enteric methane emission from dairy cows with no negative effect on mil production. *Proc. Natl. Acad. Sci. USA* **2015**, *112*, 10663–10668. [CrossRef] [PubMed]
12. *NY/T816-2021*; Nutritional Requirements of Mutton. Ministry of Agriculture and Rural Affairs: Beijing, China, 2021.
13. Wang, J.; Liu, T.; Xu, J.F.; Liu, H.H.; Sun, K.; Fan, H.Y.; Zhu, S.X.; Ran, H.; Zheng, C. Mannan oligosaccharides reduce the carbo footprint by decreasing methane emission and nitrogen excretion in Xiangdong black goats. *Anim. Feed Sci. Technol.* **2025** *328*, 116464. [CrossRef]
14. Wang, R.; Si, H.B.; Wang, M.; Lin, B.; Deng, J.P.; Tan, L.W.; Liu, W.X.; Sun, X.Z.; Teklebrhan, T.; Tan, Z.L. Effects of elementa magnesium and magnesium oxide on hydrogen, methane and volatile fatty acids production in in vitro rumen batch culture *Anim. Feed Sci. Technol.* **2019**, *252*, 74–82. [CrossRef]
15. AOAC International. *Official Methods of Analysis of AOAC International*, 18th ed; Oxford University Press: Oxford, UK, 2012.
16. Soest, P.J.V.; Robertson, J.B.; Lewis, B.A. Methods for dietary fiber, neutral detergent fiber, and nonstarch polysaccharides i relation to animal nutrition. *J. Dairy Sci.* **1991**, *74*, 83–97. [CrossRef]
17. National Research Council. *Nutrient Requirements of Small Ruminants: Sheep, Goats, Cervids, and New World Camelids*; The Nationa Academies Press: Washington, DC, USA, 2007.
18. Broderick, G.A.; Kang, J.H. Automated simultaneous determination of ammonia and total amino acids in ruminal fluid an in vitro media. *J. Dairy Sci.* **1980**, *63*, 64–75. [CrossRef]
19. Yang, X.H.; Liu, T.; Zhou, J.W.; An, L.J.; Pan, F.M.; Zhang, H.; Wang, X.J.; Xu, G.Y.; Zheng, C. Effects of Bacillus subtilis addition t milk replacer on growth performance, nutrient digestibility, intestinal crobiota, and short-chain fatty acid concentration of H lambs. *Anim. Feed Sci. Technol.* **2025**, *319*, 116175. [CrossRef]
20. Tang, L.; Li, Y.; Srivathsan, A.; Gao, Y.; Li, K.; Hu, D.; Zhang, D. Gut microbiomes of endangered przewalski's horse population in short- and long-term captivity: Implication for species reintroduction based on the soft-release strategy. *Front. Microbiol.* **2020** *11*, 363. [CrossRef]
21. Li, X.; Brejnrod, A.D.; Ernst, M.; Rykær, M.; Herschend, J.; Olsen, N.M.C.; Dorrestein, P.C.; Rensing, C.; Sørensen, S.J. Heavy meta exposure causes changes in the metabolic health-associated gut microbiome and metabolites. *Environ. Int.* **2019**, *126*, 454–467 [CrossRef]
22. Kozich, J.J.; Westcott, S.L.; Baxter, N.T.; Highlander, S.K.; Schloss, P.D. Development of a dual-index sequencing strategy an curation pipeline for analyzing amplicon sequence data on the miseq illumina sequencing platform. *Appl. Environ. Microbio* **2013**, *79*, 5112–5120. [CrossRef]
23. Magoc, T.; Salzberg, S.L. Flash: Fast length adjustment of short reads to improve genome assemblies. *Bioinformatics* **2011**, *27* 2957–2963. [CrossRef]

https://doi.org/10.3390/ani1608115

24. Caporaso, J.G.; Kuczynski, J.; Stombaugh, J.; Bittinger, K.; Bushman, F.D.; Costello, E.K.; Fierer, N.; Pena, A.G.; Goodrich, J.K.; Gordon, J.I.; et al. Qiime allows analysis of high-throughput community sequencing data. *Nat. Methods* **2010**, *7*, 335–336. [CrossRef]
25. Rognes, T.; Flouri, T.; Nichols, B.; Quince, C.; Mahe, F. Vsearch: A versatile open source tool for metagenomics. *PeerJ* **2016**, *4*, e2584. [CrossRef] [PubMed]
26. Wang, Z.; Yu, Y.; Li, X.; Xiao, H.; Zhang, P.; Shen, W.; Wan, F.; He, J.; Tang, S.; Tan, Z.; et al. Fermented soybean meal replacement in the diet of lactating holstein dairy cows: Modulated rumen fermentation and ruminal microflora. *Front. Microbiol.* **2021**, *12*, 625857. [CrossRef] [PubMed]
27. Wang, Z.; Elekwachi, C.; Jiao, J.; Wang, M.; Tang, S.; Zhou, C.; Tan, Z.; Forster, R.J. Changes in metabolically active bacterial community during rumen development, and their alteration by rhubarb root powder revealed by 16s rrna amplicon sequencing. *Front. Microbiol.* **2017**, *8*, 159. [CrossRef] [PubMed]
28. Fomenky, B.E.; Chiquette, J.; Bissonnette, N.; Talbot, G.; Chouinard, P.Y.; Ibeagha-Awemu, E.M. Impact of saccharomyces cerevisiae boulardii cncmi-1079 and lactobacillus acidophilus bt1386 on total lactobacilli population in the gastrointestinal tract and colon histomorphology of holstein dairy calves. *Anim. Feed Sci. Technol.* **2017**, *234*, 151–161. [CrossRef]
29. Abhauer, K.P.; Wemheuer, B.; Daniel, R.; Meinicke, P. Tax4fun: Predicting functional profiles from metagenomic 16s rrna data. *Bionformatics* **2015**, *31*, 2882–2884.
30. Liu, H.; Puchala, R.; LeShure, S.; Gipson, T.A.; Flythe, M.D.; Goetsch, A.L. Effects of lespedeza condensed tannins alone or with monensin, soybean oil, and coconut oil on feed intake, growth, digestion, ruminal methane emission, and heat energy by yearling alpine doelings. *J. Anima Sci.* **2019**, *97*, 885–899. [CrossRef]
31. Greening, C.; Geier, R.; Wang, C.; Woods, L.C.; Morales, S.E.; McDonald, M.J.; Rushton-Green, R.; Morgan, X.C.; Koike, S.; Leahy, S.C.; et al. Diverse hydrogen production and consumption pathways influence methane production in ruminants. *ISME J.* **2019**, *13*, 2617–2632. [CrossRef]
32. Li, J.; Zhao, S.; Meng, Z.; Gao, Y.; Miao, J.; Mao, S.; Jin, W. Effects of fumarate and nitroglycerin on in vitro rumen fermentation, methane and hydrogen production, and on microbiota. *Biology* **2023**, *7*, 1011. [CrossRef]
33. Olijhoek, D.W.; Hellwing, A.L.F.; Brask, M.; Weisbjerg, M.R.; Højberg, O.; Larsen, M.K.; Dijkstra, J.; Erlandsen, E.J.; Lund, P. Effect of dietary nitrate level on enteric methane production, hydrogen emission, rumen fermentation, and nutrient digestibility in dairy cows. *J. Dairy Sci.* **2016**, *99*, 6191–6205. [CrossRef]
34. Veneman, J.B.; Muetzel, S.; Hart, K.J.; Faulkner, C.L.; Moorby, J.M.; Perdok, H.B.; Newbold, C.J. Does dietary mitigation of enteric methane production affect rumen function and animal productivity in dairy cows? *PLoS ONE* **2015**, *10*, e104282. [CrossRef]
35. Asanuma, N.; Yokoyama, S.; Hino, T. Effects of nitrate addition to a diet on fermentation and microbial populations in the rumen of goats, with special reference to selenomonas ruminantium having the ability to reduce nitrate and nitrite. *Anim. Sci. J.* **2014**, *86*, 378–384. [CrossRef] [PubMed]
36. Meller, R.A.; Wenner, B.A.; Ashworth, J.; Gehman, A.M.; Lakritz, J.; Firkins, J.L. Potential roles of nitrate and live yeast culture in suppressing methane emission and influencing ruminal fermentation, digestibility, and milk production in lactating jersey cows. *J. Dairy Sci.* **2019**, *102*, 6144–6156. [CrossRef] [PubMed]
37. Rebelo, L.R.; Luna, I.C.; Messana, J.D.; Araujo, R.C.; Simioni, T.A.; Granja-Salcedo, Y.T.; Vito, E.S.; Lee, C.; Teixeira, I.A.M.A.; Rooke, J.A.; et al. Effect of replacing soybean meal with urea or encapsulated nitrate with or without elemental sulfur on nitrogen digestion and methane emissions in feedlot cattle. *Anim. Feed Sci. Techol.* **2019**, *257*, 114293. [CrossRef]
38. van Zijderveld, S.M.; Gerrits, W.J.J.; Apajalahti, J.A.; Newbold, J.R.; Dijkstra, J.; Leng, R.A.; Perdok, H.B. Nitrate and sulfate: Effective alternative hydrogen sinks for mitigation of ruminal methane production in sheep. *J. Dairy Sci.* **2010**, *93*, 5856–5866. [CrossRef]
39. Seyyedsalehi, M.S.; Mohebbi, E.; Tourang, F.; Sasanfar, B.; Boffetta, P.; Zendehdel, K. Association of dietary nitrate, nitrite, and n-nitroso compounds intake and gastrointestinal cancers: A systematic review and meta-analysis. *Toxics* **2023**, *11*, 190. [CrossRef]
40. Yin, H.; Li, B.; Wang, X.; Xi, Z. Effect of ammonium and nitrate supplies on nitrogen and sucrose metabolism of cabernet sauvignon. *J. Sci. Food Agric.* **2020**, *100*, 5239–5250. [CrossRef]
41. Almeida, K.V.; Santos, G.T.; Daniel, J.L.P.; Nino-de-Guzman, C.; Amaro, F.X.; Sultana, H.; Arriola, K.G.; Araujo, R.C.; Vyas, D. Effects of nitrate sources on in vitro methane production and ruminal fermentation parameters in diets differing in starch degradability. *Anim. Feed Sci. Technol.* **2022**, *292*, 115437. [CrossRef]
42. Santos, K.C.; Carvalho, F.F.R.; Carriero, M.M.; Magalhães, A.L.R.; Batista, A.M.V.; Fagundes, G.M.; Bueno, I.C.S. Use of different carbohydrate sources associated with urea and implications for in vitro fermentation and rumen microbial populations. *Anim. Prod. Sci.* **2020**, *60*, 1028–1038. [CrossRef]
43. Park, T.; Wijeratne, S.; Meulia, T.; Firkins, J.L.; Yu, Z. The macronuclear genome of anaerobic ciliate entodinium caudatum reveals its biological features adapted to the distinct rumen environment. *Genomics* **2021**, *113*, 1416–1427. [CrossRef]
44. Tan, Z.; Liu, J.; Wang, L. Factors affecting the rumen fluid foaming performance in goat fed high concentrate diet. *Front. Vet. Sci.* **2024**, *11*, 1299404. [CrossRef]

https://doi.org/10.3390/ani16081150

45. Bokharaeian, M.; Ghoorchi, T.; Toghdory, A.; Esfahani, I.J. The dose-dependent role of sage, clove, and pine essential oils in modulating ruminal fermentation and biohydrogenation of polyunsaturated fatty acids: A promising strategy to reduce methane emissions and enhance the nutritional profile of ruminant products. *Appl. Sci.* **2023**, *13*, 11605. [CrossRef]
46. Mahmoudi-Abyane, M.; Alipour, D.; Moghimi, H.R. Effects of different sources of nitrogen on performance, relative population of rumen microorganisms, ruminal fermentation and blood parameters in male feedlotting lambs. *Animal* **2019**, *14*, 1438–1446. [CrossRef] [PubMed]
47. Ungerfeld, E.M.; Aedo, M.F.; Martínez, E.D.; Saldivia, M. Inhibiting methanogenesis in rumen batch cultures did not increase the recovery of metabolic hydrogen in microbial amino acids. *Microorganisms* **2019**, *7*, 115. [CrossRef] [PubMed]
48. Hassan, F.U.; Guo, Y.; Li, M.; Tang, Z.; Peng, L.; Liang, X.; Yang, C. Effect of methionine supplementation on rumen microbiota, fermentation, and amino acid metabolism in in vitro cultures containing nitrate. *Microorganisms* **2021**, *9*, 1717. [CrossRef]
49. Zhou, X.; Stevens, M.J.A.; Neuenschwander, S.; Schwarm, A.; Kreuzer, M.; Bratus-Neuenschwander, A.; Zeitz, J.O. The transcriptome response of the ruminal methanogen methanobrevibacter ruminantium strain m1 to the inhibitor lauric acid. *BMC Res. Notes* **2018**, *11*, 135. [CrossRef]
50. Clemens, E.T.; Johnson, R.R. Influence of dietary nitrogen source, concentrate level and biuret level in sheep on the adaptation of rumen microorganisms to biuret as a nonprotein nitrogen source. *J. Nutr.* **2023**, *103*, 1406–1413. [CrossRef]
51. Patra, A.K.; Yu, Z. Effects of garlic oil, nitrate, saponin and their combinations supplemented to different substrates on in vitro fermentation, ruminal methanogenesis, and abundance and diversity of microbial populations. *J. Appl. Microbiol.* **2015**, *119*, 127–138. [CrossRef]
52. Guyader, J.; Tavendale, M.; Martin, C.; Muetzel, S. Dose-response effect of nitrate on hydrogen distribution between rumen fermentation end products: An in vitro approach. *Anim. Prod. Sci.* **2016**, *56*, 224–330. [CrossRef]
53. Guyader, J.; Ungerfeld, E.M.; Beauchemin, K.A. Redirection of metabolic hydrogen by inhibiting methanogenesis in the rumen simulation technique (rusitec). *Front. Microbiol.* **2017**, *8*, 393. [CrossRef]
54. Villar, M.L.; Hegarty, R.S.; Clay, J.W.; Smith, K.A.; Godwin, I.R.; Nolan, J.V. Dietary nitrate and presence of protozoa increase nitrate and nitrite reduction in the rumen of sheep. *J. Anim. Physiol. Anim. Nutr.* **2020**, *104*, 1242–1255. [CrossRef] [PubMed]
55. Orskov, E.R. Starch digestion and utilization in ruminants. *J. Anim. Sci.* **1986**, *63*, 1624–1633. [CrossRef] [PubMed]
56. Nozhevnikova, A.N.; Simankova, M.V.; Parshina, S.N.; Kotsyurbenko, O.R. Temperature characteristics of methanogenic archaea and acetogenic bacteria isolated from cold environments. *Water Sci. Technol.* **2018**, *44*, 41–48. [CrossRef]
57. Feng, X.Y.; Dijkstra, J.; Bannink, A.; van Gastelen, S.; France, J.; Kebreab, E. Antimethanogenic effects of nitrate supplementation in cattle: A meta-analysis. *J. Dairy Sci.* **2020**, *103*, 11375–11385. [CrossRef]
58. Zhao, L.; Meng, Q.; Ren, L.; Liu, W.; Zhang, X.; Huo, Y.; Zhou, Z. Effects of nitrate addition on rumen fermentation, bacterial biodiversity and abundance. *Anim. Biosci.* **2015**, *28*, 1433–1441. [CrossRef]
59. Romero-Pérez, A.; Alemu, A.; Araujo, R.; Beauchemin, K. 98 effect of slow release nitrate and essential oil on animal performance and methane emissions from feedlot cattle fed high-grain finishing diets. *J. Anim. Sci.* **2018**, *96*, 409. [CrossRef]
60. Troy, S.M.; Duthie, C.A.; Hyslop, J.J.; Roehe, R.; Ross, D.W.; Wallace, R.J.; Waterhouse, A.; Rooke, J.A. Effectiveness of nitrate addition and increased oil content as methane mitigation strategies for beef cattle fed two contrasting basal diets. *J. Anim. Sci.* **2015**, *93*, 1815–1823. [CrossRef]
61. Bharanidharan, R.; Tomple, B.M.; Lee, J.; Nirmal Athauda, A.A.K.; Huh, S.; Hong, W.; Kim, N.Y.; Lim, D.H.; Kim, J.G.; Kim, K.H.; et al. Effects of dietary supplementation of nitrate on enteric methane production, performance, and rumen microbiome of hanwoo steers. *J. Anim. Sci.* **2025**, *103*, skaf109. [CrossRef]
62. Ripple, W.J.; Smith, P.; Haberl, H.; Montzka, S.A.; McAlpine, C.; Boucher, D.H. Ruminants, climate change and climate policy. *Nat. Clim. Change* **2013**, *4*, 2–5. [CrossRef]
63. Marais, B.J.P.; Therion, J.J.; Mackie, R.I.; Kistner, A.; Dennison, C. Effect of nitrate and its reduction products on the growth and activity of the rumen microbial population. *Br. J. Nutr.* **1988**, *59*, 301–313. [CrossRef]
64. Zhou, Z.; Yu, Z.; Meng, Q. Effects of nitrate on methane production, fermentation, and microbial populations in in vitro ruminal cultures. *Bioresour. Technol.* **2012**, *103*, 173–179. [CrossRef]
65. Morgavi, D.P.; Forano, E.; Martin, C.; Newbold, C.J. Microbial ecosystem and methanogenesis in ruminants. *Animal* **2010**, *4*, 1024–1036. [CrossRef]
66. Iwamoto, M.; Asanuma, N.; Hino, T. Ability of selenomonas ruminantium, veillonella parvula, and wolinella succinogenes to reduce nitrate and nitrite with special reference to the suppression of ruminal methanogenesis. *Anaerobe* **2002**, *8*, 209–215. [CrossRef]

7. Nolan, J.V.; Hegarty, R.S.; Hegarty, J.; Godwin, I.R.; Woodgate, R. Effects of dietary nitrate on fermentation, methane production and digesta kinetics in sheep. *Anim. Prod. Sci.* **2010**, *50*, 801–806. [CrossRef]
8. Patra, A.K.; Yu, Z. Combinations of nitrate, saponin, and sulfate additively reduce methane production by rumen cultures in vitro while not adversely affecting feed digestion, fermentation or microbial communities. *Bioresour. Technol.* **2014**, *155*, 129–135. [CrossRef]

https://doi.org/10.3390/ani16081150

animals

MDPI

Article

The Seasonal Dietary Shift and Niche Resilience of Yaks on the Qinghai–Tibetan Plateau

Shuai Zheng [1,2], Yuning Ru [2], Mengyuan Xu [2], Yushou Ma [1], Yuan Ma [1,*] and Na Guo [1,2,*]

1 Qinghai Provincial Key Laboratory of Adaptive Management on Alpine Grassland, Academy of Animal Science and Veterinary Medicine, Qinghai University, Xining 810016, China
2 School of Life Sciences, Shandong University, Qingdao 266237, China
* Correspondence: mayuan608@163.com (Y.M.); guon@sdu.edu.cn (N.G.)

Simple Summary

Ecological theory posits that seasonal food availability drives dramatic temporal variation in dietary niche breadth, yet alternative paradigms offer contrasting predictions about its effect. Traditional Optimal Foraging Theory predicts that herbivores should broaden their dietary niche when resources become scarce. However, it remains unclear how high-altitude specialists like the yak adjust their fine-scale foraging strategies to navigate the extreme transition from resource-rich summers to resource-limited autumns. By sequencing plant DNA from fecal samples, we analyzed the seasonal diet dynamics of grazing yaks on the Qinghai–Tibetan Plateau. We found a dramatic dietary shift, as instead of expanding their niche under autumn scarcity, yaks significantly contracted a dietary niche, switching from diverse high-quality forbs to specialized abundant graminoids. These results suggest that yaks function as adaptive mixed feeders, employing a selective energy-maximization strategy to persist in one of the world's harshest environments.

Abstract

Understanding how herbivores adjust their foraging strategies to cope with seasonal resource fluctuations has been central to the nutritional ecology. Optimal Foraging Theory (OFT) predicts that generalists should broaden their dietary niche when high-quality resources are scarce, but empirical evidence in extreme environments remains poorly understood. We used *trn*L-P6 metabarcoding of fecal samples (n = 10/season) and a local reference library of 120 plant species to quantify diet composition and niche metrics of free-ranging yaks (*Bos grunniens*) on the Qinghai–Tibetan Plateau in June (summer) and October (autumn) 2024. Yaks shifted from a diverse, forb-dominated diet (e.g., Polygonaceae, Rosaceae) in summer to a specialized diet dominated by grasses in autumn. Although dietary richness and total niche width (TNW) decreased in autumn, phylogenetic diversity remained stable, indicating a strategic shift to distinct evolutionary lineages to ensure functional redundancy. Furthermore, food network analyses demonstrated a transformation from a flexible, modular foraging pattern in summer to a highly integrated, synchronized network in autumn. These findings suggest that under the distinct quality–quantity trade-off of high-altitude ecosystems, yaks adopt an energy-maximization strategy by minimizing search costs, aligning with the opportunity cost constraints of OFT, rather than randomly expanding their niche. This insight into selective foraging dynamics is critical for developing sustainable grazing practices that accommodate the natural adaptive behaviors of alpine herbivores.

Keywords: foraging strategy; dietary selection; niche breadth; yak; Qinghai–Tibetan Plateau

 https://doi.org/10.3390/ani1604061

1. Introduction

Understanding the foraging strategies of large herbivores is fundamental to ecology, as food resources are often the primary limiting factor for ungulate populations [1,2]. While the physiological functions of nutrient intake, such as body maintenance, thermoregulation, and reproduction, are well established [3], the specific drivers behind food choice remain complex. Herbivores do not simply consume available biomass; they must navigate a landscape filled with heterogeneous plant species, varying phenological states that influence forage palatability [4,5]. This challenge is particularly acute in seasonal environments, where animals must adapt to drastic fluctuations in forage properties and availability. Consequently, how herbivores balance their dietary needs against these environmental constraints directly dictates their body condition, survival, and ultimately, population dynamics [6,7].

Optimal Foraging Theory (OFT) and niche theory provide contrasting frameworks for predicting how herbivores adjust their dietary niche width in response to seasonal resource fluctuations. The classic "prey model" of OFT posits that diet breadth is inversely related to the availability of high-quality resources; as preferred foods become depleted during lean seasons, foragers are predicted to broaden their diet to include lower-quality items [8,9]. Conversely, niche-based models suggest that resource scarcity intensifies competition, forcing individuals or species to specialize on narrower subsets of resources to minimize overlap, whereas abundant resources during "fat seasons" allow for dietary generalization and overlap [10,11]. Consistent with niche-based models, meta-analyses have found that dietary partitioning was generally greatest when food abundance was low, suggesting that competition for limited food drives partitioning [12]. Despite these theoretical advances, empirical evidence remains conflicting [13,14]. Recent research on ungulates (e.g., bushbucks) in heterogeneous landscapes implies that these adjustments are more nuanced which is driven by energetic trade-offs, herbivores may narrow their realized niches to maximize energy intake under fluctuating conditions [15]. However, it remains unknown foraging strategies for ruminant prioritize in extreme environments where resource transitions are rapid and severe.

Tibetan Plateau, with a mean elevation exceeding 4000 m above sea level, is the world's largest and highest alpine grassland [16]. It represents one of the world's harshest grazing environments, characterized by an exceptionally short plant growing season of only 90 to 120 days [17]. Despite these environmental challenges, the plateau supports unique biodiversity. The yak (*Bos grunniens*) is an iconic symbol of high-altitude, providing the essential resources, including meat, milk, and fuel for millions of nomadic pastoralists in high-altitude [18,19]. For these large herbivores, the growing season is critical for survival, and previous studies have emphasized the general impact of seasonality on alpine herbivores [20]; the fine-scale adaptive strategies occurring during the phenological transition from summer to autumn remains poorly understood. Unlike the growing season in summer, autumn presents a unique quality–quantity trade-off, while food biomass remains relatively high, nutritional quality declines rapidly due to plant senescence [21]. This transitional period is decisive, as yaks must maximize energy intake to accumulate sufficient reserves for the upcoming non-growing season. Therefore, assessing whether yaks adopt a niche-expansion strategy or maintain selective segregation in response to these autumnal resource fluctuations is critical for understanding their survival mechanisms, with potential applications in resource management and environmental monitoring.

Traditional methods in determining herbivores diet include direct observation, stomach or fecal microbiology, infrared spectrophotometry, and stable isotope analyses [22].

https://doi.org/10.3390/ani16040613

These methods are effort-intensive and often lack the resolution to construct the fine-scale diet composition and the subtle patterns of co-occurrence among plant items, for grazing yaks, which are highly mobile and feed on diverse plant species [23–26]. Compared to the traditional methods, DNA metabarcoding may reveal more detailed information of herbivores' diet and offers an opportunity to reconstruct herbivore diets webs [1,20].

Therefore, we quantified the seasonal dynamics of dietary niches and food web of grazing yaks in summer and autumn using DNA metabarcoding. We hypothesized that (1) yak diet composition and food web exhibit seasonal turnover, and (2) the seasonal shifts drive an expansion of niche width or niche differentiation. Results from this study could provide insights into the behavioral mechanisms yaks employ to navigate the trade-off between resource quality and quantity.

2. Materials and Methods

2.1. Study Site

The study was conducted in Huangyuan County (3100 m a.s.l.; 36.9133° N, 100.9346° E), located on the Qinghai–Tibetan Plateau. The region is characterized by a typical continental monsoon climate, with a mean annual temperature of −0.4 °C (ranging from −11.9 to 11.7 °C) and mean annual precipitation of 505 mm in 2024. Approximately 80% of the rainfall occurs during the plant growing season from May to October. The dominant vegetation is alpine meadow. The dominant and associated species included *Elymus nutans, Festuca rubra, Poa crymophila, Stipa aliena, Carex przewalskii, Kobresia humilis, Saussurea superba, Oxytropis kansuensis, Ajania tenuifolia, Potentilla nivea, Tibetia himalaica*.

2.2. Sample Collection

Fecal samples were used as a non-invasive proxy to determine the diet of free-ranging adult yaks. To ensure a representative sampling, samples were collected from different herds grazing within the study area. Sampling was conducted in June (summer, growing season, n = 10) and October (autumn, senescent season, n = 10) in 2024. Fecal samples were collected immediately after observing different individuals defecate, with each defecation event sampled only once, using sterile spatulas to minimize ground contamination, placed into sterile tubes, and flash-frozen in liquid nitrogen. All samples were transported to the laboratory and stored at −80 °C until DNA extraction.

2.3. Local DNA Reference Library Construction

To ensure accurate taxonomic identification of dietary plants, we constructed a local DNA reference library. A comprehensive botanical survey was conducted across the study area, and voucher specimens of all encountered plant species (n = 120) were collected. Plant identification was confirmed by expert botanists [27]. For each species, total genomic DNA was extracted from silica-dried leaf material. The *trn*L (UAA) intron P6 loop region was amplified using the c/d primers [28] and sequenced via the Sanger sequencing. All curated sequences were compiled into a local, high-quality reference database for subsequent taxonomic assignment. These associated DNA data were archived in the National Library of Medicine (NCBI).

2.4. DNA Extraction, Amplification, and Sequencing

Total genomic DNA was extracted from 0.2 g of each fecal sample using the QIAamp® Fast DNA Stool Mini Kit (Qiagen, Valencia, CA, USA), following the manufacturer's protocol [1,20]. Extraction blanks were included throughout the process to monitor for potential contamination. DNA concentration and quality were assessed using a NanoDrop 2000 spectrophotometer (Thermo Scientific, Wilmington, DE, USA) and 1.5% agarose gel electrophoresis, respectively [1]. The P6 loop of the chloroplast trnL (UAA) intron was amplified using the primers *trn*L-g (5′-GGGCAATCCTGAGCCAA-3′) and *trn*L-h (5′-CCATTGAGTCTCTGCACCTATC-3′) [28]. PCR amplicons were purified, pooled, and sequenced on an Illumina NovaSeq 6000 platform, generating 2 × 300 bp paired-end reads.

2.5. Bioinformatic and Statistical Analyses

The raw sequence data was processed using the OBITools4 pipeline [29]. Reads were assigned to their respective samples using *ngsfilter*. Low-quality sequences (mean quality score <30) and those with ambiguous nucleotides were discarded. Identical reads were dereplicated into unique sequences using *obiuniq*. A stringent filtering process was then applied using *obiclean* to remove putative PCR and sequencing errors, excluding variants with low abundance (≤1000 reads total), short length (≤10 bp), or those differing by only one nucleotide from a much more abundant sequence (≤0.05%). Sequences with low identity (≤95%) to our local reference database were also removed. Taxonomic assignment of the final, high-quality molecular operational taxonomic units (mOTUs) was performed against local reference library with a 100% identity criterion. The relative read abundance (RRA) of each plant taxon was used to quantify dietary composition. To standardize sequencing depth, the mOTU table was rarefied to 2000 reads per sample. This process was repeated 10 times, and the averaged RRA values were used for all subsequent ecological analyses to ensure robustness [1,30].

Dietary diversity was assessed using Hill numbers of order q = 0 (species richness) and q = 1 (the exponential of the Shannon index), calculated with the *hilldiv* v1.5.3 R package. Dietary phylogenetic diversity was calculated using the *picante* v1.8.2 package. A phylogenetic tree of identified plant taxa was constructed by aligning the *trn*L sequences with *MAFFT* v7.525 and inferring the tree with *FastTree* v2.1.11 [31,32]. We calculated the Mean Phylogenetic Distance (MPD) to evaluate the phylogenetic breadth of the diet. The diet-plant network was visualized using a bipartite graph with the *bipartite* v2.23 R package [33].

We constructed a Bray–Curtis dissimilarity matrix based on the dietary mOTU data and performed a non-metric multidimensional scaling (NMDS) analysis using the *vegan* v2.7.2 package in R. Pairwise differences in dietary composition between seasons were tested using PERMANOVA [34]. We constructed and visualized plant co-occurrence networks to explore potential associations among dietary components. Based on the relative abundance matrix, pairwise Spearman correlations were calculated using the *Hmisc* v5.2.3 package. Significant correlations ($|r| > 0.6$, $p < 0.05$) were retained to build undirected networks with the *igraph* v 2.1.4 package, and the networks were visualized using *ggraph* v2.2.2 [35–37].

3. Results

3.1. Seasonal Dietary Diversity and Composition

Total 20 samples included 510 unique trnL-P6 sequences (mOTUs) from 44 plant families. The diet dietary of yaks exhibited a dramatic seasonal shift between summer and autumn. Dietary richness (Hill number q = 0) was significantly higher in summer than in autumn (144.8 ± 1.5 vs. 109 ± 2.0, $p < 0.05$), dietary diversity (q = 1) and phylogenetic diversity (MPD) had no significant difference between summer and autumn (Figure 1A–C), while aboveground biomass of alpine grassland was higher in autumn than in summer (Figure 1D).

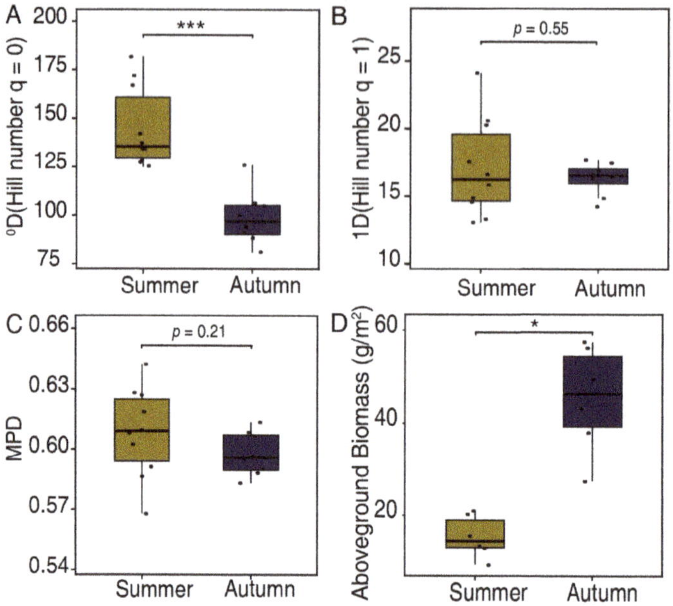

Figure 1. Dietary diversity of yaks and above-ground biomass of alpine grassland vegetation in both summer and autumn. (**A**) dietary richness (0D). (**B**) dietary richness and evenness (1D). (**C**) mean pairwise phylogenetic distance (MPD, overall clustering of food plant taxa on the phylogenetic tree). (**D**) aboveground biomass of the study area where yak foraging. Statistical significance was assessed using the Wilcoxon test (* $p < 0.05$, *** $p < 0.0001$).

Examination of quantitative food-web interactions revealed seasonal shifts regarding main dietary plant families between summer and autumn. At the family level, the top 10 plant taxa in terms of relative read abundance (RRA) across both seasons included Rosaceae, Polygonaceae, Fabaceae, Poaceae, Cyperaceae, Asteraceae, Caryophyllaceae, Ranunculaceae, Liliaceae, and Primulaceae in both summer and autumn. The diet composition in summer was dominated by forbs, with Rosaceae (36%), Polygonaceae (21%), and Fabaceae (12%) being the most abundant families. In contrast, mean RRA of Poaceae (grasses) was greatest (39%), less for Asteraceae (13%), lesser for Cyperaceae (3.6%) and Polygonaceae (2.4%) in autumn (Figure 2). To determine which plant taxa most contributed to the dietary niche between summer and winter in yaks, LEfSe analysis was performed. Rosaceae and Polygonaceae were the significant diet biomarkers in summer, whereas Poaceae and Apiaceae characterized in autumn (Figure 3). Overall, the yaks transitioned from a forb-dominated diet in summer to grass dominated diet in autumn.

https://doi.org/10.3390/ani16040613

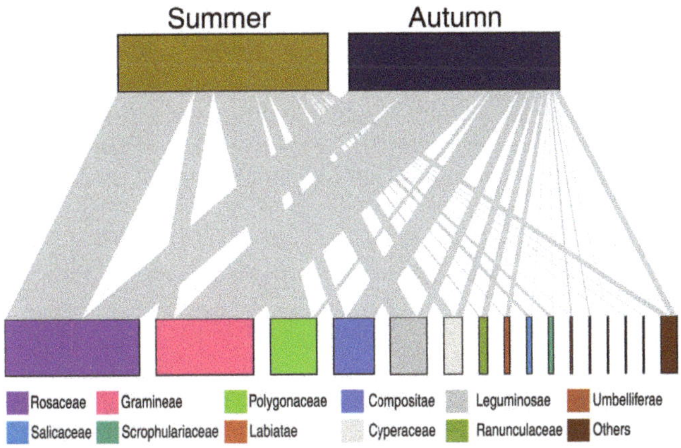

Figure 2. Dietary composition of yak in both summer and autumn. Bipartite networks depicting the dietary composition of yaks. Lines connect yaks (**upper bar**) to dietary plant sequences (**lower bar**), which are colored by plant family. The width of the upper bar represents the overall dietary composition, the links represent the proportional contribution of each plant family, and the width of the lower bar represents the total intake proportion of each plant family.

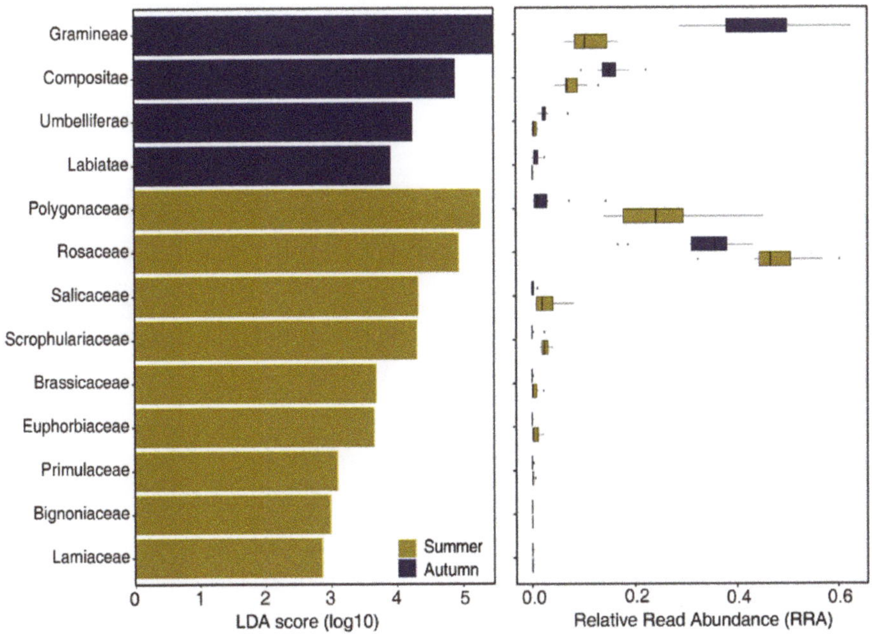

Figure 3. Key plant taxa between month and its relative read abundance (RRA). The diagram in the left was the results of LEfSe analysis on dietary data of yak in two seasons, showing the linear discriminant analysis (LDA) scores (threshold = 2) of food components consumed by yaks in different seasons, along with boxplots of the relative read abundance (RRA) for the corresponding plant families.

3.2. Seasonal Food Plant Partitioning

Nonmetric multidimensional scaling (NMDS) based on dietary dissimilarity revealed clear seasonal partitioning of plant taxa consumed between summer and autumn among individual fecal samples (PERMANOVA: F = 23.85, R^2 = 0.57, p = 0.001) (Figure 4A).

https://doi.org/10.3390/ani16040613

These results are consistent with the pairwise Pianka niche-overlap index (Figure 4B), which showed a mean value of 0.8730 ± 0.0122 in summer compared to 0.9135 ± 0.0107 in autumn. The total niche breadth (TNW) was significantly higher in summer than in autumn (1.87 vs. 1.67, $p = 0.034$) (Figure 5A), consistent with the higher dietary species richness observed. Within-individual diversity (WID) was lower in autumn than in summer (Figure 5B), whereas among-individual diversity (AID) showed the opposite trend, increasing in autumn (Figure 5C). Together with the Pianka index results, the higher AID/TNW ratio (degree of specialization) in autumn indicates that as food resource quantity and quality declined, individual differentiation became more pronounced despite the increase in overall dietary overlap.

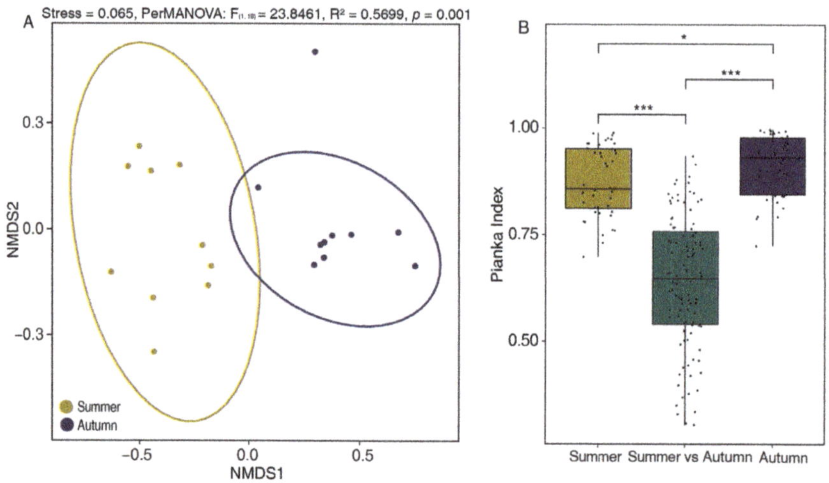

Figure 4. Seasonal variation in yak dietary composition. (**A**) Non-metric multidimensional scaling (NMDS) analysis based on Bray–Curtis similarity of dietary data. (**B**) Pianka index values representing dietary niche overlap among months (* $p < 0.05$, *** $p < 0.0001$); values closer to 1 indicate higher similarity, while values closer to 0 indicate lower overlap. The colors represent paired Pianka index values within/among months.

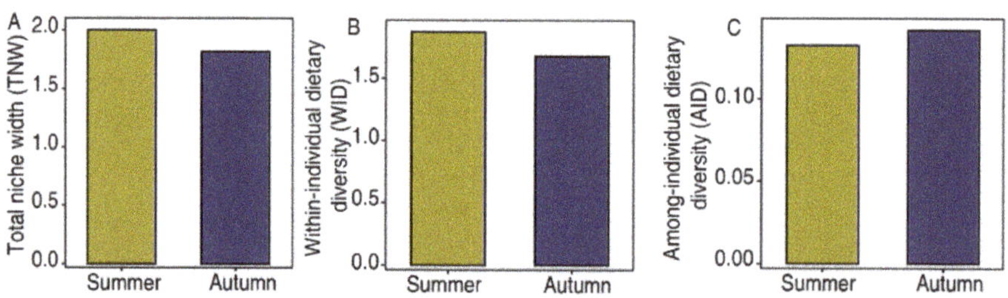

Figure 5. Variation of niche breadth in yaks between summer and autumn. (**A**) Total Niche Width (TNW, reflecting the overall niche width). (**B**) Within Individual Diversity (WID, reflecting the proportion of individual niche width in the total niche width of the population). (**C**) Among Individual Diversity (AID, reflecting the proportion of among-individual differences in total niche breadth).

3.3. Food-Webs Roles of Plant Taxa

The co-occurrence networks of dietary plant composition revealed a distinct seasonal shift in foraging strategy between summer and autumn (Figure 6). The dietary network in summer was sparser and more modular, characterized by fewer edges, higher modularity

https://doi.org/10.3390/ani16040613

and a higher proportion of positive correlations. Although Rosaceae was the most abundant family, Polygonaceae served as a key hub species, connecting different dietary modules. This modular pattern indicates that yaks evolved a more opportunistic and flexible foraging strategy, switching between different patches of preferred forbs, rather than consuming resources in a synchronized manner. Food web in autumn was denser and more integrated, exhibiting increased edge density, a higher clustering coefficient, and greater assortativity. It was characterized by a large, interconnected core of co-consumed plants, suggesting that yaks consistently consume a core set of available grasses and forbs together. Notably, a negative correlation between Poaceae and Fabaceae within food network indicates avoidance foraging pattern.

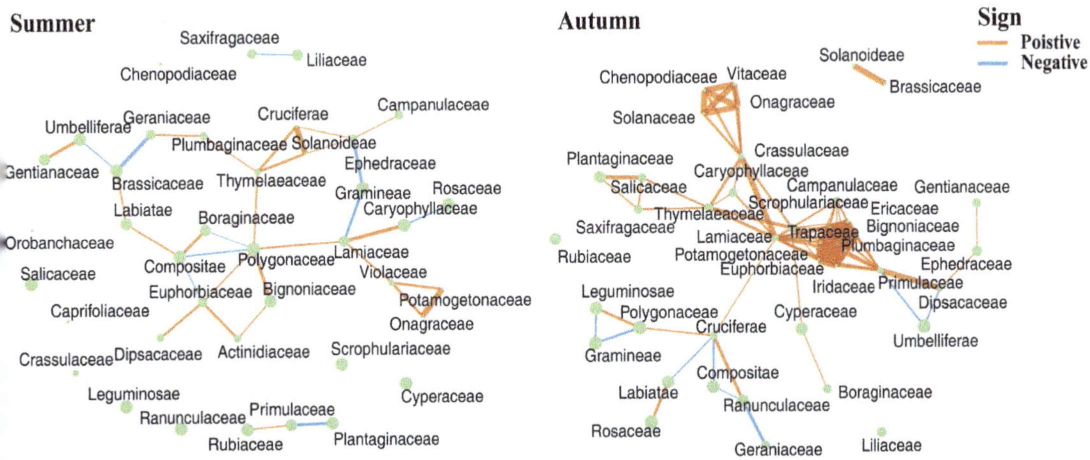

Figure 6. Co-occurrence network of dietary components in the yaks. Each node represents a distinct plant taxon detected in the yak diet. Node size indicates the relative read abundance of the corresponding plant family (nonlinear scaling). Edge width reflects the strength of the co-occurrence relationship, and edge color denotes the sign of the correlation (positive or negative).

4. Discussion

Although seasonal dietary plasticity is well-documented in generalist herbivores, empirical investigation into how population-level niche dynamics emerge from individual foraging strategies during critical phenological transitions remains limited. This lack of knowledge is particularly critical for alpine ungulates, which are compelled to optimize their foraging strategies amidst the severe seasonal trade-off between declining forage quality and abundant standing biomass. By leveraging the high taxonomic resolution of DNA metabarcoding, we demonstrate that free-grazing yaks exhibit a fundamental reconfiguration of their dietary niche between summer and autumn. We show that yaks transition from a diverse, forb-dominated diet (e.g., Rosaceae and Polygonaceae) in summer to a specialized, grass- and sedge-dominated diet (e.g., Poaceae) in autumn. This shift is driven by two complementary mechanisms: a contraction of the total niche breadth (TNW) coupled with an unexpected increase in individual specialization (AID/TNW) and network connectivity. Our results suggest that yaks adopt a context-dependent energy-maximization strategy, with narrowing their niche to focus on abundant and high-fiber resources, while maintaining distinct individual foraging patterns to mitigate competition. These findings suggest a potentially generalizable framework for understanding how the realized dietary niches of alpine herbivores are constrained by the dynamic interplay of seasonal phenology and physiological plasticity.

https://doi.org/10.3390/ani16040613

4.1. Seasonal Dietary Diversity and Niche Breadth

The results reveal a distinct seasonal pattern in dietary richness and diversity driven by the phenological trade-off between resource quality and quantity. Although above-ground biomass was found to be higher in autumn than in summer, likely due to the accumulation of lignified graminoid stems, yaks displayed a contraction in their dietary niche in autumn. Specifically, dietary taxonomic richness peaked in summer, whereas autumn was characterized by narrowing niche, despite phylogenetic diversity showing no significant difference. This seasonal contraction can be interpreted through the framework of OFT, particularly regarding opportunity costs [8,38]. In the resource-rich summer, the abundance of high-quality vegetation allows yaks to be selective, foraging on a wide richness of plants. However, in autumn, as diverse forbs senesce and become scarce, the opportunity cost of searching for these rare, nutrient-dense items becomes prohibitively high. Consequently, yaks minimize search effort by shifting their focus to the most reliable and abundant food sources such as grasses. Although the metabolic energy of grasses declines in autumn, their high biomass makes them the most economical energy source in helps of gut microbiome [21,39]. This strategy supports the theoretical prediction that animals facing higher starvation risk or search costs will prioritize intake over nutritional quality [40], resulting in the observed reduction in dietary richness and niche breadth in autumn.

Variations in dietary profiles often reflect underlying disparities in nutritional quality, driven by factors such as individual physiological requirements and environmental heterogeneity [41]. Although direct nutritional assessment was not conducted in the present study, our previous investigation into the annual nutritional dynamics of alpine forage [20], together with established ecological principles suggest that summer diets provide peak nutrition, whereas high-quality resources become progressively depleted in winter [42]. In this study, yaks prioritized ephemeral, nutrient-dense resources in summer. This foraging behavior closely parallels that of the North American bison, which similarly concentrates on high-quality graminoids and forbs during the productive season to meet the metabolic demands of a large body size [43,44]. Crucially, by exploiting fast-growing vegetation in summer, herbivores can accumulate somatic energy reserves while preserving persistent forage stocks for the winter range. This mechanism, often termed the 'storage effect', serves to mitigate seasonal resource bottlenecks and alleviate intraspecific competition [45,46]. These findings underscore the morphophysiological plasticity of large herbivores in response to environmental variability [10]. Deciphering the limits of this flexibility is vital for adaptive management in the face of rapid climate change [47,48], where scalable, data-driven approaches are essential for revealing previously obscured patterns of dietary adaptation in wildlife populations.

4.2. Seasonal Shifts in Diet Composition

Yaks diet vary seasonally, and available evidence suggests that our results are likely to conservatively describe dietary separation. In summer, yaks heavily utilized nutritious forbs, particularly from the Polygonaceae and Rosaceae families (e.g., *Polygonum sibiricum*, *P. viviparum*, *Potentilla anserina*). These species green up quickly and provide high protein content, facilitating rapid compensatory growth and fat accumulation before winter. However, the consumption of these diverse forbs reflects a nutritional strategy, yaks prioritize digestible energy and protein over the metabolic cost of detoxification. Many of these plants are rich in plant secondary metabolites (PSMs) [21,49–52]. The ability to tolerate these compounds is likely attributed to the yak's symbiotic gut microbiota, which functions as a 'biological shield' by metabolizing potentially harmful PSMs (e.g., tannins and alkaloids) into non-toxic compounds [53]. This co-evolved capacity enables yaks to exploit nutrient-

https://doi.org/10.3390/ani1604061

dense resources in the growing season that would otherwise be unpalatable to less adapted herbivores. In autumn, grasses and sedges became most common in yak diet, which are common throughout alpine grassland and provide lower-quality resources in autumn [21] that yaks switch to when higher-quality resources are most limited. The seasonal shifts in diet composition further highlight remarkable physiological adaptations to the quality–quantity trade-off. As ruminants indigenous to the Qinghai–Tibetan Plateau, yaks harbor a specialized and highly efficient gut microbiome capable of degrading complex structural carbohydrates (i.e., cellulose and hemicellulose) [54]. These microbial communities ferment high-fiber substrates into volatile fatty acids (VFAs), serving as the primary energy source for the host [55,56]. This enhanced fermentation capacity allows yaks to maximize nutrient extraction from abundant fiber-rich plants, thereby mitigating nutritional limitations and maintaining high intake rates to support fat reserves before the onset of the harsh winter.

Large herbivorous mammals are typically categorized along a functional continuum, ranging from grazers that specialize on monocots (mainly grasses) to browsers that consume dicots, with mixed feeders consuming substantial amounts of both [57,58]. While yaks are commonly described as grazers within traditional foraging guild, our result suggests mixed-feeding strategies may be common in high-altitude large herbivores. This dietary plasticity reflects an adaptive response to specific biogeographical variables of the regions, particularly the intense seasonal fluctuations and winter snow cover [59]. Consistent with recent insights from Yellowstone ecosystem [60], our data indicate that in the extreme environments, extrinsic environmental forcing often exerts a stronger influence on diet. Consequently, strict dietary specialization may be less common than previously assumed in highly seasonal habitats. Instead, dietary generalization serves as a dominant survival strategy, allowing herbivores to navigate the complex spatiotemporal dynamics of plant phenology and nutritional availability [61].

4.3. Seasonal Dietary Niche Separation and Food Webs

Seasonality clearly influenced the predominance of cluster assignments for both the plants and animals in food network. Ecologists have long debated the drivers of population niche width, contrasting the Niche Variation Hypothesis (NVH), which posits that population niche expansion stems from individual specialization [10], with Optimal Foraging Theory (OFT), which assumes conspecifics broaden their diets uniformly under resource limitation [8]. Typically, both theories predict niche expansion under scarcity. Unlike previous findings in moose (Alces alces), which tend to maintain a generalist strategy or expand their niche under scarcity [46], food scarcity did not drive niche expansion but rather contraction in yaks. The divergent trends observed in TNW and AID between seasons suggest a distinct foraging strategy. In summer, the high TNW and modular network structure reflect an opportunistic generalist strategy consistent with OFT in resource-rich environments However, the autumn dynamics present a fascinating deviation. While TNW contracted due to resource senescence, the increase in AID suggests that individuals differentiated their foraging paths within this constrained niche. This aligns with the NVH's premise that intraspecific competition drives individual specialization, likely mediated by searching and handling disparate food items (e.g., scarce forbs vs. abundant grasses).

This individual partitioning is mirrored in the transformation of the food co-occurrence network. The modular structure observed in summer reflects a flexible foraging strategy where yaks exploit various high-quality patches without strong coupling. In contrast, the dense, integrated network in autumn suggests a constrained synchronization, where the population is forced to converge on a core set of grasses and sedges. Crucially, the negative correlation between Poaceae and Fabaceae in the autumn network highlights a stark functional trade-off. As suggested by foraging theory [62], individuals are constrained

https://doi.org/10.3390/ani16040613

by cognition or physiology from maximizing intake of all food types simultaneously. In autumn, yaks focus on maximizing biomass intake from abundant but low-quality grasses and sedges and may lack the time or foraging efficiency to seek out rare, high-protein legumes, and vice versa. This trade-off drives the observed niche partitioning, preventing complete dietary overlap even when resources are scarce.

These findings have implications for grassland management. Since the autumn diet shifts towards fibrous grasses and sedges, the primary nutritional bottleneck becomes protein limitation [63,64]. Therefore, management interventions may shift from generic supplementation to strategic nitrogen delivery during the phenological transition. Providing catalytic protein sources (e.g., urea-molasses blocks) specifically in autumn would maintain microbial fermentation efficiency, allowing yaks to better utilize the abundant dry grass biomass and minimize body weight loss during the cold season.

Low sample size may not sufficiently uncover the full range of food items that yaks consumed. The dominant dietary components and main feeding interactions in the communities were most likely captured, allowing robust characterization of broad seasonal shifts in yak dietary niches and food-web structure. Although the dietary niche of yaks exhibited clear shifts between summer and autumn, the limited temporal sampling restricts inference about fine-scale transition dynamics, such as early versus late summer or early versus late autumn dietary variation. Further incorporating larger sample sizes and finer temporal sampling across the Qinghai–Tibetan Plateau will refine our understanding of yak foraging dynamics and trophic interactions.

5. Conclusions

This study characterizes the seasonal diet niche and network dynamic of yaks on the Qinghai–Tibetan Plateau. We observed that yaks shifted from a forb-dominated diet in summer to a diet primarily composed of grasses and sedges in autumn. Instead of expanding their dietary niche in response to reduced diet resources in the autumn, yaks appeared to narrow their niche while maintain stable dietary phylogenetic diversity. Yaks exhibited consistent foraging patterns, characterized by distinct plant co-occurrence networks. These findings provide insights into the dietary plasticity of large herbivores in alpine grassland and highlight the importance of aligning grazing management with plant phenology.

Author Contributions: N.G. conceived and designed the study. N.G., Y.R., Y.M. (Yushou Ma); Y.M. (Yuan Ma); and M.X. conducted DNA analyses. N.G. and S.Z. performed bioinformatic analyses and analyses the data. N.G. and S.Z. drafted the manuscript. All authors have read and agreed to the published version of the manuscript.

Funding: This research was funded by Chief Scientist Program of Qinghai Province (grant no. 2024-SF-101), the Open Project of Qinghai Provincial Key Laboratory of Adaptive Management on Alpine Grassland (grant no. 2023-GHSYS-KF-02), the State Key Laboratory of Hulless Barley and Yak Germplasm Resources and Genetic Improvement (grant no. XZNKY-CZ-2022-016-10), Natural Science Foundation Youth Program of Shandong Province (grant no. ZR2024QC346), Gansu Youth Science and Technology Fund (grant no. 22JR5RA521).

Institutional Review Board Statement: All procedures on yaks in this study were approved by the Committee of Animal Ethics of Shandong University (Protocol number SYDWLL-2024-060).

Informed Consent Statement: Not applicable.

Data Availability Statement: The raw data supporting the conclusions of this article will be made available by the authors on request.

Conflicts of Interest: The authors declare no conflicts of interest.

https://doi.org/10.3390/ani16040613

References

1. Pansu, J.; Hutchinson, M.C.; Anderson, T.M.; te Beest, M.; Begg, C.M.; Begg, K.S.; Bonin, A.; Chama, L.; Chamaillé-Jammes, S.; Coissac, E.; et al. The generality of cryptic dietary niche differences in diverse large-herbivore assemblages. *Proc. Natl. Acad. Sci. USA* **2022**, *119*, e2204400119. [CrossRef]
2. Ogutu, J.O.; Owen-Smith, N. ENSO, rainfall and temperature influences on extreme population declines among African Savanna ungulates. *Ecol. Lett.* **2003**, *6*, 412–419. [CrossRef]
3. Spitzer, R.; Coissac, E.; Cromsigt, J.P.; Felton, A.M.; Fohringer, C.; Landman, M.; Neumann, W.; Raubenheimer, D.; Singh, N.J.; Taberlet, P. Macro-nutritional balancing in a circumpolar boreal ruminant under winter conditions. *Funct. Ecol.* **2023**, *37*, 1256–1268. [CrossRef]
4. Parikh, G.L.; Forbey, J.S.; Robb, B.; Peterson, R.O.; Vucetich, L.M.; Vucetich, J.A. The influence of plant defensive chemicals, diet composition, and winter severity on the nutritional condition of a free-ranging, generalist herbivore. *Oikos* **2017**, *126*, 196–203. [CrossRef]
5. Shipley, L.A.; Blomquist, S.; Danell, K. Diet choices made by free-ranging moose in northern sweden in relation to plant distribution, chemistry, and morphology. *Can. J. Zool.* **1998**, *76*, 1722–1733. [CrossRef]
6. Couturier, S.; Côte, S.D.; Huot, J.; Otto, R.D. Body-condition dynamics in a northern ungulate gaining fat in winter. *Can. J. Zool.* **2009**, *87*, 367–378. [CrossRef]
7. White, R.G. Foraging patterns and their multiplier effects on productivity of Northern ungulates. *Oikos* **1983**, *40*, 377–384. [CrossRef]
8. Stephens, D.W.; Krebs, J.R. *Foraging Theory*; Princeton University Press: Princeton, NJ, USA, 1986; p. 6.
9. Schoener, T.W. The controversy over interspecific competition: Despite spirited criticism, competition continues to occupy a major domain in ecological thought. *Am. Sci.* **1982**, *70*, 586–595.
10. Bolnick, D.I.; Svanbäck, R.; Fordyce, J.A.; Yang, L.H.; Davis, J.M.; Hulsey, C.D.; Forister, M.L. The ecology of individuals: Incidence and implications of individual specialization. *Am. Nat.* **2003**, *161*, 1–28. [CrossRef]
11. Robinson, B.W.; Wilson, D.S. optimal foraging, specialization, and a solution to Liem's paradox. *Am. Nat.* **1998**, *151*, 223–235. [CrossRef]
12. Porter, C.K.; Golcher-Benavides, J.; Benkman, C.W. Seasonal patterns of dietary partitioning in vertebrates. *Ecol. Lett.* **2022**, *25*, 2463–2475. [CrossRef]
13. Dostine, P.L.; Franklin, D.C. A comparison of the diet of three finch species in the Yinberrie Hills Area, Northern Territory. *Emu—Austral Ornithol.* **2002**, *102*, 159–164. [CrossRef]
14. Smith, J.M. Optimization theory in evolution. *Annu. Rev. Ecol. Syst.* **1978**, *9*, 31–56. [CrossRef]
15. Walker, R.H.; Hutchinson, M.C.; Potter, A.B.; Becker, J.A.; Long, R.A.; Pringle, R.M. Mechanisms of individual variation in large herbivore diets: Roles of spatial heterogeneity and state-dependent foraging. *Ecology* **2023**, *104*, e3921. [CrossRef]
16. Zhu, Q.; Chen, H.; Peng, C.; Liu, J.; Piao, S.; He, J.-S.; Wang, S.; Zhao, X.; Zhang, J.; Fang, X. An early warning signal for grassland degradation on the Qinghai-Tibetan Plateau. *Nat. Commun.* **2023**, *14*, 6406. [CrossRef]
17. Speakman, J.R.; Chi, Q.; Ołdakowski, Ł.; Fu, H.; Fletcher, Q.E.; Hambly, C.; Togo, J.; Liu, X.; Piertney, S.B.; Wang, X.; et al. Surviving winter on the Qinghai-Tibetan Plateau: Pikas suppress energy demands and exploit yak feces to survive winter. *Proc. Natl. Acad. Sci. USA* **2021**, *118*, e2100707118. [CrossRef] [PubMed]
18. FAO Statistical. *FAO Statistical Yearbook 2014: Asia and the Pacific Food and Agriculture*; Food and Agriculture Organization of the United Nations Regional Office for Asia and the Pacific: Bangkok, Thailand, 2014.
19. Qiu, Q.; Zhang, G.; Ma, T.; Qian, W.; Wang, J.; Ye, Z.; Cao, C.; Hu, Q.; Kim, J.; Larkin, D.M.; et al. The yak genome and adaptation to life at high altitude. *Nat. Genet.* **2012**, *44*, 946–949. [CrossRef]
20. Guo, N.; Wu, Q.; Shi, F.; Niu, J.; Zhang, T.; Degen, A.A.; Fang, Q.; Ding, L.; Shang, Z.; Zhang, Z. Seasonal dynamics of diet-gut microbiota interaction in adaptation of yaks to life at high altitude. *npj Biofilms Microbiomes* **2021**, *7*, 38. [CrossRef] [PubMed]
21. Long, R. Feed value of native forages of the Tibetan Plateau of China. *Anim. Feed Sci. Tech.* **1999**, *80*, 101–113. [CrossRef]
22. Papachriston, T.G.; Dziba, L.E.; Provenza, F.D. Botanical composition determination of range herbivore diets: A review. *Small Rumin. Res.* **2005**, *59*, 141–156.
23. Soininen, E.M.; Valentini, A.; Coissac, E.; Miquel, C.; Gielly, L.; Brochmann, C.; Brysting, A.K.; Sønstebø, J.H.; Ims, R.A.; Yoccoz, N.G. Analysing diet of small herbivores: The efficiency of dna barcoding coupled with high-throughput pyrosequencing for deciphering the composition of complex plant mixtures. *Front. Zool* **2009**, *6*, 16. [CrossRef]
24. Bonin, M.; Dussault, C.; Taillon, J.; Lecomte, N.; Côté, S.D. Combining stable isotopes, morphological, and molecular analyses to reconstruct the diet of free-ranging consumers. *Ecol. Evol.* **2020**, *10*, 6664–6676. [CrossRef] [PubMed]
25. Kartzinel, T.R.; Chen, P.A.; Coverdale, T.C.; Erickson, D.L.; Kress, W.J.; Kuzmina, M.L.; Rubenstein, D.I.; Wang, W.; Pringle, R.M. DNA metabarcoding illuminates dietary niche partitioning by African large herbivores. *Proc. Natl. Acad. Sci. USA* **2015**, *112*, 8019–8024. [CrossRef]

https://doi.org/10.3390/ani16040613

26. Boyse, E.; Robinson, K.P.; Carr, I.M.; Valsecchi, E.; Beger, M.; Goodman, S.J. Inferring species interactions from co-occurrence networks with environmental dna metabarcoding data in a coastal marine food web. *Mol. Ecol.* **2025**, *34*, e17701. [CrossRef]
27. Feng, H.Y.; Pan, J.B. *Field Guide to Wild Plants of China (Qilian Mountains)*; The Commercial Press: Beijing, China, 2015.
28. Pierre, T.; Eeic, C.; François, P.; Lydovic, G.; Christian, M.; Alice, V.; Thierry, V.; Gérard, C.; Christian, B.; Eske, W. Power and limitations of the chloroplast trnL (UAA) intron for plant DNA barcoding. *Nucleic Acids Res.* **2007**, *35*, e14.
29. Caporaso, J.G.; Kuczynski, J.; Stombaugh, J.; Bittinger, K.; Bushman, F.D.; Costello, E.K.; Fierer, N.; Peña, A.G.; Goodrich, J.K Gordon, J.I.; et al. QIIME allows analysis of high-throughput community sequencing data. *Nat. Methods* **2010**, *7*, 335–336 [CrossRef]
30. Guyton, J.A.; Pansu, J.; Hutchinson, M.C.; Kartzinel, T.R.; Potter, A.B.; Coverdale, T.C.; Daskin, J.H.; da Conceição, A.G.; Pee M.J.; Stalmans, M.E. Trophic rewilding revives biotic resistance to shrub invasion. *Nat. Ecol. Evol.* **2020**, *4*, 712–724. [CrossRef]
31. Kuraku, S.; Zmasek, C.M.; Nishimura, O.; Katoh, K. aLeaves facilitates on-demand exploration of metazoan gene family trees on MAFFT sequence alignment server with enhanced interactivity. *Nucleic Acids Res.* **2013**, *41*, 22–28. [CrossRef]
32. Price, M.N.; Dehal, P.S.; Arkin, A.P. FastTree: Computing large minimum evolution trees with profiles instead of a distance matrix *Mol. Biol. Evol.* **2009**, *26*, 1641–1650. [CrossRef] [PubMed]
33. Dormann, C.; Frund, J.; Bluthgen, N.; Gruber, B. Indices, graphs and null models: Analyzing bipartite ecological networks. *J. Eco* **2009**, *2*, 7–24. [CrossRef]
34. Oksanen, J.; Blanchet, F.G.; Kindt, R.; Legendre, P.; Minchin, P.R.; O'hara, R.B.; Simpson, G.L.; Solymos, P.; Stevens, M.H.H Wagner, H. *Package 'Vegan.' Community Ecology Package*, version 2.0-10; R Foundation for Statistical Computing: Vienna, Austria 2013.
35. Harrell, F.E.; Dupont, C. *Hmisc: Harrell miscellaneous*, R package version 4.1-1; R Foundation for Statistical Computing: Vienna Austria, 2018.
36. Csárdi, G.; Nepusz, T. The Igraph software package for complex network research. *J. Complex. Syst.* **2006**, *1695*, 1–9.
37. Pedersen, T.L. *Ggraph: An Implementation of Grammar of Graphics for Graphs and Networks*, R Package Version 2.0.4; R Foundation for Statistical Computing: Vienna, Austria, 2020.
38. Nonacs, P. State dependent behavior and the marginal value theorem. *Behav. Ecol.* **2001**, *12*, 71–83. [CrossRef]
39. He, J.; Zhou, Y.; Kong, L.; Li, B.; Yu, J.; Gong, H.; Sun, X.; Cui, G. Effects of altitude change on nutritional quality and elymus nutans regularity in Qinghai-Tibet Plateau. *J. Northeast Agric. Univ. (Engl. Ed.)* **2021**, *28*, 69–76.
40. Barnett, C.A.; Bateson, M.; Rowe, C. State-dependent decision making: Educated predators strategically trade off the costs and benefits of consuming aposematic prey. *Behav. Ecol.* **2007**, *18*, 645–651. [CrossRef]
41. Gill, B.A.; Wittemyer, G.; Cerling, T.E.; Musili, P.M.; Kartzinel, T.R. Foraging history of individual elephants using dna metabarcoding. *Roy. Soc. Open Sci.* **2023**, *10*, 230337. [CrossRef]
42. DelGiudice, G.D.; Singer, F.J.; Seal, U.S.; Bowser, G. Physiological responses of yellowstone bison to winter nutritional deprivation *J. Wildl. Manag.* **1994**, *58*, 24–34. [CrossRef]
43. Geremia, C.; Merkle, J.A.; Eacker, D.R.; Wallen, R.L.; White, P.J.; Hebblewhite, M.; Kauffman, M.J. Migrating bison engineer the green wave. *Proc. Natl. Acad. Sci. USA* **2019**, *116*, 25707–25713. [CrossRef]
44. Letten, A.D.; Ke, P.-J.; Fukami, T. Linking modern coexistence theory and contemporary niche theory. *Ecol. Monogr.* **2017**, *87* 161–177. [CrossRef]
45. Chesson, P.L.; Warner, R.R. Environmental variability promotes coexistence in lottery competitive systems. *Am. Nat.* **1981**, *117* 923–943. [CrossRef]
46. Durant, S.M. Competition refuges and coexistence: An example from serengeti carnivores. *J. Anim. Ecol.* **1998**, *67*, 370–386 [CrossRef]
47. Potter, A.B.; Hutchinson, M.C.; Pansu, J.; Wursten, B.; Long, R.A.; Levine, J.M.; Pringle, R.M. Mechanisms of dietary resource partitioning in large herbivore assemblages: A plant-trait-based approach. *J. Ecol.* **2022**, *110*, 817–832. [CrossRef]
48. Radloff, F.G.; der Waal, C.V.; Bond, A.L. Extensive browsing by a conventional grazer? Stable carbon isotope analysis reveals extraordinary dietary flexibility among sanga cattle of north central namibia. *Isot. Environ. Health Stud.* **2013**, *49*, 318–324 [CrossRef]
49. Steinkellner, S.; Lendzemo, V.; Langer, I.; Schweiger, P.; Khaosaad, T.; Toussaint, J.P.; Vierheilig, H. Flavonoids and strigolactones in root exudates as signals in symbiotic and pathogenic plant-fungus interactions. *Molecules* **2007**, *12*, 1290–1306. [CrossRef [PubMed]
50. Zhang, X.; Long, R.; Dan, R.; Ding, X. Analysis of phenols content of three species of polygonum in alpine meadows. *J. Gansu Agric. Univ.* **2008**, *43*, 126–129.
51. Poutaraud, A.; Michelot-Antalik, A.; Plantureux, S. Grasslands: A source of secondary metabolites for livestock health. *J. Agric Food Chem.* **2017**, *65*, 6535–6553. [CrossRef]
52. Sun, P.; Hao, R.; Fan, F.; Wang, Y.; Zhu, F. Adaptation of high-altitude plants to plateau abiotic stresses: A case study of the Qinghai-Tibet Plateau. *Int. J. Mol. Sci.* **2025**, *26*, 2292. [CrossRef] [PubMed]

https://doi.org/10.3390/ani16040613

53. Mcsweeney, C.; Palmer, B.; McNeill, D.; Krause, D. Microbial interactions with tannins: Nutritional consequences for ruminants. *Anim. Feed Sci. Tech.* **2001**, *91*, 83–93. [CrossRef]
54. Mi, J.; Jing, X.; Ma, C.; Shi, F.; Cao, Z.; Yang, X.; Yang, Y.; Kakade, A.; Wang, W.; Long, R. A metagenomic catalogue of the ruminant gut Archaeome. *Nat. Commun.* **2024**, *15*, 9609. [CrossRef]
55. Gharechahi, J.; Vahidi, M.F.; Bahram, M.; Han, J.-L.; Ding, X.-Z.; Salekdeh, G.H. Metagenomic analysis reveals a dynamic microbiome with diversified adaptive functions to utilize high lignocellulosic forages in the cattle rumen. *ISME J.* **2021**, *15*, 1108–1120. [CrossRef]
56. Zhang, Z.; Xu, D.; Wang, L.; Hao, J.; Wang, J.; Zhou, X.; Wang, W.; Qiu, Q.; Huang, X.; Zhou, J.; et al. Convergent evolution of rumen microbiomes in high-altitude mammals. *Curr. Biol.* **2016**, *26*, 1873–1879. [CrossRef]
57. Gordon, I.J.; Prins, H.H.; Mallon, J.; Puk, L.D.; Miranda, E.B.; Starling-Manne, C.; van der Wal, R.; Moore, B.; Foley, W.; Lush, L. The ecology of browsing and grazing in other vertebrate taxa. In *The Ecology of Browsing and Grazing II*; Springer: Berlin/Heidelberg, Germany, 2019; pp. 339–404.
58. Pringle, R.M.; Abraham, J.O.; Anderson, T.M.; Coverdale, T.C.; Davies, A.B.; Dutton, C.L.; Gaylard, A.; Goheen, J.R.; Holdo, R.M.; Hutchinson, M.C.; et al. Impacts of large herbivores on terrestrial ecosystems. *Curr. Biol.* **2023**, *33*, 584–610. [CrossRef]
59. Codron, D.; Brink, J.S.; Rossouw, L.; Clauss, M.; Codron, J.; Lee-Thorp, J.A.; Sponheimer, M. Functional differentiation of african grazing ruminants: An example of specialized adaptations to very small changes in diet. *Biol. J. Linn. Soc.* **2008**, *94*, 755–764. [CrossRef]
60. Hoff, H.K.; Littleford-Colquhoun, B.L.; Kartzinel, R.Y.; Anderson, H.M.; Geremia, C.; McGarvey, L.M.; Segal, C.; Kartzinel, T.R. The apportionment of dietary diversity in wildlife. *Proc. Natl. Acad. Sci. USA* **2025**, *122*, e2502691122. [CrossRef] [PubMed]
61. Littleford-Colquhoun, B.L.; Geremia, C.; McGarvey, L.M.; Merkle, J.A.; Hoff, H.K.; Anderson, H.; Segal, C.R.; Kartzinel, R.Y.; Maywar, I.J.; Nantais, N. Body size modulates the extent of seasonal diet switching by large mammalian herbivores in yellowstone national park. *Roy. Soc. Open Sci.* **2024**, *11*, 240136. [CrossRef]
62. Araújo, M.S.; Bolnick, D.I.; Layman, C.A. The ecological causes of individual specialisation. *Ecol. Lett.* **2011**, *14*, 948–958. [CrossRef]
63. Wang, H.; Long, R.; Liang, J.B.; Guo, X.; Ding, L.; Shang, Z. Comparison of nitrogen metabolism in yak (*Bos grunniens*) and indigenous cattle (*Bos taurus*) on the Qinghai-Tibetan Plateau. *Asian-Australas. J. Anim. Sci.* **2011**, *24*, 766–773. [CrossRef]
64. Long, R.J.; Dong, S.K.; Hu, Z.Z.; Shi, J.J.; Dong, Q.M.; Han, X.T. Digestibility, Nutrient balance and urinary purine derivative excretion in dry yak cows fed oat hay at different levels of intake. *Lives. Prod. Sci.* **2004**, *88*, 27–32. [CrossRef]

https://doi.org/10.3390/ani16040613

animals

MDPI

Article

Multi-Omics Analysis Reveals Concentrate Supplementation Alleviates Body Weight Loss by Regulating Rumen Function in Lactating Tibetan Sheep During the Cold Season

Chao Yang [1], Qingling Ma [2], Jiancui Wang [2], Zhiyou Wang [2] and Shengzhen Hou [1,*]

1 State Key Laboratory of Plateau Ecology and Agriculture, Qinghai University, Xining 810016, China; yangchaocas@126.com

2 College of Agriculture and Animal Husbandry, Qinghai University, Xining 810016, China; qinglingma@126.com (Q.M.); wangjiancui2025@126.com (J.W.); wangzhiyou52@163.com (Z.W.)

* Correspondence: qhdxhsz@163.com

Simple Summary

Systematic studies on the effects of concentrate supplementation on the rumen function and microbiology alterations in Tibetan ewes during their lactation period are limited. Our multi-omics analysis showed that higher concentrate levels significantly alleviated body weight loss by regulating rumen histomorphology, enhancing cellulase activity, and altering the abundance of key bacteria (*Lachnospiraceae_XPB1014_group*, *Anaerovibrio*, *Ruminococcus*, and *Pseudobutyrivibrio*). Furthermore, concentrate supplementation levels altered the expression of genes (*TRPA1*, *EPHB1*, *GATA3*, *C4*, *ABCG2*, *THBS4*, and *TNFRSF11B*) involved in immune regulation, signal transduction, and nutrient digestion. Correlation analysis confirmed that these changes are interconnected, as indicated by significant associations between rumen bacteria, fermentation parameters, and digestive enzyme activities. Our findings clarify that concentrate supplementation mitigates weight loss by orchestrating functional improvements across the rumen ecosystem.

Abstract

The parturition season of grazing Tibetan ewes spans from October to March, a period that exacerbates the adverse impacts of nutrient-deficient herbage on milk yield, body condition, and postpartum recovery. To alleviate the weight loss of ewes during the cold seasons, we provided concentrate supplements at four levels (dry matter (DM) basis), 260 g (C1), 440 g (C2), 520 g (C3), and 610 g (C4), alongside a basal diet of grazed pasture. A total of 96 multiparous Tibetan ewes (third parity, body weight: 45.17 ± 3.69 kg (body weight (BW) were enrolled within 12–18 h postpartum and randomly allocated to four dietary groups ($n = 24$ ewes per group). We measured growth performance, ruminal histomorphology, fermentation parameters, and digestive enzymes. A multi-omics technique (16S rRNA gene sequencing and RNA-seq) was employed to investigate the mechanisms underlying alterations in ruminal function. The results showed that increasing the concentrate level decreased body weight loss and increased average dry matter intake ($p < 0.05$). Rumen morphology was significantly altered: papilla width and muscle layer thickness were greatest in the C4 group, whereas submucosal thickness was highest in the C1 group ($p < 0.05$). Cellulase activity was lowest in the C1 group ($p < 0.05$). Papilla width of lactating Tibetan ewes in the C4 group was higher ($p < 0.05$) than that in the C1 and C3 groups. Concentrate supplementation altered ruminal microbiota composition and diversity. Each group exhibited a distinct microbial signature: the C1 group was characterized by *Lachnospiraceae_XPB1014_group*, *Candidatus_Omnitrophus*, *Paenibacillus*, and *unclassified_Oligoflexaceae*; the C2 group was enriched in *Papillibacter*,

Anaerovibrio, *V9D2013_group*, and *unclassified_Peptococcaceae*; the C3 group was characterized by *unclassified_Bacteroidales_RF16_group*; and the C4 group was characterized by *Ruminococcus*, *Pseudobutyrivibrio*, and *Mitsuokella* ($p < 0.05$). Transcriptomic analysis identified differentially expressed genes (*TRPA1*, *EPHB1*, *GATA3*, *C4*, *ABCG2*, *THBS4*, and *TNFRSF11B*) that are predominantly involved in immune regulation, signal transduction, and nutrient digestion. The results of Spearman correlation analysis showed that *Anaerovibrio* was negatively correlated with propionate (r = −0.565, $p < 0.05$). However, it was positively correlated with the ratio of acetate and propionate (r = 0.579, $p < 0.05$). Moreover, *Lachnospiraceae_XPB1014_group* was negatively correlated with cellulase (r = −0.699, $p < 0.05$) and α-amylase (r = −0.514, $p < 0.05$). These findings suggest that the increasing concentrate supplementation alleviates body weight loss in lactating Tibetan sheep by orchestrating improvements in rumen histomorphology, digestive function, altering bacteria composition, and ruminal immune and modulating host epithelial gene expression.

Keywords: histomorphology; digestive enzymes; rumen fermentation; microbiota; lactating Tibetan sheep

1. Introduction

The Qinghai–Tibet Plateau—recognized as the "Roof of the World" and the "Third Pole"—is the highest and most extensive plateau on earth. Its extreme environmental conditions—including hypobaric hypoxia, severe cold, high UV-B radiation, and a low partial pressure of oxygen—impose stringent selective pressures on resident organisms [1]. Tibetan sheep (*Ovis aries*), an endemic livestock species on the plateau, are phenotypically and ecotypically stratified into three distinct topographical ecotypes: highland-, valley-, and oula-types [2]. These animals represent a critical indigenous genetic resource, having endured millennia of exposure to high-altitude, cryogenic, hypoxic, and high-irradiance conditions. Through long-term natural and artificial selection, they have evolved a suite of integrated physiological, morphological, and molecular adaptations. These include augmented erythropoiesis, blunted hypoxic pulmonary vasoconstrictive response, dense double fleece, and enhanced DNA-repair capacity, which collectively confer exceptional fitness within this harsh cryosphere environment [3].

The abbreviated pasture growing season typically leads to an insufficient nutrient supply for grazing ruminants, particularly acute from November to May, a period that consistently induces live-weight loss in Tibetan sheep within extensive grazing systems [4]. Crucially, the parturition season for grazing Tibetan ewes occurs from October to March of the subsequent year. This time exacerbates the adverse impacts of herbage nutrient deficits on both lactation performance and postpartum recovery [5]. Consequently, mitigating nutrient deficiency during lactation is a pivotal constraint on the production efficiency of the Tibetan sheep industry. Although herders routinely provide concentrate diets to lactating ewes to maximize nutrient intake, formulating the most cost-effective supplementation regimen remains a significant challenge. Moreover, the digestive efficiency of nutrients from pasture and concentrate supplements in the rumen critically impacts postpartum recovery in Tibetan ewes.

As the hallmark digestive organ of ruminants, the rumen underpins their adaptation to high-altitude ecosystems. Ruminal feed fermentation and hydrolysis are governed almost exclusively by a complex and dynamic microbial consortium [6]. This pregastric microbial ecosystem comprises bacteria, methanogenic archaea, ciliate protozoa, and anaerobic fungi, collectively mediating the degradation of dietary substrates [7]. Functionally, rumen microbes can be categorized by their primary metabolic roles, such as amylolytic, fibrolytic, and proteolytic [8]. Their catabolism yields volatile fatty acids (VFAs) and ammonia-N—key metabolites absorbed and utilized by the host. The composition and metabolic activity of this microbiome are modulated by diet composition, feeding regimen, ambient environment, health status, and host genotype [9]. Inter-species and inter-breed divergence in microbial profiles may therefore arise during environmental adaptation. For example, elevated ambient temperature increases the abundance of *Streptococcus bovis* and other amylolytic bacterial populations in Holstein cattle [10], while the same stressor reduces amylolytic populations in small ruminants [11]. In plateau-adapted herbivores, the gut—especially the rumen—harbors a highly diverse, cellulolytic microbiota that facilitates the utilization of low-quality forage and sustains short-chain fatty acid (SCFA)-dependent energy homeostasis under chronic nutrient scarcity [12]. However, the responsiveness of the rumen microbiome to graded levels of concentrate supplementation during the cold season, and the concomitant shifts in ruminal nutrient absorption and post-ruminal metabolism, remain to be elucidated.

In the current study, we used RNA-sequencing and 16S rDNA amplicon sequencing to investigate the ruminal function and the associated shifts in bacterial community structure and function in response to different levels of concentrate supplementation in lactating Tibetan sheep. Specifically, our objectives were 3-fold: (1) to quantify the extent to which incremental concentrate allowances mitigate body weight loss in lactating ewes; (2) to evaluate treatment-induced alterations in rumen histomorphometry, fermentation end-products, and the activity of key digestive enzymes; and (3) to elucidate the regulatory mechanisms by which concentrate level reshapes rumen microbial community structure and associated metabolic pathways. Our study identified the optimal concentrate inclusion rate for minimizing body weight loss and delineated the corresponding adaptive responses of rumen function and microbiota. This work aims to provide an evidence-based nutritional strategy to enhance the productivity of lactating Tibetan ewes during the cold seasons.

2. Materials and Methods

2.1. Animal and Experimental Design

The experiment was conducted at Jinzang Tibetan Sheep Breeding Professional Cooperative from October 2023 to January 2024, and all protocols were approved by the Animal Ethics Committee of Qinghai University (permit No. QUA-2023-0112).

A total of 96 multiparous Tibetan ewes (third parity, body weight (BW): 45.17 ± 3.69 kg were enrolled within 12–18 h postpartum and randomly allocated to four dietary groups (n = 24 ewes per group). Each group comprised three pens (replicates) with eight ewes per pen. To simulate voluntary dry matter intake (DMI) under natural grazing conditions, the metabolizable energy (ME) requirement was calculated using metabolic BW (BW0.75) according to the method described by Liu H. [13]. Grazing Tibetan sheep have been reported to consume 62.8 g dry matter (DM) kg/BW0.75 daily. Based on the mean metabolic BW of 17.37 kg recorded in the present study, the baseline DMI was estimated to be 1.10 kg DM/d per ewe, corresponding to 73.4 g crude protein (CP), 15 g ether extract (EE), and 8.18 MJ metabolic energy daily from native pasture. Accounting for the elevated nutrient demands of lactation, roughage allowance (DM basis) was increased to 1.40 kg per ewe per day (50:50 oat hay: wheat straw), isonitrogenous and isoenergetic with the above-mentioned pasture. On this basal roughage, ewes received daily concentrate supplements (DM basis) of 260 g (C1), 440 g (C2), 520 g (C3), and 610 g (C4) per ewe, respectively; ingredient composition and nutritional levels are presented in Table 1, and the ingredient composition and chemical composition of diets in each group was presented in Table S1.

Table 1. Ingredient composition and nutritional levels in concentrates for Tibetan ewes.

Item	Content
Ingredient, % DM basis	
Corn	32.00
Wheat	7.00
Palm meal	30.00
Soybean meal	4.00
Rapeseed meal	18.00
Cottonseed meal	2.00
Glucose	2.00
Medical stone	1.20
Sodium chloride	1.10
Sodium bicarbonate	0.35
$CaHPO_4$	0.35
Premix [1]	2.00
Total	100
Nutritional level [2]	
Dry matter, %	88.00
Digestible energy, MJ/kg	10.44
Crude protein, %	14.63
Ether extract, %	2.97
Neutral detergent fiber, %	26.66
Acid detergent fiber, %	13.53
Calcium, %	0.95
Phosphorus, %	0.55

[1] Prvided per kilogram of premix: $CuSO_4$ 0.25 g, $CoSO_4$ 0.08 g, $MnSO_4$ 0.67 g, ZnO 0.83 g, Na_2SeO_3 0.17 g, $Ca(IO_3)_2$ 0.25 g, Vitamine A 145,000 IU, Vitamine D_3 34,000 IU, Vitamine E 41.67 mg, antioxidant iron 0.83 g, ammonium nicotinate 83.33 mg; [2] Digestible energy is calculated based on the ingredients of the diet and their digestible energy content; other values of nutrients are determined by the Association of Official Analytical Chemists (AOAC) method.

Prior to the trial, all pens were thoroughly cleaned, disinfected, and equipped with creep barriers to separate lambs during concentrate feeding, enabling accurate individual recording of feed intake of the ewes. Ewes were offered feed in two equal meals at 08:30 and 16:40 h and had ad libitum access to fresh water. Lambs were temporarily removed from dams before the reception of roughage and concentrate; they were returned to the ewes only after the ewes completed consumption of the respective diets. Throughout the experimental period, lambs received exclusively the dam's milk and were not provided creep feed or milk replacer.

2.2. Sample Collection

At the end of animal feeding experiment, six experimental Tibetan ewes were selected from each group (two ewes from each pen within one group) for slaughter after a fast of 12 h. Anesthetization of the selected ewes was precisely carried out via an intravenous injection of sodium pentobarbital administered at a dosage of 50 mg/kg BW. Following the induction of anesthesia, exsanguination was performed until the ewes attained a surgical plane of anesthesia. Subsequently, the abdominal cavity was carefully opened, and the rumen was dissected and separated from the reticulum. The ruminal digesta was then filtered through four layers of gauze, and the resultant rumen fluid was collected in 5 mL cryogenic vials (Corning®, Corning, NY, USA). The collected rumen fluid was immediately subjected to freezing using liquid nitrogen and subsequently maintained at −80 °C for subsequent determination of volatile fatty acids and digestive enzymes. Ruminal tissue of the abdominal sac was dissected and washed in sterile culture dishes (Corning®, NY, USA) with ice-cold DNase/RNase-free ddH_2O (R1600, Solarbio LIFE SCIENCES, Beijing, China) until clean (3–4 times). The tissues were cut into small pieces, then a part of the tissues was transferred into 10% neutral buffered formalin for morphology observation; the other tissues were placed in a sterile sample bag (Whirl-Pak®, Pleasant Prairie, WI, USA), which was immediately snap-frozen in liquid nitrogen and stored at −80 °C for RNA extraction.

2.3. Histomorphology Analysis

The ruminal tissue samples fixed in the buffered formalin were washed and trimmed into small square pieces (0.5 mm × 0.5 mm), dehydrated in gradient ethanol, and finally embedded in paraffin wax. Five sections for each sample were sliced, installed on glass slides, and stained with eosin and hematoxylin. Three random straight papillae of each slide were selected to measure the papilla height, width, and surface area using a fluorescence microscope (Olympus, Tokyo, Japan) with a calibrated 10-fold eyepiece graticule. Furthermore, the thickness of the submucosal layer and muscle layer were measured.

2.4. Determination of Rumen Rermentation

The pH of ruminal fluid was assayed using a portable pH meter (PHS-1701, Shanghai Boqu Instrument Co., Ltd., Shanghai, China). Prior to measurement, meticulous calibration of the pH meter was carried out utilizing a standard buffer solution to ensure the accuracy and reliability of the pH readings. For each ruminal fluid sample, three repeated measurements were performed to obtain robust data. Regarding the determination of volatile fatty acids (VFAs) in ruminal fluid, a gas chromatography system (Agilent 7890A, Agilent Inc., Palo Alto, CA, USA) equipped with a capillary column (AE-FFAP, specifications: 30 m × 0.25 mm × 0.33 µm; ATECH Technologies Co., Lanzhou, China) was employed, following the procedures detailed in a previous study [14]. Specifically, ruminal fluid samples were thawed and subjected to centrifugation at 16,000× *g* for 10 m at 4 °C. Subsequently, 2 mL of the supernatant was carefully collected and mixed with 400 µL of 25% metaphosphoric acid. The mixture was left to stand for 4 h at 4 °C to allow for adequate precipitation of proteins and other macromolecules. After this incubation period, the sample underwent a second round of centrifugation at 16,000× *g* for 10 min at 4 °C. To an aliquot of 200 µL of the resulting supernatant, 200 µL of crotonic acid (10 g/L) was added as an internal standard. The sample was then filtered through a 0.45-µm filter to eliminate any residual particulates that might interfere with the chromatographic analysis. In terms of the gas chromatography operating conditions, the injector and detector temperatures were precisely set at 200 °C and 250 °C, respectively. The column temperature was programmed to increase from 45 °C to 150 °C at a heating rate of 20 °C/min and was maintained at 150 °C for an additional 5 min to ensure complete elution of all VFA components.

2.5. Measurement of Digestive Enzymes

Ruminal fluid samples were thawed at 4 °C in a refrigerator. Subsequently, the samples were centrifuged at 10,000× *g* for 5 min at 4 °C to obtain the supernatant. The activities of several key digestive enzymes in the supernatant, including cellulase, hemicellulase, α-amylase, and β-amylase, were assayed using corresponding biochemical reagent kits (AIDISHENG, Jiangsu Aidisheng Biological Technology Co., Ltd., Yancheng, Jiangsu, China). All operations were conducted in strict accordance with the manufacturer's instructions. Finally, the optical density values were measured with a microplate reader to determine the enzyme activities.

2.6. 16s rDNA Sequencing and Data Analysis

Total DNA was extracted from rumen fluid samples of 24 Tibetan ewes using the *SteadyPure* Stool DNA Extraction Kit (AG21036, Accurate Biology, Changsha, Hunan, China), following the manufacturer's protocol. The DNA concentration was determined with a Nanodrop-2000 spectrophotometer (Thermo Fisher Scientific, Waltham,

MA, USA), and the quality was evaluated by 2% agarose gel electrophoresis. The V3-V4 regions of the bacterial 16S rRNA gene were amplified via PCR using the forward primer 338F (5′-ACTCCTACGGGAGGCAGCA-3′) and reverse primer 806R (5′-GGACTACHVGGGTWTCTAAT-3′) [15]. The sequencing library was constructed using the Illumina Next®Ultra™ DNA Library Prep Kit (New England Biolabs, Ipswich, MA, USA). Both Qubit@ 2.0 Fluorometer and Agilent Bioanalyzer 2100 systems were employed to assess the library quality. Finally, the qualified amplicon libraries were sequenced on the Illumina NovaSeq PE250 platform (Illumina, San Diego, CA, USA) to generate 150 bp paired-end reads.

The processing of raw data commenced with quality filtering using Trimmomatic (v0.33) [16]. Following this, primer sequences were identified and excised via Cutadapt (v1.9.1). Subsequently, paired-end (PE) reads were assembled using USEARCH (v10), after which UCHIME (v8.1) was employed to eliminate chimeric sequences, resulting in high-quality clean reads [17]. These clean reads were then subjected to analysis through the Divisive amplicon denoising algorithm 2 (DADA2) method within the QIIME2 framework to generate a feature table and produce de-noised sequences [18]. Each representative feature sequence was annotated against the SILVA database (v138), and feature abundance normalization was performed based on the relative abundance of each sample across various taxonomic ranks [19]. QIIME2 was utilized to compute alpha diversity indices, including ACE, Chao1, Shannon, and Simpson, as well as beta diversity values [20]. Beta diversity was quantified using the unweighted or weighted Unifrac dissimilarity metric, Bray–Curtis, and Binary Jaccard distance, and Adonis analysis was conducted to evaluate the similarity of bacterial communities among the 24 samples. Finally, the linear discriminant analysis effect size (LEfSe) tool was applied to identify the effect size of species contributing to differences between samples using the Kruskal–Wallis test with a significance level of $p < 0.05$ and a linear discriminant analysis (LDA) score threshold of >3.5.

2.7. RNA Isolation, Transcriptomic Sequencing, and Bioinformatics Analysis

Total RNA was extracted from approximately 100 mg of frozen tissue ground in liquid nitrogen using the SteadyPure Universal RNA Extraction Kit (AG21017, Accurate Biology, Hunan, China), following the manufacturer's procedures. The integrity of the extracted RNA was assessed using the Agilent 2100 Bioanalyzer system (Agilent Technologies, Santa Clara, CA, USA), with an RNA Integrity Number (RIN) greater than 7.0 indicating sufficient quality for library construction. Subsequently, 1 μg of RNA per sample was utilized for cDNA library construction using the NEBNext UltraTM RNA Library Prep Kit (Illumina, San Diego, CA, USA) in accordance with the manufacturer's recommendations. The libraries were then amplified via PCR and validated using the Agilent 2100 Bioanalyzer system (Agilent Technologies, CA, USA). Finally, the cDNA libraries were sequenced on the Illumina HiSeq2000 platform (Illumina, CA, USA) to generate high-quality, 150 bp paired-end reads.

Raw data were subjected to quality control by FAST-QC (version 0.18.0) to obtain clean data after removing adapters and low-quality reads [21]. Clean data were then mapped to the ribosome RNA (rRNA) database to remove rRNA using Bowtie2 (version 2.2.8) [22]. We used HISAT2 (version 2.1.0) to align the clean reads to the sheep reference genome (ARS-UI_Ramb_v3.0) with default parameters [23]. The mapped reads of each sample were assembled using StringTie (version 1.3.1), and the expression of genes was estimated by transcripts per kilobase of exon model per million mapped reads (TPM) [24]. Principal Component Analysis (PCA) was conducted based on expressed gene data using R (R version 3.4.2) to assess the difference among the four groups. Differentially expressed genes (DEGs) were identified by DESeq2 package using the following parameters: fold change > 2 and false discovery rate (FDR) < 0.05 [25]. The Kyoto Encyclopedia of Genes and Genomes (KEGG) pathway enrichment for DEGs was performed on KOBAS software (version 3.0) [26], and pathways on level 3 underwent variance analysis in which adjusted $p < 0.05$ was considered as a significant difference. A Short Time-series Expression Miner (STEM) analysis was conducted to capture coordinated transcriptional responses across concentrate levels, and the analytical parameters were referred to the previous study [27].

2.8. Statistical Analysis

Raw data of rumen histomorphology, fermentation parameters, and the levels of digestive enzymes were preprocessed using Excel 2022 software, and statistical analyses were performed by the univariate analysis of variance (ANOVA), followed by Tukey's HSD test for multiple comparison in IBM SPSS Statistics 25.0. All data are shown as the mean ± standard error of the mean (SEM), and $p < 0.05$ was considered to be statistically significant.

3. Results

3.1. Growth Performance of Lactating Tibetan Sheep

According to the data in Table 2, the initial BW differed significantly ($p < 0.05$) among the experimental groups, and the BW on D34 and D68 did not affect by concentrate supplementation. The extent of average BW loss during the study period was decreased ($p < 0.05$) with increasing concentrate supplementation levels. Average dry matter intake was increased ($p < 0.05$) with increasing concentrate supplementation levels.

Table 2. Growth performance of lactating Tibetan sheep.

Item	Group				SEM	*p*-Value
	C1	C2	C3	C4		
Initial BW, kg	45.75 ^{a}	45.26 ^{a}	44.95 ^{b}	43.92 ^{b}	5.542	0.013
D34 BW, kg	41.57	44.78	44.86	44.68	4.641	0.543
D68 BW, kg	37.39	44.30	44.81	45.25	4.457	0.381
Average BW loss, kg	−0.123 ^{b}	−0.014 ^{a}	−0.002 ^{a}	0.020 ^{a}	0.083	<0.001
Average dry matter intake, kg	1.19 ^{d}	1.37 ^{c}	1.56 ^{b}	1.72 ^{a}	0.164	<0.001

BW, body weight; SEM, standard error of the mean. ^{a,b,c,d} Means of a row with no common superscript are significantly different ($p < 0.05$).

3.2. Histomorphological Analysis of Rumen

As shown in Table 3 and Figure 1, the level of concentrate supplementation significantly affected ($p < 0.05$) the rumen histomorphology, papilla width, submucosal thickness, and muscle layer thickness in lactating Tibetan ewes. Specifically, the papilla width was higher ($p < 0.05$) in the C4 group than that in the C1 and C2 groups. Conversely, submucosal thickness was higher ($p < 0.05$) in the C1 group than other groups. For muscle layer thickness, the C4 group exhibited a greater value ($p < 0.05$) than both the C1 and C2 groups, while the value in the C3 group was also higher ($p < 0.05$) than that in the C1 group.

Table 3. Rumen histological and morphological parameters of lactating Tibetan ewes.

Items	Group				SEM	p-Value
	C1	C2	C3	C4		
Papilla height, μm	930.01	947.49	1065.39	1054.08	461.70	0.612
Papilla width, μm	456.26 [b]	510.35 [b]	530.53 [ab]	646.35 [a]	226.09	0.008
Papilla surface area, mm^2	423.26	520.76	574.66	569.71	353.64	0.339
Submucosal thickness, μm	423.37 [a]	355.92 [b]	354.44 [b]	365.25 [b]	113.19	0.027
Muscle layer thickness, μm	1193.13 [c]	1219.41 [bc]	1398.37 [ab]	1579.55 [a]	435.72	<0.001

[a,b,c] Means of a row with no common superscript are significantly different ($p < 0.05$).

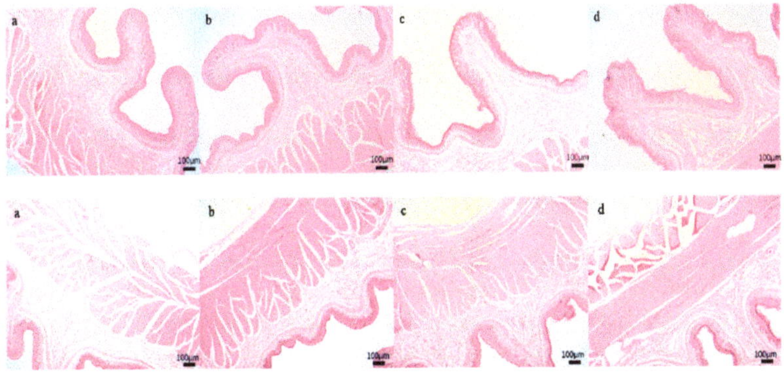

Figure 1. Rumen histomorphology of lactating Tibetan ewes (4×). (**a–d**) Rumen of lactating Tibetan ewes in the C1, C2, C3, and C4 group, respectively.

3.3. Rumen Fermentation Parameters

The level of concentrate supplementation had no significant effect ($p > 0.05$) on ruminal pH, molar ratio of acetate, propionate, butyrate, isobutyrate, valerate, isovalerate, total VFA, or the ratio of acetate and propionate (Table 4).

Table 4. Rumen fermentation parameters of lactating Tibetan ewes.

Item	Group				SEM	p-Value
	C1	C2	C3	C4		
pH	7.12	7.22	6.95	7.12	0.048	0.254
Acetate, %	68.80	68.30	66.70	65.99	0.505	0.158
Propionate, %	17.19	17.04	19.17	16.72	0.364	0.060
Butyrate, %	8.62	9.27	9.13	11.48	0.486	0.163
Isobutyrate, %	2.15	2.18	1.99	2.24	0.080	0.747
Valerate, %	0.86	0.81	0.78	0.91	0.026	0.341
Isovalerate, %	2.38	2.39	2.23	2.62	0.094	0.569
Total VFA, mmol/L	33.87	44.75	46.79	40.08	1.932	0.074
Acetate/Propionate	4.01	4.04	3.52	3.97	0.411	0.087

3.4. Ruminal Digestive Enzyme of Lactating Tibetan Sheep

The activity of ruminal digestive enzymes, including hemicellulose, α-amylase, and β-amylase, was not significantly affected ($p > 0.05$) by the levels of concentrate supplementation (Table 5). However, the activity of cellulase in the C1 group was lower than that in the C2 group ($p < 0.05$).

Table 5. The activity of digestive enzymes in the rumen of lactating Tibetan sheep.

Item	Group				SEM	*p*-Value
	C1	C2	C3	C4		
Cellulase, μg/h/mL	4.19 ^{b}	8.66 ^{a}	7.39 ^{ab}	7.78 ^{ab}	0.613	0.042
Hemicellulase, nmol/min/mL	60.38	78.36	55.24	59.69	3.634	0.107
α-amylase, μg/min/mL	316.26	328.17	349.96	336.29	9.278	0.655
β-amylase, mg/min/mL	2.16	2.70	4.19	5.96	0.602	0.101

^{a,b} Means of a row with no common superscript are significantly different ($p < 0.05$).

3.5. Microbial Composition of Rumen in Lactating Tibetan Sheep

16S rRNA sequencing of ruminal fluid samples from the four ewe groups produced 1,876,149 raw reads. Following quality control, 1,723,117 clean reads were obtained, with an average of 71,797 clean reads per sample. A total of 36,068 amplicon sequence variants (ASVs) were generated following quality control. The rarefaction curves confirmed that the sequencing data coverage was adequate to describe the ruminal microbial composition of lactating Tibetan ewes (Figure S1). To evaluate the effects of concentrate supplementary levels on ruminal microbiota, we assessed α-diversity and β-diversity. The α-diversity analysis revealed that concentrate supplementation levels did not affect ACE, Chao1, and Simpson indices. However, the Shannon index was higher ($p < 0.05$) in the C1 group than that in the C3 group (Figure 2A). Principal coordinate analysis (PCoA) based on unweighted and weighted UniFrac, Bray–Curtis, and Binary Jaccard distances revealed distinct clustering among the four treatment groups. This separation was confirmed to be statistically significant by ANOSIM analysis ($p = 0.007$, $p = 0.022$, $p = 0.001$, and $p = 0.001$, respectively; Figure 2B).

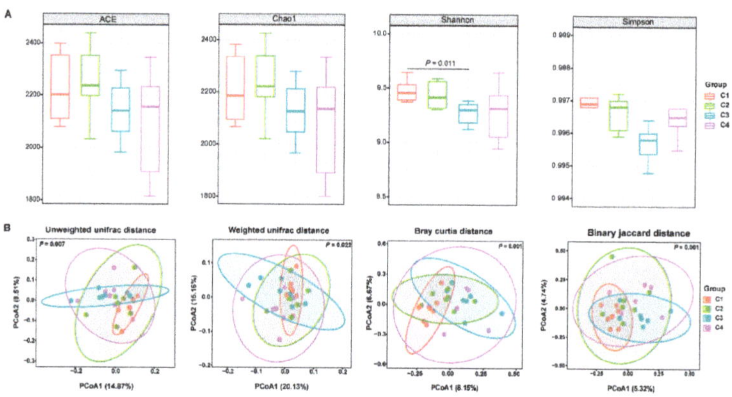

Figure 2. Rumen bacteria diversity of lactating Tibetan ewes. (**A**) Alpha diversity indices, including ACE, Chao1, Shannon, and Simpson. (**B**) Principal coordinates analysis (PCoA) based on unweighted or weighted Unifrac distance metric, Bray–Curtis, and Binary Jaccard distance metric, and $p < 0.05$ represents the significant difference of analysis of similarities (ANOISM).

Taxonomic analysis identified a total of 22 phyla across the treatment. Six of these had an average abundance greater than 1%, with *Bacteroidota*, *Firmicutes*, and *Patescibacteria* being the most dominant phyla (Figure 3A). At the genus level, 24 genera exhibited an average relative abundance above 1%. The dominant genera included *Prevotella*, *Rikenellaceae_RC9_gut_group*, *unclassified_F082*, *Christensenellaceae_R_7_group*, *Succiniclasticum*, *Prevotellaceae_UCG_003*, and *Prevotellaceae_UCG_001* (Figure 3B). LEfSe analysis identified specific taxa enriched in each treatment group. The C1 group was characterized by a higher relative abundance of *Lachnospiraceae_XPB1014_group*,

Candidatus_Omnitrophus, *Paenibacillus*, and *unclassified_Oligoflexaceae*. The C2 group was enriched in *Papillibacter*, *Anaerovibrio*, *V9D2013_group*, and *unclassified_Peptococcaceae*. The C3 group showed a higher relative abundance of *unclassified_Bacteroidales_RF16_group*. The C4 group was characterized by a higher relative abundance of *Ruminococcus*, *Pseudobutyrivibrio*, and *Mitsuokella* ($p < 0.05$ and LDA > 3.5; Figure 3C).

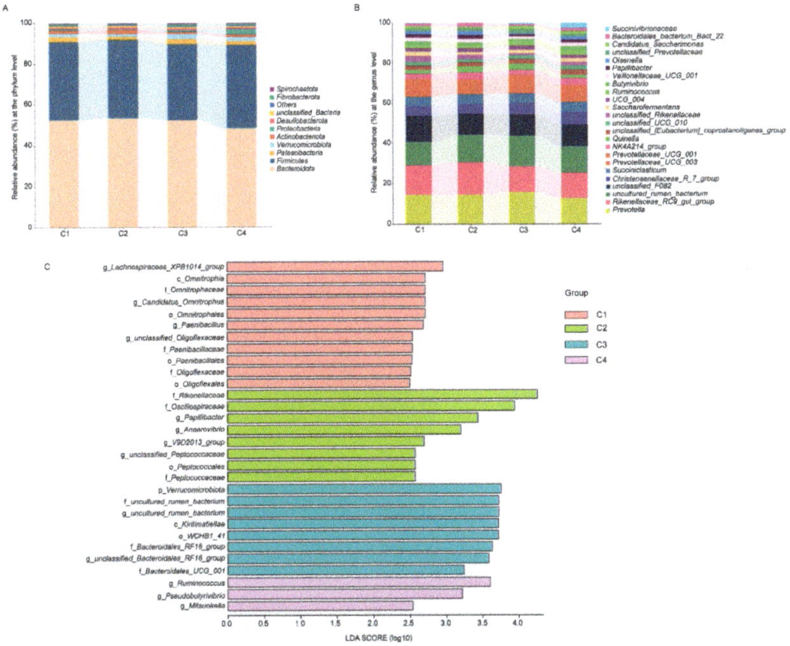

Figure 3. Rumen bacteria composition of lactating Tibetan ewes supplemented different concentrate levels. (**A**) Bacteria composition and the relative abundance at the phylum level. (**B**) Bacteria composition and the relative abundance at the genus level. (**C**) The representative bacteria in each group analyzed by the linear discriminant analysis effect size (LEfSe) analysis ($p < 0.05$ and linear discriminant analysis (LDA) > 3.5).

3.6. Transcriptomic Profile of Rumen in Lactating Tibetan Sheep

To elucidate the impact of graded levels of concentrate supplementation on the ruminal transcriptome, we performed high-throughput RNA sequencing, yielding 1.026 billion high-quality paired-end reads (4.28 million reads per sample). Of these, 89.28% were successfully mapped to the sheep reference genome. We defined expressed genes as those with transcripts per million (TPM) > 1 in at least 50% of the samples within any treatment group; this criterion retained 12,858 genes for downstream analyses. Principal component analysis (PCA) revealed no clear segregation among the four dietary groups (Figure 4A). Differential expression analysis identified 3, 17, and 17 differentially expressed genes (DEGs) in the C2 vs. C1, C3 vs. C1, and C4 vs. C1 comparisons, respectively (Figure 4B–D). The limited number of DEGs precluded robust functional enrichment of these individual gene sets.

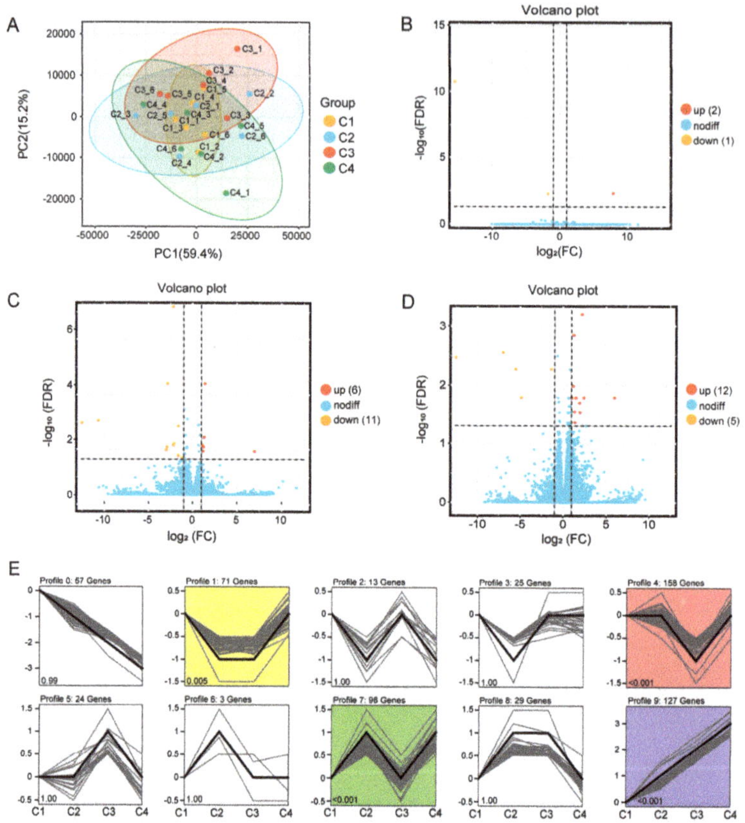

Figure 4. Rumen transcriptome profile of lactating Tibetan ewes supplemented different concentrate levels. (**A**) Principal component analysis (PCA) of rumen transcriptome profile. (**B–D**) Differential expressed gene (adjusted $p < 0.05$ and $|\log_2(\text{Fold Change})| > 1$) identification between C2 vs. C1, C3 vs. C1, and C4 vs. C1. (**E**) Short Time-series Expression Miner (STEM) analysis of rumen transcriptome.

To capture coordinated transcriptional responses across concentrate levels, we performed a STEM analysis, yielding nine distinct gene cluster profiles (Figure 4E). Among these, four exhibited statistically significant temporal patterns ($p < 0.05$): profile 1 ($n = 71$), profile 4 ($n = 158$), profile 7 ($n = 98$), and profile 9 ($n = 127$). Kyoto Encyclopedia of Genes and Genomes (KEGG) pathway enrichment showed that genes in profile 4 were significantly enriched in hematopoietic cell lineage ($p < 0.05$; Figure 5B), whereas genes in profile 7 were enriched in complement and coagulation cascades ($p < 0.05$; Figure 5C). No significant enrichments were detected for profiles 1 or 9 ($p > 0.05$; Figure 5A,D). Notably, representative DEGs—including TRPA1 and EPHB1 (profile 1), GATA3 (profile 4), C4 and ABCG2 (profile 7), and THBS4 and TNFRSF11B (profile 9)—are predominantly implicated in immune regulation, signal transduction, and nutrient digestion, suggesting that concentrate supplementation modulates ruminal function through these biological processes.

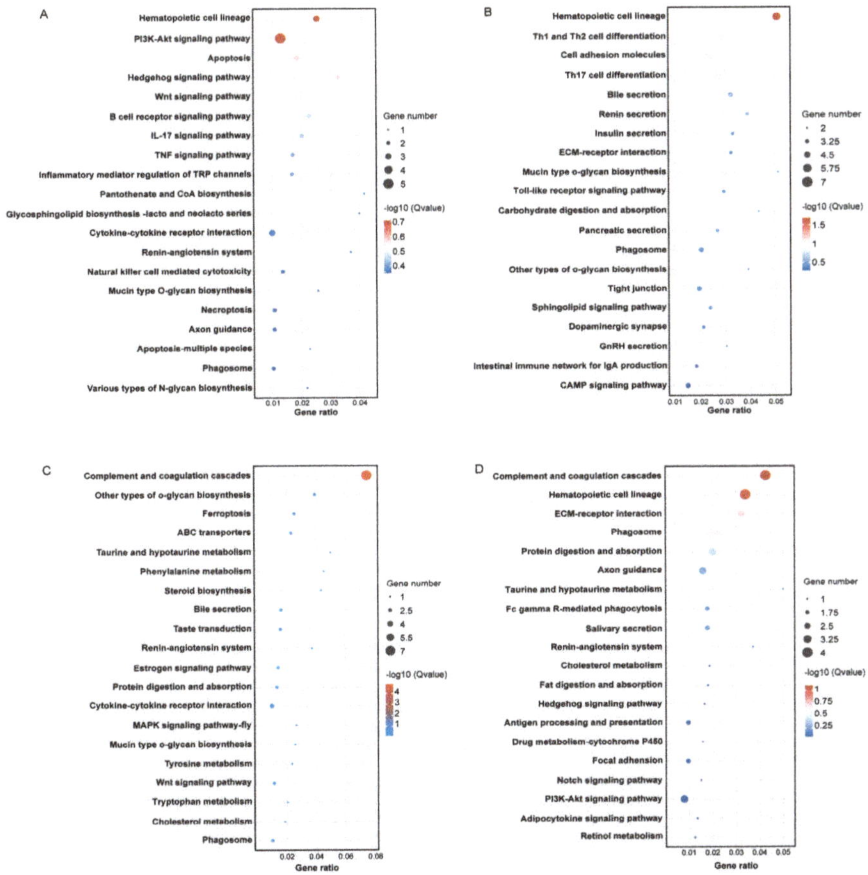

Figure 5. Kyoto Encyclopedia of Genes and Genomes (KEGG) pathway enrichment analysis of the genes in the profile 1 (**A**), profile 4 (**B**), profile 7 (**C**), and profile 9 (**D**).

3.7. Correlation Analysis Between Phenotype and Bacteria

We performed Spearman correlation analysis to investigate the relationship between the phenotype (rumen histomorphometric indicators, fermentation parameters, and digestive enzyme activity) and ruminal bacteria (representative bacteria in each group generated by LEfSe analysis) on OmicStudio tools https://www.omicstudio.cn/tool (accessed on 13 August 2025). The results showed that the relative abundance of *Anaerovibrio* was negatively correlated with propionate (r = −0.565, $p < 0.05$) but positively correlated with the ratio of acetate and propionate (r = 0.579, $p < 0.05$; Figure 6). Moreover, *Lachnospiraceae_XPB1014_group* was negatively correlated with cellulase (r = −0.699, $p < 0.05$) and α-amylase (r = −0.514, $p < 0.05$).

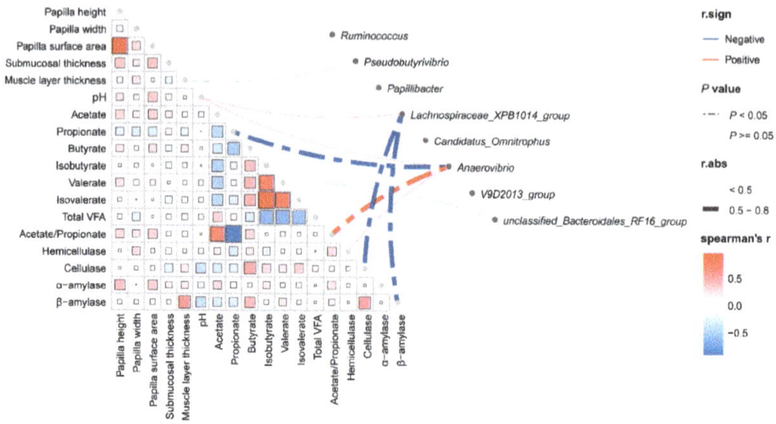

Figure 6. Spearman correlation analysis of the phenotype (rumen histomorphology indicators, fermentation parameters, and digestive enzyme activity) and ruminal bacteria. $|r| > 0.5$ and $p < 0.05$ are considered significant correlations. Read dashed line represents positive correlation and blue dashed line represents negative correlation.

4. Discussion

Extreme temperature is considered one of the main environmental stresses affecting animal production, reproduction, physiology, and health [28]. On the Qinghai Tibetan Plateau, cold season temperatures range from −10 to −30 °C. These harsh conditions commonly induce a negative energy balance in Tibetan sheep, especially in lactating ewes, thereby seriously impairing lactation performance and postpartum recovery [29]. Thus, identifying effective nutritional strategies to ensure adequate energy intake for Tibetan sheep during the cold season is crucial.

Several studies have focused on dietary intervention, such as protein levels [30], concentrate-to-forage ratios [31], and dietary types [32], on the growth performance of growing Tibetan sheep during the cold season. However, few studies have explored nutritional regulation to alleviate body weight loss in lactating Tibetan ewes or their underlying regulatory mechanism. Previous research indicates that Tibetan sheep supplemented with oat hey (800 g DM/d) exhibited lower body weight loss and higher nitrogen retention and energy utilization during the winter season than those fed only native herbage [33]. Similarly, supplementing with 250 g/d of concentrate has been shown to reduce body weight loss and increase milk yield of Tibetan ewes during the cold season [34]. In this study, the increase in concentrate level significantly decreased body weight loss of lactating Tibetan ewes; the BW of those on D34 and D68 was similar. However, the BW of lambs on D60 significantly increased with increasing concentrate levels (Table S2), indicating that higher concentrate intake enhanced the availability of protein, fat, and energy, which may promote the utilization of energy and other nutrients and ameliorate negative energy balance, thereby alleviating body weight loss and supporting lactation and postpartum recovery.

The ruminal epithelium is organized into leaf-like papillae that constitute both the primary absorptive interface and a critical barrier against translocation of luminal microbes or their metabolites [35]. Quantitative morphology of papillae (height, width, and absorptive surface area) is widely accepted as a proxy for the epithelial absorptive capacity for volatile fatty acids (VFAs) and other fermentation end-products. Previous studies have demonstrated that the level of concentrate supplementation modulates papillar morphogenesis [36]; the present observation that papillar width in lactating Tibetan sheep increased with concentrate level corroborates this relationship. Histologically, the ruminal epithelium comprises, from lumen to serosa, the stratum corneum, stratum granulosum, stratum spinosum, and stratum basale. This stratified structure collectively governs VFA metabolism, vectorial transport, and barrier function [37]. The thickness of the epithelium directly influences nutrient absorption. Furthermore, the observed decrease in the thickness of the submucosal layer may be attributed to the increased production of VFAs and other metabolites, which can drive adaptive changes in morphology in rumen adaptation to improve the efficiency of the absorption of nutrients and maintain the ruminal homeostasis. A diet of coarse, withered grass typically promotes development of the rumen muscle layer, as increased ruminal motility is required to degrade and digest its high cellulose and hemicellulose content [36]. In this study, although roughage intake was similar across the treatments, the muscle layer thickness increased with increasing concentrate levels. It suggests that the greater energy availability from concentrate fermentation may promote the need for the vigorous ruminal motility required to digest forage, potentially leading to a relative enhancement in muscular development.

The rumen microbiota plays a crucial role in food digestion, growth and development, energy balance, and immune regulation in ruminant animals [38]. Ruminal fermentation parameters are key indicators underlying differences in ruminant growth performance. Among them, VFAs are fermentation products of microorganisms in the rumen, and their production depends on diet and the rumen microbiota [39]. In the previous study, the ruminal fluid samples were obtained 3–4 h after the morning feeding using a stomach tube attached to an electric pump in Tibetan sheep under barn feeding, and Tibetan sheep that received different concentrate-to-forage ratio diets had similar ruminal pH, VFA concentration, and acetate-to-propionate ratio [40], which was inconsistent with our results. In this study, the molar ratio of VFAs and pH of ruminal fluid were not affected by concentrate supplementation. This lack of significant difference may be attributed to the extended interval between feeding and sample collection (fasting 12 h before sample collection), during which VFAs were absorbed by the ruminal epithelium. In addition, the above results might be due to the ability of the rumen system that can adapt to appropriate dietary concentrate levels through the self-adjustment of rumen microorganisms [41]. *Lachnospiraceae_XPB1014_group* is an important bacterial taxon within the gut microbiota. This group can decompose pectin-like substances and degrade complex polysaccharides such as galacturonic acid, producing short-chain fatty acids (SCFAs) such as acetic acid and butyric acid that provide energy for the host [42]. In this study, *Lachnospiraceae_XPB1014_group* was negatively correlated with cellulase and α-amylase. This is consistent with its specialized role in pectin degradation, rather than in breaking down cellulose or starch. The changes in bacterial genera such as *Papillibacter* in the rumen of dairy cows are closely related to the rumen acidification environment. Under high-energy diet conditions, the rumen pH of dairy cows will decrease, which then causes a significant reduction in rumen *Papillibacter* abundance [43]. *Papillibacter* is a known butyrate producer [44], and its relative abundance is often indicative of the butyrate production levels. However, in our study, the relative abundance of *Papillibacter* was higher in the C2 group than other groups, yet the molar ratio of butyrate and pH in the C2 group did not differ from the other groups. *Anaerovibrio* is a well-known rumen lipolytic bacterium whose fermentation end-products are propionate and succinic acid, and its relative abundance was higher in the sheep fed a mixed ration than in those fed a forage-only diet [45]. The higher relative abundance of *Anaerovibrio* in the C2 group indicates that Tibetan ewes in this group may possess a greater capacity for fat degradation. Furthermore, *Anaerovibrio* was negatively correlated with propionate and positively correlated with the ratio of acetate and propionate. These results are contradictory to its function and warrant further study. The *V9D2013_group*, a unique bacterial genus in the gastrointestinal tract, degrades cellulose to produce butyrate and inhibits inflammation and oxidative stress [46]. The abundance of *V9D2013_group* was higher in the C2 group than other groups. Meanwhile, the activity of cellulose was higher in the C2 group, indicating that *V9D2013_group* may enhance cellulolytic activity to facilitate butyrate production. The relative abundance of family *Peptococcaceae* is possibly correlated with energy intake and conversion, as well as butyrate production [47]. The increased abundance of *unclassified Peptococcaceae* in the C2 group suggests that members of this taxon likely contribute to enhanced butyrate production.

In the current study, *Ruminococcus*, *Pseudobutyrivibrio*, and *Mitsuokella* were identified as key bacterial genera influenced by the dietary treatments. The genus *Ruminococcus* comprises obligate anaerobic, Gram-positive fibrolytic symbionts that ubiquitously inhabit the mammalian gut and enzymatically depolymerize diverse plant cell-wall polysaccharides [48]. Moreover, *Ruminococcus gnavus* cross-feeds on the malto-oligosaccharides and transient glucose released by *Ruminococcus bromii* during starch hydrolysis, channeling these metabolites to acetate [49]. These results indicate that *Ruminococcus* plays a comprehensive role in carbohydrate metabolism, contributing to the degradation of both structure and starch. *Pseudobutyrivibrio*, a genus within the Firmicutes phylum, effectively degrades hemicellulose and produces butyrate [50]. The major fermentation products of *Mitsuokella* are lactate, succinate, and acetate, and it is unlikely that this genus directly contributed to the butyrate production in the rumen [51]. In the C4 group, Tibetan ewes received 620 g of concentrate per day in addition to 1.4 kg of roughage. This increased concentrate likely provides more fermentable substrates, thereby directly facilitating the proliferation of *Ruminococcus*, *Pseudobutyrivibrio*, and *Mitsuokella*.

The gene expression pattern of rumen epithelium is largely affected by dietary transition. The rumen epithelial transcriptome showed a high number of genes that were differentially expressed genes in cattle exposed to rumen acidosis when feeding a high-starch diet. These genes impacted critical biological pathways, specifically those responsible for cell signaling and morphogenesis [52]. When ruminants experience malnutrition during the cold season, the expression of genes involved in carbohydrate metabolism and energy conversion was upregulated [53]. In the current study, the increased concentrate supplementation mainly affected the genes involved in immune regulation, signal transduction, and nutrient digestion. This suggests that greater nutrient intake provides abundant fermentable substrates, leading to increased production of VFAs and other end-products. These metabolites subsequently support rumen development and maintain homeostasis by modulating immune responses, signal transduction pathways, and nutrient metabolism.

5. Conclusions

Concentrate supplementation effectively alleviated the body weight loss in lactating Tibetan sheep experiencing the extreme temperatures of the cold season. Increased concentrate levels induced adaptive rumen morphological changes, specifically increased papilla width, and decreased submucosal and muscle layer thickness, which collectively enhance nutrient digestion and absorption. Meanwhile, concentrate supplementation altered the abundance of key rumen bacteria (*Lachnospiraceae_XPB1014_group*, *Candidatus_Omnitrophus*, *Paenibacillus*, *Papillibacter*, *Anaerovibrio*, *V9D2013_group*, *unclassified_Bacteroidales_RF16_group*, *Ruminococcus*, *Pseudobutyrivibrio*, and *Mitsuokella*), modulating the activity of digestive enzymes to optimize rumen fermentation and energy harvest for maintaining normal metabolism. These dietary changes also triggered adaptive shifts in rumen epithelium, promoting homeostasis through regulation of immune, signal transduction, and nutrient metabolism. We conclude that supplementing with 610 g (DM basis) of concentrate is an effective strategy to reduce weight loss, improving rumen morphology and improving fermentation efficiency in lactating Tibetan ewes during the cold season.

Supplementary Materials: The following supporting information can be downloaded at: https://www.mdpi.com/article/10.3390/ani15192791/s1. Table S1: Ingredient composition and chemical composition of diets in each group; Table S2: Body weight changes in lambs in each group; Figure S1: The rarefaction curves and Shannon curves of each Tibetan sheep rumen fluid sample.

Author Contributions: Conceptualization, C.Y. and S.H.; methodology, C.Y.; software, Q.M.; validation, C.Y., Z.W. and S.H.; formal analysis, J.W.; investigation, J.W.; resources, Z.W.; data curation, C.Y. and J.W.; writing—original draft preparation, C.Y.; writing—review and editing, C.Y. and Q.M.; visualization, Q.M.; supervision, S.H.; project administration, S.H.; funding acquisition, C.Y. All authors have read and agreed to the published version of the manuscript.

Funding: This research was funded by the Construction of Standardized Production System for Improving Quality and Efficiency of Tibetan Sheep Industry (2022-NK-169).

Institutional Review Board Statement: All animal experiments were approved by the Animal Ethics Committee of Qinghai University (permit No. QUA-2023-0112).

Informed Consent Statement: Not applicable.

Data Availability Statement: The datasets presented in this study can be found in online repositories. The transcriptomics and 16s rDNA sequence data of this study have been submitted to the Sequence Read Archive (SRA) database, and the data are accessible through SRA Series accession number PRJNA1305533 and PRJNA1305115.

Acknowledgments: All authors thank the efforts of students Fengshuo Zhang, Zhenling Wu, Yu Zhang, Qiurong Ji, Qvyangangmao Su, and Kaina Zhu in Qinghai University for animal trial and sample collecting.

Conflicts of Interest: The authors declare no conflicts of interest.

References

1. Guo, T.; Wang, X.; Zhang, Q.; Wei, L.; Liu, H.; Zhao, N.; Hu, L.; Xu, S. Comparative Analysis of the Composition of Fatty Acids and Metabolites between Black Tibetan and Chaka Sheep on the Qinghai—Tibet Plateau. *Animals* **2022**, *12*, 2745. [CrossRef] [PubMed]
2. Song, Y.; Yuan, C.; An, X.; Guo, T.; Zhang, W.; Lu, Z.; Liu, J. Genome-Wide Selection Signals Reveal Candidate Genes Associated with Plateau Adaptation in Tibetan Sheep. *Animals* **2024**, *14*, 3212. [CrossRef]
3. Xin, G.; Ge, C.; Gao, Q.; Zhang, J.; Nie, Y.; Yang, Y.; Zhang, D.; Li, H.; Ren, Y. Effects of soil ingestion on nutrient digestibility and rumen bacterial diversity of Tibetan sheep. *Chemosphere* **2022**, *308*, 136000. [CrossRef]
4. Shang, Z.; Gibb, M.; Leiber, F.; Ismail, M.; Ding, L.; Guo, X.; Long, R. The sustainable development of grassland-livestock systems on the Tibetan plateau: Problems, strategies and prospects. *Rangel. J.* **2014**, *36*, 267–296. [CrossRef]
5. Zhang, S. Research on the Value-Added Process of the Value Chain of Tibetan Sheep Based on the Rotational Grazing. *J. Adv. Agric. Technol.* **2023**, *10*, 42–51. [CrossRef]
6. Weimer, P.J. Degradation of cellulose and hemicellulose by ruminal microorganisms. *Microorganisms* **2022**, *10*, 2345. [CrossRef]
7. Ibrahim, N.A.; Alimon, A.R.; Yaakub, H.; Samsudin, A.A.; Candyrine, S.C.L.; Wan Mohamed, W.N.; Md Noh, A.; Fuat, M.A.; Mookiah, S. Effects of vegetable oil supplementation on rumen fermentation and microbial population in ruminant: A review. *Trop. Anim. Health Prod.* **2021**, *53*, 422. [CrossRef]
8. Russell, J.B.; Muck, R.E.; Weimer, P.J. Quantitative analysis of cellulose degradation and growth of cellulolytic bacteria in the rumen. *FEMS Microbiol. Ecol.* **2009**, *67*, 183–197. [CrossRef]
9. Henderson, G.; Cox, F.; Ganesh, S.; Jonker, A.; Young, W.; Janssen, P.H. Rumen microbial community composition varies with diet and host, but a core microbiome is found across a wide geographical range. *Sci. Rep.* **2015**, *5*, 14567. [CrossRef]
10. Chen, Q.; Sha, Y.; Liu, X.; Gao, M.; Chen, X.; Yang, W.; Huang, W.; Wang, J.; He, Y.; Gao, X. A Study on the Differences in Rumen Microbiota–Liver Gluconeogenesis–Mitochondrial Interaction Between Tibetan Sheep and Hu Sheep in the Qinghai–Tibet Plateau. *Animals* **2025**, *15*, 1603. [CrossRef]
11. Tajima, K.; Nonaka, I.; Higuchi, K.; Takusari, N.; Kurihara, M.; Takenaka, A.; Mitsumori, M.; Kajikawa, H.; Aminov, R.I. Influence of high temperature and humidity on rumen bacterial diversity in Holstein heifers. *Anaerobe* **2007**, *13*, 57–64. [CrossRef]

12. Wu, X.; Zhang, G.; Zhang, W.; Zhou, J.; Cong, H.; Yang, G.; Liu, G. Rumen microbiota helps Tibetan sheep obtain energy more efficiently to survive in the extreme environment of the Qinghai–Tibet Plateau. *Front. Microbiol.* **2024**, *15*, 1431063. [CrossRef]
13. Liu, H. Effect of Seasonal Dynamics on Feed Intake, Blood Parameters, Rumen Microbiota and Metabolites in Grazing Tibetan Sheep. Master's Thesis, Lanzhou University, Lanzhou, China, 2023.
14. Li, F.; Yang, X.; Cao, Y.; Li, S.; Yao, J.; Li, Z.; Sun, F. Effects of dietary effective fiber to rumen degradable starch ratios on the risk of sub-acute ruminal acidosis and rumen content fatty acids composition in dairy goat. *Anim. Feed Sci. Technol.* **2014**, *189*, 54–62. [CrossRef]
15. Zhang, Y.; Ji, Q.; Gui, L.; Gebeyew, K.; Hou, S.; Wang, Z.; Han, L.; Yang, C. The influence of resveratrol and β-Hydroxy-β-methyl butyric acid supplementation alone or in combination on the development and health of the duodenum in Tibetan sheep. *Front. Microbiol.* **2025**, *16*, 1612102. [CrossRef] [PubMed]
16. Bolger, A.M.; Marc, L.; Bjoern, U. Trimmomatic: A flexible trimmer for Illumina sequence data. *Bioinformatics* **2014**, *30*, 2114–2120. [CrossRef] [PubMed]
17. Edgar, R.C.; Haas, B.J.; Clemente, J.C.; Quince, C.; Knight, R. UCHIME improves sensitivity and speed of chimera detection. *Bioinformatics* **2011**, *27*, 2194–2200. [CrossRef]
18. Callahan, B.J.; McMurdie, P.J.; Rosen, M.J.; Han, A.W.; Johnson, A.J.A.; Holmes, S.P. DADA2: High-resolution sample inference from Illumina amplicon data. *Nat. Methods* **2016**, *13*, 581–583. [CrossRef]
19. Quast, C.; Pruesse, E.; Yilmaz, P.; Gerken, J.; Schweer, T.; Yarza, P.; Peplies, J.; Glöckner, F.O. The SILVA ribosomal RNA gene database project: Improved data processing and web-based tools. *Nucleic Acids Res.* **2013**, *41*, D590–D596. [CrossRef]
20. Caporaso, J.G.; Kuczynski, J.; Stombaugh, J.; Bittinger, K.; Bushman, F.D.; Costello, E.K.; Fierer, N.; Peña, A.G.; Goodrich, J.K.; Gordon, J.I. QIIME allows analysis of high-throughput community sequencing data. *Nat. Methods* **2010**, *7*, 335–336. [CrossRef]
21. Chen, S.; Zhou, Y.; Chen, Y.; Gu, J. fastp: An ultra-fast all-in-one FASTQ preprocessor. *Bioinformatics* **2018**, *34*, i884–i890. [CrossRef]
22. Langmead, B.; Salzberg, S.L. Fast gapped-read alignment with Bowtie 2. *Nat. Methods* **2012**, *9*, 357–359. [CrossRef] [PubMed]
23. Kim, D.; Langmead, B.; Salzberg, S.L. HISAT: A fast spliced aligner with low memory requirements. *Nat. Methods* **2015**, *12*, 357–360. [CrossRef] [PubMed]
24. Pertea, M.; Pertea, G.M.; Antonescu, C.M.; Chang, T.-C.; Mendell, J.T.; Salzberg, S.L. StringTie enables improved reconstruction of a transcriptome from RNA-seq reads. *Nat. Biotechnol.* **2015**, *33*, 290–295. [CrossRef] [PubMed]
25. Love, M.I.; Huber, W.; Anders, S. Moderated estimation of fold change and dispersion for RNA-seq data with DESeq2. *Genome Biol.* **2014**, *15*, 550. [CrossRef]
26. Bu, D.; Luo, H.; Huo, P.; Wang, Z.; Zhang, S.; He, Z.; Wu, Y.; Zhao, L.; Liu, J.; Guo, J.; et al. KOBAS-i: Intelligent prioritization and exploratory visualization of biological functions for gene enrichment analysis. *Nucleic Acids Res.* **2021**, *49*, W317–W325. [CrossRef]
27. Ernst, J.; Bar-Joseph, Z. STEM: A tool for the analysis of short time series gene expression data. *BMC Bioinform.* **2006**, *7*, 191. [CrossRef]
28. Rojas-Downing, M.M.; Nejadhashemi, A.P.; Harrigan, T.; Woznicki, S.A. Climate change and livestock: Impacts, adaptation, and mitigation. *Clim. Risk Manag.* **2017**, *16*, 145–163. [CrossRef]
29. Fan, Q.; Cui, X.; Wang, Z.; Chang, S.; Wanapat, M.; Yan, T.; Hou, F. Rumen microbiota of Tibetan sheep (Ovis aries) adaptation to extremely cold season on the Qinghai-Tibetan Plateau. *Front. Vet. Sci.* **2021**, *8*, 673822. [CrossRef]
30. Wang, X.; Xu, T.; Zhang, X.; Geng, Y.; Kang, S.; Xu, S. Effects of dietary protein levels on growth performance, carcass traits, serum metabolites, and meat composition of Tibetan sheep during the cold season on the Qinghai-Tibetan plateau. *Animals* **2020**, *10*, 801. [CrossRef]
31. Wang, S.; Tang, W.; Jiang, T.; Wang, R.; Zhang, R.; Ou, J.; Wang, Q.; Cheng, X.; Ren, C.; Chen, J. Effect of dietary concentrate-to-forage ratios during the cold season on slaughter performance, meat quality, rumen fermentation and gut microbiota of tibetan sheep. *Animals* **2024**, *14*, 3305. [CrossRef]
32. Xu, T.; Xu, S.; Hu, L.; Zhao, N.; Liu, Z.; Ma, L.; Liu, H.; Zhao, X. Effect of dietary types on feed intakes, growth performance and economic benefit in Tibetan sheep and yaks on the Qinghai-Tibet Plateau during cold season. *PLoS ONE* **2017**, *12*, e0169187.
33. Abdelraheem, N.; Li, F.; Guo, P.; Sun, Y.; Liu, Y.; Cheng, Y.; Hou, F. Oat hay as winter feed improves digestibility, nitrogen balance and energy utilization of Tibetan sheep (Ovis aries) in the Qinghai Tibetan Plateau. *Livest. Sci.* **2019**, *230*, 103854. [CrossRef]
34. Gui, L.-S.; Raza, S.H.A.; Allam, F.A.E.A.; Zhou, L.; Hou, S.; Khan, I.; Kakar, I.U.; Abd El-Aziz, A.H.; Jia, J.; Sun, Y. Altered milk yield and rumen microbial abundance in response to concentrate supplementation during the cold season in Tibetan sheep. *Electron. J. Biotechnol.* **2021**, *53*, 80–86. [CrossRef]
35. Graham, C.; Simmons, N.L. Functional organization of the bovine rumen epithelium. *Am. J. Physiol.-Regul. Integr. Comp. Physiol.* **2005**, *288*, R173–R181.
36. Liu, X.; Sha, Y.; Lv, W.; Cao, G.; Guo, X.; Pu, X.; Wang, J.; Li, S.; Hu, J.; Luo, Y. Multi-omics reveals that the rumen transcriptome, microbiome, and its metabolome co-regulate cold season adaptability of Tibetan sheep. *Front. Microbiol.* **2022**, *13*, 859601.
37. Steele, M.A.; Penner, G.B.; Chaucheyras-Durand, F.; Guan, L.L. Development and physiology of the rumen and the lower gut: Targets for improving gut health. *J. Dairy Sci.* **2016**, *99*, 4955–4966. [CrossRef]

8. Grinberg, I.R.; Yin, G.; Borovok, I.; Miller, M.E.B.; Yeoman, C.J.; Dassa, B.; Yu, Z.; Mizrahi, I.; Flint, H.J.; Bayer, E.A. Functional phylotyping approach for assessing intraspecific diversity of Ruminococcus albus within the rumen microbiome. *FEMS Microbiol. Lett.* **2015**, *362*, 1–10. [CrossRef]
9. Sha, Y.; Hu, J.; Shi, B.; Dingkao, R.; Wang, J.; Li, S.; Zhang, W.; Luo, Y.; Liu, X. Supplementary feeding of cattle-yak in the cold season alters rumen microbes, volatile fatty acids, and expression of SGLT1 in the rumen epithelium. *PeerJ* **2021**, *9*, e11048. [CrossRef]
0. Liu, H.; Xu, T.; Xu, S.; Ma, L.; Han, X.; Wang, X.; Zhang, X.; Hu, L.; Zhao, N.; Chen, Y. Effect of dietary concentrate to forage ratio on growth performance, rumen fermentation and bacterial diversity of Tibetan sheep under barn feeding on the Qinghai-Tibetan plateau. *PeerJ* **2019**, *7*, e7462. [CrossRef]
1. Zhang, T.; Zhuang, S.; Dong, W.; Chang, G.; Yan, R.; Wang, T. Effects of different dietary concentrate to forage ratios on rumen fluid pH and VFA levels and blood VFA levels in dairy goats. *Anim. Husb. Vet. Med.* **2013**, *4*, 5–10.
2. Liu, R.; Peng, C.; Jing, D.; Xiao, Y.; Zhu, W.; Zhao, S.; Zhang, J.; Chen, X.; Li, J. Lachnospira is a signature of antihistamine efficacy in chronic spontaneous urticaria. *Exp. Dermatol.* **2022**, *31*, 242–247. [CrossRef] [PubMed]
3. Mao, S.; Zhang, R.; Wang, D.; Zhu, W. Impact of subacute ruminal acidosis (SARA) adaptation on rumen microbiota in dairy cattle using pyrosequencing. *Anaerobe* **2013**, *24*, 12–19. [CrossRef]
4. Xue, Y.; Lin, L.; Hu, F.; Zhu, W.; Mao, S. Disruption of ruminal homeostasis by malnutrition involved in systemic ruminal microbiota-host interactions in a pregnant sheep model. *Microbiome* **2020**, *8*, 138. [CrossRef] [PubMed]
5. Liu, K.; Xu, Q.; Wang, L.; Wang, J.; Guo, W.; Zhou, M. The impact of diet on the composition and relative abundance of rumen microbes in goat. *Asian-Australas. J. Anim. Sci.* **2016**, *30*, 531. [CrossRef] [PubMed]
6. Hao, Y.; Ji, Z.; Shen, Z.; Xue, Y.; Zhang, B.; Yu, D.; Liu, T.; Luo, D.; Xing, G.; Tang, J. Increase dietary fiber intake ameliorates cecal morphology and drives cecal species-specific of short-chain fatty acids in white pekin ducks. *Front. Microbiol.* **2022**, *13*, 853797. [CrossRef]
7. Xue, Y.; Hu, F.; Guo, C.; Mei, S.; Xie, F.; Zeng, H.; Mao, S. Undernutrition shifted colonic fermentation and digest-associated bacterial communities in pregnant ewes. *Appl. Microbiol. Biotechnol.* **2020**, *104*, 5973–5984. [CrossRef]
8. Christopherson, M.R.; Dawson, J.A.; Stevenson, D.M.; Cunningham, A.C.; Bramhacharya, S.; Weimer, P.J.; Kendziorski, C.; Suen, G. Unique aspects of fiber degradation by the ruminal ethanologen Ruminococcus albus 7 revealed by physiological and transcriptomic analysis. *BMC Genom.* **2014**, *15*, 1066. [CrossRef]
9. Crost, E.H.; Le Gall, G.; Laverde-Gomez, J.A.; Mukhopadhya, I.; Flint, H.J.; Juge, N. Mechanistic insights into the cross-feeding of Ruminococcus gnavus and Ruminococcus bromii on host and dietary carbohydrates. *Front. Microbiol.* **2018**, *9*, 2558. [CrossRef]
0. Louis, P.; Flint, H.J. Formation of propionate and butyrate by the human colonic microbiota. *Environ. Microbiol.* **2017**, *19*, 29–41. [CrossRef]
1. Lan, G.; Ho, Y.; Abdullah, N. Mitsuokella jalaludinii sp. nov., from the rumens of cattle in Malaysia. *Int. J. Syst. Evol. Microbiol.* **2002**, *52*, 713–718. [CrossRef]
2. Li, W.; Gelsinger, S.; Edwards, A.; Riehle, C.; Koch, D. Transcriptome analysis of rumen epithelium and meta-transcriptome analysis of rumen epimural microbial community in young calves with feed induced acidosis. *Sci. Rep.* **2019**, *9*, 4744. [CrossRef]
3. Guo, X.; Sha, Y.; Pu, X.; Xu, Y.; Yao, L.; Liu, X.; He, Y.; Hu, J.; Wang, J.; Li, S. Coevolution of rumen epithelial circRNAs with their microbiota and metabolites in response to cold-season nutritional stress in Tibetan sheep. *Int. J. Mol. Sci.* **2022**, *23*, 10488. [CrossRef]

animals

MDPI

Article

Effects of Partial Replacement of Alfalfa Hay with Alfalfa Silage in Dairy Cows: Impacts on Production Performance and Rumen Microbiota

Tian Xia [1,2,3,†], Zixin Liu [1,2,3,†], Ziyan Yang [3,4], Aoyu Jiang [1,2,3], Chuanshe Zhou [1,2,3,*] and Zhiliang Tan [1,2,3,5]

1 State Key Laboratory of Forage Breeding-by-Design and Utilization, Institute of Subtropical Agriculture, Chinese Academy of Sciences, Changsha 410125, China; xiat9797@163.com (T.X.); lzx@isa.ac.cn (Z.L.); jiangaoyu21@mails.ucas.ac.cn (A.J.); zltan@isa.ac.cn (Z.T.)
2 University of Chinese Academy of Sciences, Beijing 101408, China
3 Hunan Provincial Key Laboratory of Animal Nutritional Physiology and Metabolic Process, National Engineering Laboratory for Pollution Control and Waste Utilization in Livestock and Poultry Production, Institute of Subtropical Agriculture, Chinese Academy of Sciences, Changsha 410125, China; 15544552854@163.com
4 College of Animal Science and Technology, Ningxia University, Yinchuan 750021, China
5 Yuelushan Laboratory, Changsha 410128, China
* Correspondence: zcs@isa.ac.cn; Tel.: +86-731-84619795; Fax: +86-731-84612685
† These authors contributed equally to this work.

Simple Summary

Alfalfa (*Medicago sativa* L.) is one of the major high-quality forages for ruminants and is an ideal feed source for high-yielding dairy cows. The currently common forms of utilization are alfalfa hay and alfalfa silage. However, there is currently limited research on the effects of alfalfa silage partially replacing alfalfa hay in dairy cows. Our research has found that partially replacing alfalfa hay with alfalfa silage feed is beneficial for improving the production performance of dairy cows and altering rumen microbial diversity. This provides a reliable source of data for future practical production.

Abstract

As an important feed source for ruminants, alfalfa's rational and efficient utilization is of great significance for the production and economic benefits of pastures. This study focuses on Sanhe dairy cows and includes a control group (CON group, alfalfa in the diet is hay) and an experimental group (AS group, alfalfa silage partially replaces alfalfa hay of equal dry weight). The feeding experiment lasted for 60 days. The results revealed that, compared with the CON group, the AS group exhibited increased milk yield, milk protein, and milk fat. There were no significant differences in apparent digestibility, serum biochemical indicators, and volatile fatty acid (VFA) levels between the two groups. However, the microbial composition of the rumen differed significantly between the two groups of cows based on β-diversity. On the genus level, compared with the CON group, the relative abundance of *Erysipelatoclostridium*, *Pseudoflavonifractor*, and *Candidatus Saccharimonas* in the AS group was significantly reduced. In summary, partially replacing alfalfa hay with alfalfa silage feed is beneficial for improving the production performance of cows and changing rumen microbial diversity. These findings provide a basis for the effective utilization of alfalfa.

Keywords: dairy cow; alfalfa silage; alfalfa hay; production performance; rumen microbiota

1. Introduction

Alfalfa (*Medicago sativa* L.) is one of the major high-quality forages for ruminants because of its high protein content, and it has desirable characteristics as a feed source for high-producing dairy cows [1]. Furthermore, alfalfa is a vital source of energy for ruminants due to its high nutrient content, which includes vitamins, minerals, and crude fiber. Alfalfa has attracted considerable attention and widespread cultivation because of its growing significance in animal husbandry [2]. The development of human health cannot be separated from high-quality dairy products. Correspondingly, the scale of dairy farming and the number of dairy cows are constantly increasing. Consequently, there is a growing need for premium alfalfa in the husbandry of ruminant animals. At the moment, alfalfa hay and silage are the most widely used types. There are some variations in the nutritional value and feeding effects of alfalfa hay and silage on cows because of the diverse production methods.

Alfalfa silage feed has high digestible energy, and nutrients such as vitamin B and vitamin C are well preserved [3]. High-quality alfalfa silage can have a crude protein level of over 18%, as well as enhanced rumen-degradable protein and organic acid contents, both of which can increase the amount of nutrients available to ruminants [4]. Research has shown that, compared with feeding alfalfa hay, alfalfa silage results in a higher degradation rate of crude protein in the rumen [5]. In addition, alfalfa silage improves the taste of the feed during fermentation, and compared with alfalfa hay, alfalfa silage has better palatability [6,7]. A study on sheep showed that feeding silage significantly increased feed intake and improved the digestibility of dry matter and crude protein [8]. Alfalfa silage can increase the financial benefits of pastures because of its comparatively easy production process, which enables greater leaf utilization and minimizes losses. Alfalfa silage, however, necessitates rigorous fermentation control. At the same time, it is impossible to overlook the severe storage and transportation limitations of silage feed.

Alfalfa hay is currently the main form of preparation for alfalfa [9]. Hay feed often has a higher content of crude fiber [10]. It has been shown in ruminants that the proportion of Firmicutes increased with the crude fiber content in hay [11]. In addition, research has shown that the dry matter intake (DMI) of cows fed alfalfa hay was greater than that of cows fed alfalfa silage [5]. Moreover, during the drying process of alfalfa hay, there is less loss of antioxidants [12]. However, the production process of hay may lead to the loss of some water-soluble vitamins [13]. Although alfalfa hay is more conducive to pasture storage and has a higher protein content when compared with alfalfa silage, alfalfa hay has disadvantages such as poor palatability, low digestion and utilization rate, and higher cost.

The rumen is the most critical organ among ruminants, serving as a primary site of feed conversion, with rumen microbiota playing a key role in this process. The digestion and absorption of hay and silage feed in the rumen mainly depend on the action of rumen microorganisms. Both feeds can be fermented to produce volatile fatty acids (VFAs), approximately 50~85% of which are directly transported and absorbed by rumen epithelial cells as the main energy source for ruminants [14,15]. Cellulose and hemicellulose are the main components of alfalfa hay and silage. Compared with hay, alfalfa silage contains more water-soluble carbohydrates, which are the main substrates for rapid rumen fermentation [16,17]. The VFAs, acetate, and β-hydroxybutyric acid produced by rumen microbial fermentation of carbohydrates are directly absorbed by the gastric mucosa into mammary tissue and serve as precursor substances for milk fat synthesis [18]. However, rapid fermentation of silage may increase the risk of rumen acidosis [16]. The decomposition rate of cellulose and hemicellulose in hay is slow, and the production rate of VFAs is consequently low—this low VFAs production rate helps avoid the rapid accumulation of acids in the rumen [19]. Since alfalfa silage and alfalfa hay are distinct rumen fermentation substrates, they can influence the fermentation processes of rumen microbes and, consequently, the productivity of dairy cows. However, dairy cow performance can be improved by strategically combining hay and silage in their diet to optimize rumen fermentation efficiency as well as nutrient absorption and utilization.

Alfalfa silage and alfalfa hay each possess unique advantages. Combining these two feeds has the potential to improve the balance of protein and energy in dairy cow diets [20]. Moreover, this combination exhibits potential benefits in enhancing feed palatability, increasing dairy cow feed intake, and optimizing rumen fermentation [21,22]. Luo et al.'s study demonstrated that when corn silage and alfalfa hay were combined, dairy cows exhibited a higher dry matter digestibility and fiber digestibility compared with those in the group using fed corn silage alone. This difference may be attributed to the larger rumen volume, which is promoted by the higher neutral detergent fiber (NDF) content of alfalfa hay [23]. Cui et al. demonstrated that when wheat silage was used to partially replace oat hay (at replacement ratios of 36% or 64%), relative to the group fed oat hay alone, the concentrations of volatile fatty acids and microbial crude protein in lambs were significantly elevated, and the abundance of cellulolytic bacteria was also enhanced. This provides additional energy for ruminants. This phenomenon may be attributed to the increased fiber digestibility and a broader spectrum of nitrogen sources [24]. Furthermore, appropriate silage-to-hay ratios and effective alfalfa processing technologies enhance the economic benefits of grasslands and promote the sustainable development of animal husbandry. However, current research on the effects of alfalfa silage partially replacing alfalfa hay on dairy cows remains limited, and studies exploring the differences in nutritional or functional properties between these two forages are also scarce. Therefore, this study aims to investigate the effects of alfalfa silage partially replacing alfalfa hay on the production performance, serum biochemical indices, and rumen microbiota of dairy cows, thereby providing reliable data support for future practical animal production.

2. Materials and Methods

2.1. Animals, Experimental Design, and Diets

All procedures for animal experimentation were carried out according to guidelines approved by the Animal Care Committee, Institute of Subtropical Agriculture, Chinese Academy of Sciences (ISA-2022-0083).

Hulunbuir Agricultural Reclamation Xiertala Farm and Ranch Co. Ltd. (Hulunbuir City, China) provided the Sanhe dairy cows used in this study. A total of 20 Sanhe dairy cows with a parity of 2 and in mid-lactation were randomly divided into two groups, with 10 cows per group. Based on previous research data, milk yield was used as the primary index; the sample size required to achieve 80% statistical power was 8 cows per group, calculated using G*Power software (3.1.9.7). Therefore, considering potential model construction failure and animal mortality, 10 cows were selected for this experiment. The experimental group (AS group) received 3.0 kg of alfalfa silage to replace 1.5 kg of alfalfa hay on an equal dry matter basis, while the control group (CON group) was fed a basal diet containing 2.5 kg of alfalfa hay. The composition and nutritional levels of the diets are presented in Table 1. A 15-day pre-feeding period and a 45-day formal feeding period were included in the 60-day experimental period. In accordance with standard production protocols, each cow was housed individually in a pen. The cowshed was equipped with an outdoor exercise area and indoor fans. The experiment was conducted from July to September. The sprinkler system was used to cool the cows when the ambient temperature became excessively high. Cows were fed three times daily (6:30, 14:00, and 20:30). The staff cleaned up the remaining feed and recorded the daily feed intake before feeding them the next morning. The cows were allowed ad libitum access to drinking water.

Table 1. Composition and nutrient levels of the basal diet (DM basis) %.

Ingredients (%)	CON	AS
Cornsilage	35.74	35.74
Concentrated feed [1]	39.73	39.73
Alfalfa hay	9.46	3.78
Alfalfa silage	0.00	5.68
Flakedcorn	8.70	8.70
Beetrootmeal	5.74	5.74
Bakingsoda	0.63	0.63
Total	100.00	100.00
Nutrient levels [2] (%)		
CP	14.89	14.83
EE	3.05	3.08
NDF	32.37	32.59
ADF	17.04	17.16
Ash	2.97	3.02

[1] The concentrated feed for dairy cows during lactation was formulated by Xiertala first farm, with the following main ingredients: soya bean meal, DDGS, calcium phosphate, and premix. Premix composition per kilogram: 4500 mg of Fe, 1600 mg of Cu, 3000 mg of Mn, 5500 mg of Zn, 30 mg of Se, 20 mg of Co, 30 mg of I, 600 KIU of vitamin A, 200 KIU of vitamin D, 200 IU of vitamin E. [2] CP = crude protein; EE = ether extract; NDF = neutral detergent fiber; ADF = acid detergent fiber.

2.2. Sample Collection

Total mixed ration (TMR) samples were collected daily throughout the experimental period. Following the completion of the experiment, the TMR samples were proportionately mixed to determine the nutritional composition.

Milk samples were collected from the dairy cows in this experiment following the completion of the experiment. In total, 40 aliquot mL of milk samples were taken from each cow in the morning, lunchtime, and evening in a 4:3:3 ratio by volume, and they were placed into a centrifuge tube containing potassium dichromate preservative. For the determination of milk composition, the samples were kept in a refrigerator at 4 °C.

The rectal sampling method was used to collect 300–500 g of fecal samples once in the morning and once in the evening, which was performed daily from the 43rd to 45th day of the formal experimental period. After evenly mixing 200 g of each cow's fecal sample, 20 mL of 10% H_2SO_4 was added for nitrogen fixation. The treated fecal samples were then stored in a refrigerator at −20 °C until subsequent testing.

Before the morning feed on the final day of the formal experimental period, 10 mL of blood was collected from the tail vein of each cow. It was then allowed to stand for 2 h, and then centrifuged at 3000 rpm for 15 min at 4 °C. The resulting supernatant was aspirated into two mL centrifuge tubes, which were then kept at −20 °C until subsequent testing.

Rumen fluid was collected from the dairy cows before the morning feed on the 45th day of the formal experimental period. A rumen fluid sampler with a gastric tube was first inserted into the dairy cow's mouth; subsequently, 100 mL of rumen fluid was aspirated. The first 50 mL of the aspirated rumen fluid was discarded to avoid saliva contamination. The remaining collected rumen fluid was filtered to remove solid residues, then aliquoted into 10 mL centrifuge tubes. These tubes were immediately transferred to a liquid nitrogen tank for rapid freezing and subsequent storage until testing.

All sample collection procedures were conducted under the premise of ensuring the health of the experimental cows. Feces from the cows were collected via rectal sampling. Blood was collected from the veins at the base of the cow's tail. The rumen fluid sampler equipped with a gastric tube was employed to collect rumen fluid via the cow's mouth. All sample collection activities complied with the relevant animal ethical standards. The specific operational procedures were consistent with those described in previous studies [25].

2.3. Analytical Procedures

2.3.1. Diets, Feces, and Apparent Digestibility

The determination methods for feed and fecal components were consistent with those previously employed by Liu et al. [26]. The apparent digestibility was determined using the method previously employed by Van Keulen et al. [27], employing acid-insoluble ash in feces and feed as an endogenous indicator. The formula was as follows:

$$Apparent\ digestibility\ (\%) = [1 - \frac{A_d \times N_f}{A_f \times N_d}] \times 100$$

In the formula:

A_d and A_f represent the acid-insoluble ash (AIA) content in the diet and feces, respectively (g/kg);

N_d and N_f represent the nutrient content of a certain nutrient in the diet and feces, respectively (g/kg).

2.3.2. Serum Biochemical Parameters

Serum biochemical parameters were measured using an automatic biochemical analyzer (Hitachi 7600; Hitachi, Tokyo, Japan), including total protein (TP, A045-2-2), albumin (ALB, A028-1-1), globulin (GLB), alanine aminotransferase (ALT, C009-3-1), aspartate aminotransferase (AST, C010-3-1), alkaline phosphatase (ALP, A059-1-1), urea nitrogen (BUN, C013-1-1), glucose (GLU, F006-1-1), triglycerides (TG, A110-1-1), and total cholesterol (T-CHO, A111-1-1). In addition, serum immunoglobulin G (IgG, H106-1-1), immunoglobulin A (IgA, H108-1-2), immunoglobulin M (IgM, H109-1-2), interferon-γ (IFN-γ, H025-1-2), interleukin-1β (IL1β, H002-1-2), interleukin-2 (IL-2, H003-1-1), interleukin-6 (IL-6, H007-1-2), interleukin-8 (IL-8, H008-1-1), and tumor necrosis factor-alpha (TNF-α, H052-1-2) concentrations were all measured using commercial reagent kits (Nanjing JianCheng Bioengineering Institute, Nanjing, China). Refer to the commercial reagent kit manual for the determination method.

2.3.3. Milk Composition

A Fossomatic FC Type 79910 equipment (Foss Electric, Hillerod, Denmark) was used to measure the somatic cell count, and a Basic Unit MilkoScan FT+200 Type 76150 device was used to assess the quality of the milk samples [28]. The 4% FCM (Fat-Corrected Milk) and ECM (Energy-Corrected Milk) yields were calculated [25,29]. The formulas were as follows:

$$4\%\ FCM\ (Fat - Corrected\ Milk) = milk\ yield\ (kg/d) \times [milk\ fat\ (\%) \times 0.15 + 0.4]$$

$$\begin{aligned} ECM\ (Energy - Corrected\ Milk) \\ = milk\ yield\ (kg/d) \times [0.0929 \times milk\ fat\ (\%) \\ +0.0547 \times milk\ protein\ (\%) + 0.0395 \times milk\ lactose\ (\%)] \end{aligned}$$

2.3.4. Measurement of Rumen Fermentation Parameters

Rumen fluid was collected and then centrifuged at 4 °C and 10,000× g for 10 min. An aliquot (1.5 mL) of the supernatant was taken and mixed with 0.15 mL of 25% metaphosphoric acid for fixation. This mixture was then used for the determination of volatile fatty acids (VFA) [30]. The concentrations of microbial crude protein (MCP), ammonia nitrogen (NH_3-N), and VFA concentrations in the rumen fluid were determined using the methods described by Wang [31], Weatherburn [32], and Zinn [33], respectively.

2.3.5. DNA Extraction and 16S rRNA Gene Sequencing

Following the manufacturer's instructions, microbial DNA was isolated from rumen fluid samples using the E.Z.N.A.® Soil DNA Kit (Omega Bio-tek, Norcross, GA, USA). The full-length segment of the 16S rRNA gene was subjected to a polymerase chain reaction (PCR) using primers 27F (5′-AGAGTTTGATCMTGGCTCAG-3′) and 1492R (5′-ACCTTGTTACGACTT-3′) [34,35]. The 25 μL PCR amplification system contained Q5 DNA Polymerase 0.25 μL, dNTP 2.5 mM 2 μL, Forwardprimer 10 μM 1 μL, Reverseprimer 10 μM 1 μL, DNA Template 2 μL, ddH_2O 8.75 μL, 5 × reaction buffer 5 μL, and 5 × GC buffer 5 μL. Following their extraction from 2% agarose gels, amplicons were purified using the AxyPrep DNA Gel Extraction Kit (Axygen Biosciences, Union City, CA, USA) [36]. As directed by the manufacturer (Pacific Biosciences, Shanghai, China), SMRTbell libraries were created from the amplified DNA using blunt ligation. Shanghai Biozeron Biotechnology Co. Ltd. (Shanghai, China) carried out all amplicon sequencing. The SMRT Link Analysis software version 9.0 was used to process PacBio raw data and produce demultiplexed circular consensus sequence (CCS) reads.

With a 98.65% similarity criterion, OTUs were grouped using UPARSE 7.1 (http://drive5.com/uparse/) (accessed on 8 December 2022), and chimeric sequences were found and removed using UCHIME. The RDP Classifier (https://sourceforge.net/projects/rdp-classifier/) (accessed on 8 December 2022) was used to analyze the phylogenetic relationship of each 16S rRNA gene sequence at a 70% confidence level against the Silva (SSU132)16S rRNA database [37]. The beta diversity was determined using weighted and unweighted UniFrac distance matrices, and alpha and beta diversity analysis were performed using the table acquired from OTU, and then using R software (V3.1) to generate a principal coordinate analysis graph (PCoA) [37]. To compare the variations in relative bacterial abundance between the two groups, linear discriminant analysis effect size (LEfSe) was used. Significant differences between the groups are indicated by $p < 0.05$ and linear discriminant analysis (LDA) > 2 [38]. For more detailed steps, refer to previous research in our laboratory [39].

2.3.6. Correlation Analysis

The linkET R package (0.0.3) [40] was then used to visualize these microbial species and examine their associations with milk yield, milk protein, milk fat, and milk lactose using the Mantel test method. We conducted Spearman correlation tests after linearly fitting the relative abundance of rumen microbes using the ggplot2 R software (version 4.1.2).

2.4. Statistical Analysis

The Shapiro–Wilk normality test of all data was conducted using SPSS 24.0 statistical software. All data conformed to a normal distribution. The analysis was performed using Student's *t*-test using SPSS 24.0 statistical software. When $p < 0.05$, the data were statistically significant. There was no significant difference in the data when $p > 0.05$. When $0.05 < p < 0.1$, a trend was observed in the data. All data were presented as means and standard error of the mean (SEM).

3. Results

3.1. Lactation Performance, Milk Quality, and Apparent Digestibility

The effect of the partial replacement of alfalfa hay with alfalfa silage on milk yield, milk quality, and apparent digestibility of nutrients in Sanhe dairy cows was presented in Table 2. There were no significant differences in milk yield, milk quality, and apparent digestibility between the CON and AS groups ($p > 0.05$). However, compared with the CON group, the AS group showed an increasing tendency in milk yield ($p = 0.092$). The 4%

FCM and ECM in the AS group were significantly higher than those in the CON group ($p < 0.05$). Compared with the CON group, the AS group showed increases in milk yield, milk protein, and milk fat levels by 9.73%, 2.71%, and 5.70%, respectively.

Table 2. The effect of the partial replacement of alfalfa hay with alfalfa silage on lactation performance, milk quality, and apparent digestibility of dairy cows.

Items	CON	AS	SEM	*p*-Value
Milk yield (kg/d)	23.33	25.60	0.897	0.092
Milk composition				
Milk protein (%)	3.39	3.49	0.040	0.146
Milk fat (%)	3.51	3.71	0.100	0.174
Milk lactose (%)	4.64	4.64	0.075	0.993
MUN (%)	13.40	14.45	0.622	0.254
Lactation performance				
4% FCM [1]	21.61	24.47	0.792	0.024
ECM [2]	16.23	18.39	0.512	0.006
Milk protein (g/d)	0.79	0.89	0.028	0.022
Milk fat (g/d)	0.82	0.95	0.037	0.031
Milk lactose (g/d)	1.09	1.19	0.050	0.194
Milk protein (g/kg)	33.93	34.85	0.400	0.146
Milk fat (g/kg)	35.06	37.05	0.999	0.174
Milk lactose (g/kg)	46.42	46.43	0.747	0.993
Apparent digestibility				
DM (%)	90.44	87.62	0.881	0.051
EE (%)	82.52	81.42	2.060	0.709
CP (%)	63.55	63.84	2.214	0.928
NDF (%)	73.89	71.25	2.198	0.429
ADF (%)	72.09	71.85	2.427	0.946

[1] 4% FCM (Fat-Corrected Milk) was calculated as milk yield (kg/d) × [milk fat (%) × 0.15 + 0.4]. [2] ECM (Energy-Corrected Milk) was calculated as milk yield (kg/d) × [0.0929 × milk fat (%) + 0.0547 × milk protein (%) + 0.0395 × milk lactose (%)]. Values are presented as mean ± SEM. $p < 0.05$ indicates a significant difference in data between two groups in the same row. Ten samples per group were used for analysis.

3.2. Serum Biochemical Indicators

The effect of the partial replacement of alfalfa hay with alfalfa silage on serum biochemical indicators is presented in Table 3. Compared with the CON group, the serum TP and TG levels in the AS group increased by 2.70% and 8.07% ($p > 0.05$), respectively, while the serum ALP level in the AS group dropped by 7.06% ($p > 0.05$). There was no significant difference in serum biochemical indicators between the two groups ($p > 0.05$).

Table 3. The effect of the partial replacement of alfalfa hay with alfalfa silage on serum biochemical indicators of dairy cows.

Items [1]	CON	AS	SEM	*p*-Value
TP (g/L)	86.62	88.96	1.600	0.302
ALB (g/L)	38.49	37.08	0.746	0.177
GLB (g/L)	50.38	49.57	1.832	0.766
ALB/GLB	0.79	0.78	0.030	0.778
ALT (U/L)	40.90	38.59	1.624	0.322
AST (U/L)	86.67	91.80	3.171	0.267
AST/ALT	2.59	2.51	0.176	0.757
ALP (U/L)	78.80	73.24	8.045	0.679
BUN (mmol/L)	5.52	5.36	0.242	0.639
GLU (mmol/L)	3.50	3.27	0.182	0.358

Table 3. *Cont.*

Items [1]	CON	AS	SEM	*p*-Value
TG (mmol/L)	0.16	0.17	0.007	0.191
T-CHO (mmol/L)	4.90	4.87	0.180	0.918

[1] TP = total protein; ALB = albumin; GLB = globulin; ALT = alanine aminotransferase; AST = aspartate aminotransferase; ALP = alkaline phosphatase; BUN = urea nitrogen; GLU = glucose; TG = triglycerides; T-CHO = total cholesterol. Values are presented as mean ± SEM. $p < 0.05$ indicates a significant difference in data between two groups in the same row. Ten samples per group were used for analysis.

3.3. Serum Immune Parameters

The effect of the partial replacement of alfalfa hay with alfalfa silage on serum immune indicators of Sanhe dairy cows is presented in Table 4. Compared with the CON group, the content of IgG in the serum of cows in the AS group was significantly reduced (13.59%, $p < 0.05$), while the content of IL-6 in the AS group was significantly higher than that in the CON group (39.18%, $p < 0.05$). There was no significant difference in other immune-related indicators between the two groups ($p > 0.05$).

Table 4. The effect of the partial replacement of alfalfa hay with alfalfa silage on serum immune parameters of dairy cows.

Items	CON	AS	SEM	*p*-Value
IgG (mg/mL)	6.75	5.84	0.269	0.026
IgA (μg/mL)	2223.63	2018.01	96.547	0.136
IgM (μg/mL)	1425.72	1323.86	62.901	0.259
IFNγ (pg/mL)	727.81	840.99	45.257	0.097
IL1β (pg/mL)	588.70	631.48	24.785	0.235
IL2 (pg/mL)	631.43	735.98	45.290	0.113
IL4 (pg/mL)	43.63	39.16	1.912	0.146
IL6 (pg/mL)	83.41	116.09	7.100	0.003
IL8 (pg/mL)	174.63	183.36	9.589	0.518
TNFα (pg/mL)	147.09	149.93	8.757	0.818

Values are presented as mean ± SEM. $p < 0.05$ indicates a significant difference in data between two groups in the same row. Ten samples per group were used for analysis.

3.4. Rumen Fermentation Parameters

The effect of the partial replacement of alfalfa hay with alfalfa silage on rumen fermentation parameters of Sanhe dairy cows was presented in Table 5. Compared with the CON group, the MCP content in the AS group was significantly reduced ($p < 0.05$). Compared with the CON group, the AS group showed a 4.12% increase in acetic acid content and a 12.45% decrease in propionic acid content ($p > 0.05$), but there was no significant change. There was no significant change in other indicators as well ($p > 0.05$).

Table 5. The effect of the partial replacement of alfalfa hay with alfalfa silage on rumen fermentation parameters of dairy cows.

Items	CON	AS	SEM	*p*-Value
MCP (mg/mL)	1.39	1.24	0.039	0.016
NH_3-N (mg/dL)	3.64	3.99	0.388	0.516
Acetate (mmol/L)	60.98	63.50	2.450	0.461
Propionate (mmol/L)	23.18	20.30	1.530	0.187
Isobutyrate (mmol/L)	0.58	0.66	0.040	0.193
Butyrate (mmol/L)	9.91	10.96	0.632	0.261
Isovalerate (mmol/L)	1.04	1.03	0.089	0.951

Table 5. *Cont.*

Items	CON	AS	SEM	*p*-Value
Valerate (mmol/L)	1.42	1.33	0.086	0.471
TVFA (mmol/L)	100.03	96.51	3.167	0.428
Acetate/Propionate	2.78	3.17	0.200	0.190

Values are presented as mean ± SEM. $p < 0.05$ indicates a significant difference in data between two groups in the same row. Ten samples per group were used for analysis.

3.5. Rumen Microbial Richness and Diversity

A total of 1985 OTUs were present in the two groups, 2085 of which were specific to the CON group and 2642 to the AS group (Figure 1A). Compared with the CON group, there were no significant differences in the ACE index, Chao 1 index, Shannon index, and Simpson index in the AS group ($p > 0.05$, Figure 1B). The plot of principal coordinates analysis revealed clear clustering of the CON group and the AS group, with obvious separation between them. And there was a significant difference in the Bray–Curtis distance between the two groups ($p < 0.05$, Figure 1C). *Bacteroidetes* and *Firmicutes* accounted for more than 90% of the microbiota in each group, making them the dominant phyla in both groups (Figure 1D). There were notable variations across the groups: *Candidatus Saccharibacteria* relative abundances were larger in the CON group ($p < 0.05$, Figure 1F).

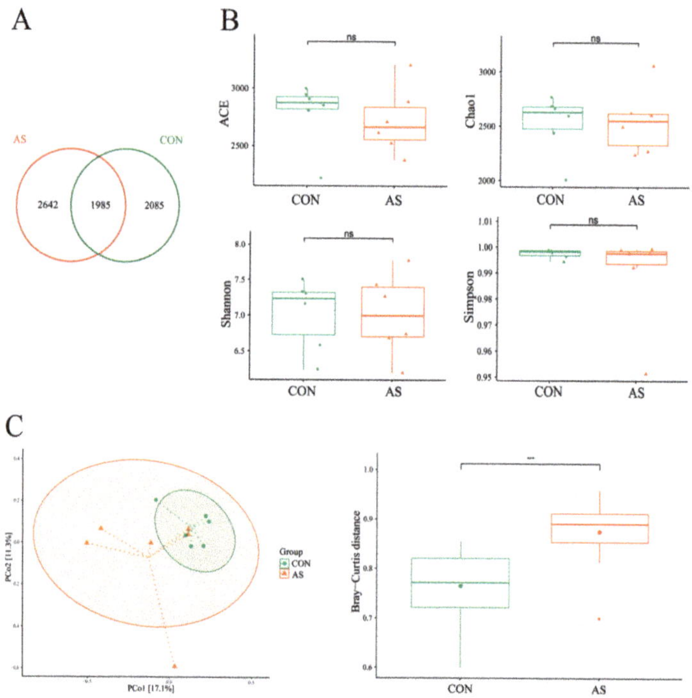

Figure 1. *Cont.*

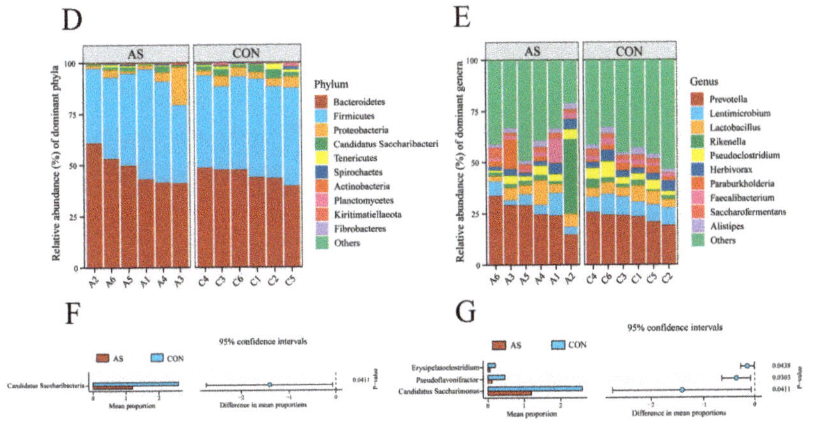

Figure 1. The effect of alfalfa silage partially replacing alfalfa hay on rumen microbiota of dairy cows. (**A**) Venn diagram of rumen microbiota on OTU level in the two groups. (**B**) Alpha diversity. (**C**) Beta diversity. (**D**) Changes and differences in microbiome on phylum level. (**E**) Changes and differences in microbiome on genus level. (**F**) The microbial communities with significant differences at the phylum level. (**G**) The microbial communities with significant differences at the genus level. "ns" indicates no significant difference between the two groups. "***" indicates that there is a highly significant difference between the two sets of data ($p < 0.001$).

In terms of genus composition, both groups mainly had *Prevotella*, *Lentimicrobium*, and *Lactobacillus* as the dominant bacterial groups (Figure 1E). Compared with the CON group, the relative abundance of *Erysipelatoclostridium*, *Pseudoflavonifractor*, and *Candidatus Saccharimonas* in the AS group was significantly reduced ($p < 0.05$, Figure 1G).

3.6. Correlation Analysis

At the phylum level, there was a significant positive correlation between rumen *Fibrobacteres* and milk yield in cows ($p < 0.01$, Figure 2A). There was a significant negative correlation between *Spirochaetes* and lactose ($p < 0.05$, Figure 2A). At the genus level, significant positive correlations were observed between *Ruminiclostridium* and milk yield, *Pectinatus* and milk fat, as well as *Lachnospira* and lactose ($p < 0.05$, Figure 2B). However, there was a negative correlation between the relative abundance of *Ottowia* and milk yield ($p < 0.05$, Figure 2B). Meanwhile, there was a significant negative correlation between *Rubripirellula* and milk protein ($p < 0.01$, Figure 2B). In addition, *Parabacteroides* and *Treponema* were negatively correlated with lactose ($p < 0.01$, Figure 2B).

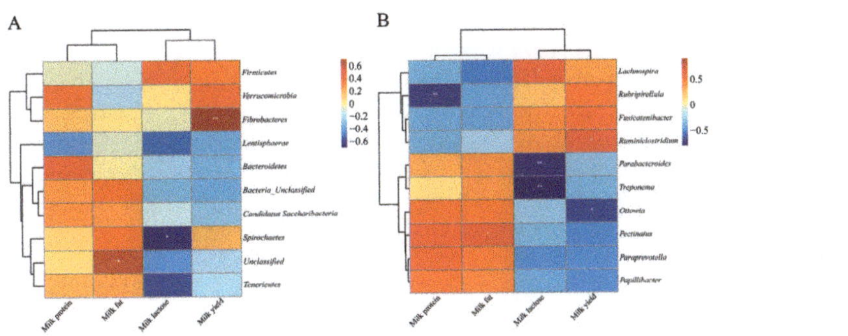

Figure 2. Correlation analysis between rumen microbiota and milk yield, milk protein, milk fat, and milk lactose. (**A**) Correlation analysis at microbial phylum level. (**B**) Correlation analysis at microbial

genus level. The red block represents positive correlation, and the blue block represents negative correlation. "*" Indicates significant difference ($p < 0.05$), "**" indicates extremely significant difference ($p < 0.01$). The mean OTUs for *Fibrobacteres* phylum were 0.5 for samples in the CON group and 1.33 for samples in the AS group. The mean OTUs for *Ruminiclostridium* genus were 61.17 for samples in the CON group and 63.33 for samples in the AS group.

4. Discussion

4.1. Lactation Performance, Milk Quality, and Apparent Digestibility

Alfalfa hay is an important feed source in dairy farming. However, the cultivation of alfalfa is constrained by climatic and soil conditions (key environmental factors) [41], alongside rising import costs [42,43] and high processing costs [44]—factors that consequently pose significant challenges to dairy farming. Alfalfa silage and alfalfa hay each possess unique advantages. Therefore, this study aims to explore the potential impacts of the partial replacement of alfalfa hay with alfalfa silage on Sanhe dairy cows.

Alfalfa was one of the main high-quality roughages in the production of dairy cows, which can provide nutrition of high quality to improve milk quality and production [45]. According to the statistical analysis results, the milk yield of the cows in the CON and AS groups was 23.33 kg/d and 25.60 kg/d, respectively, at the end of the experimental period. Although there was no significant difference in milk yield between the two groups, the milk yield in the AS group showed an upward trend than that in the CON group, and 4% FCM production was significantly increased. The 4% FCM and ECM in the AS group were significantly higher than those in the CON group. This indicates an improvement in milk quality and an increase in feed utilization efficiency [46]. Silage retains more digestible nutrients during its fermentation process, particularly soluble sugars and starch. These substances were more easily degraded by ruminal microorganisms, thereby providing the dairy cow's body with more energy and nutrients [47]. This may be the reason for the increase in 4% FCM and ECM in dairy cows. At the same time, the milk protein and fat content in the AS group also increased. The two groups' apparent digestibility and milk quality, however, did not differ significantly. In contrast, previous studies have demonstrated that replacing chopped alfalfa hay with alfalfa silage tended to increase milk fat percentage and milk fat yield, with the milk protein content also increasing [4,48]. Compared with dairy cows given alfalfa hay, those fed alfalfa silage had a 15.45% higher milk fat content and a 12.31% higher milk fat output [20]. These were similar to the results of our study. According to earlier research, silage's digestible energy value (per kilogram of dry matter) was almost 1.24 times higher than hay's [49]. Therefore, one of the reasons for the relatively higher milk protein and milk fat contents in the AS group in this study may be that alfalfa silage had higher digestible energy—particularly ruminal digestible energy—compared with alfalfa hay, which in turn provided more energy for the synthesis of milk protein and milk fat in dairy cows. In addition, the fiber digestibility of silage is higher, which can better stimulate rumen fermentation and produce more precursors for milk fat synthesis [20]. Silage feed has good palatability [50] and can effectively preserve nutritional components [51,52], especially protein and energy, in the raw materials during the production process, which may partly explain the improved milk yield in cows.

4.2. Serum Biochemistry and Immune Parameters

Blood is an important medium for the transportation of nutrients in living organisms [53]. In ruminant animals, the blood system is closely involved in rumen fermentation, the absorption of rumen fermentation products, and milk synthesis. In this study, there was no significant difference in serum biochemical indicators between the two groups. A study of Holstein calves showed that changing the proportion of hay or silage in the diet did not significantly affect blood glucose, blood urea nitrogen, total protein, and globulin [54]. This was similar to our research. Silage had stable nutritional content, and its fermentation process can lower the feed's pH value—thereby inhibiting the proliferation of pathogenic bacteria and reducing the production of potentially harmful substances [55]. This avoided additional metabolic burdens on the liver and blood system. This may partially explain why the partial replacement of alfalfa hay with alfalfa silage had no significant effect on serum biochemical parameters in the present study. In addition, determinations of immune-related indicators showed a significant decrease in IgG content and a significant increase in IL-6 content in the serum of the AS group. This observation suggested that the AS group may face potential risks of immunosuppressive or inflammation. Research has shown that improper storage of silage feed can lead to fungal spoilage and mycotoxin contamination [56]. Mycotoxins can activate the inflammatory response of the immune system, leading to the release of pro-inflammatory cytokines such as IL-6. Meanwhile, it can inhibit the differentiation of B lymphocytes and the production of antibodies, leading to a decrease in IgG synthesis [57].

4.3. Rumen Fermentation and Microorganisms

VFA, a crucial energy substrate created by microbial fermentation in the rumen, can supply 80% of ruminants' energy needs [58]. The partial substitution of alfalfa silage for alfalfa hay exerted no significant effect on VFA. However, changes in the concentrations of acetic acid and propionic acid suggest that the rumen fermentation types of the CON and AS groups may be changing. Specifically, compared with the CON group, the AS group exhibited an increasing trend in acetic acid concentration and a decreasing trend in propionic acid concentration. Acetic acid is a precursor substance for milk fat synthesis [59]. This may be the reason why the milk fat levels in the AS group were relatively high. In addition, the MCP content in the AS group significantly decreased. Previous studies have shown that partially replacing alfalfa hay with corn silage decreases ruminal MCP production. This is because corn silage has a high organic acid content—particularly its lactic acid concentration—which may inhibit the activity and proliferation of rumen microorganisms, thereby subsequently reducing the digestibility of nutrients [60]. Therefore, the synthesis of MCP was reduced. Notably, the lower antioxidant content of the silage may also upset the redox balance in ruminal microbes, decreasing the percentage of nutrients directed towards microbial protein synthesis, since rumen bacteria were known to have a lower antioxidant capacity than aerobic bacteria [61].

Multi-omics research has shown that the rumen microbiota is essential for dairy cow performance. For ruminants, the rumen microbiota acts as a bioreactor that allows dairy cows to obtain nutrients from plant matter that was indigestible to humans [62]. In the present study, 16S rRNA gene sequencing revealed a significant difference in beta diversity analysis between the CON and the AS group. At the phylum taxonomic level, the *Bacteroidetes* and *Firmicutes* accounted for the largest proportion of the ruminal microbiota. This composition pattern was consistent with many relevant previous studies [63,64]. Numerous fiber-degrading bacteria found in the *Bacteroidetes* and *Firmicutes* may break down cellulose in the feed to create volatile fatty acids and are crucial to the rumen's breakdown of carbohydrates, which supports ruminant growth and development [65]. There was no significant difference in the relative abundance of *Bacteroidetes* and *Firmicutes* between the CON and AS groups. This finding supports the normal digestion and utilization of silage or hay feed by cows. *Prevotella* was the primary prevalent bacteria in the CON and AS groups at the genus level. As a prominent genus in the ruminant digestive system, *Prevotella* is known to play crucial roles in the rumen ecology, including the synthesis of amino acids and short-chain fatty acids—such as propionate and acetate [66]. The abundance of *Prevotella* in the AS group (25.3%) was higher than that in the CON group (21.6%). Studies have shown that feeding black goats with cassava silage can increase the abundance of *Prevotella* in the rumen. This is similar to our research findings. During the ensiling process of alfalfa, its cell walls decompose, releasing more soluble sugars and thereby providing more fermentable substrate for *Prevotella*. This could explain the increased relative abundance of *Prevotella* in the AS group [67]. Partially replacing alfalfa hay with alfalfa silage increased the abundance of *Prevotella* and enhanced the degradation ability of the feed. In addition, the AS group significantly reduced the abundance of *Erysipelatoclostridium*, *Pseudoflavonifractor*, and *Candidatus Saccharimonas*. Studies have shown that *Erysipelatoclostridium* can cause severe infection in the host [68], while an increase in the abundance of *Pseudoflavonifractor* [69] and *Candidatus Saccharimonas* [70,71] was associated with host inflammation. The low content of fermentable carbohydrates (such as starch and water-soluble carbohydrates) in alfalfa hay leads to insufficient fermentation substrates for beneficial bacteria such as lactic acid bacteria in the rumen [72]. Reduced activity of beneficial ruminal bacteria may induce an imbalance in the rumen microbial community. This could explain the higher relative abundance of pathogenic bacteria in the CON group. Supplementing alfalfa silage in the AS group's diet alleviated this phenomenon. Furthermore, these changes in specific microbial taxa may account for the significant difference in β-diversity between the two groups. In contrast, since the overall ruminal microbiota remained stable and the dominant bacterial phylum was similar between groups, this could explain the absence of a significant difference in α-diversity. These two phenomena are not contradictory. These results indicated that alfalfa silage alters the composition of rumen microbiota and is beneficial for the healthy development of dairy cows. According to Cui et al.'s study, the diet containing 100% wheat silage had a lower relative abundance of *Proteobacteria*, which are Gram-negative bacteria that include pathogens like *Salmonella* and *Escherichia coli*, compared with the group that used oaten hay as their only source of fodder [24]. Similarly, Yin et al. found that the inclusion of whole-plant corn silage in the feed can reduce the abundance of some species of pathogenic bacteria in pigs, thereby reducing the incidence of disease in the animals [73]. These findings from previous studies are consistent with the results of the present study. One plausible explanation for this consistency is that the anaerobic fermentation process of alfalfa silage produces large quantities of lactic acid and other organic acids, which exert an inhibitory effect on bacterial growth. In contrast, compared with alfalfa silage, alfalfa hay is more susceptible to contamination by microorganisms (e.g., molds) during storage, thereby reducing its nutritional value and increasing the risk of diseases in Sanhe dairy cows.

4.4. Correlation Analysis

The correlation analysis results demonstrate a statistically significant positive correlation between the phylum *Fibrobacteres* and the milk yield of Sanhe dairy cows. *Fibrobacteres* has been shown in previous studies to be the principal cellulolytic active bacteria in the rumen [74,75]. The fiber is broken down in the rumen and can be used to generate VFA, providing an energy source for ruminants [76]. This may explain the statistically significant positive correlation between *Fibrobacteres* and milk yield. In addition, previous research has found that a decrease in the relative abundance of the phylum *Fibrobacteres* is observed in ewes at parturition, indicating that *Fibrobacteres* may be highly sensitive to changes in rumen ecological conditions [77]. In this study, the partial substitution of alfalfa silage for alfalfa hay changed the microbial fermentation substrate and similarly altered rumen ecological conditions. This further explains the correlation between the changes in *Fibrobacteres* and milk yield in the AS group. Specifically, the average number of OTUs for *Fibrobacteres* in the AS group was higher than that in the CON group, representing a relative increase of 166%. Further studies are needed to clarify the specific mechanisms underlying the link between *Fibrobacteres* (e.g., its OTU diversity) and milk yield. At the genus level, a statistically significant positive correlation was observed between the relative abundance of *Ruminiclostridium* and milk yield. It has been shown that the function of *Ruminiclostridium* is related to the degradation of flavonoids and xyloglucans, among others [78]. Coumarin and vanillic acid, produced by the decomposition of flavonoids, have antioxidant and anti-inflammatory effects [79]. Xyloglucan decomposition produces xylose, and glucose can provide energy for the animal body [80]. They are all effective in improving ruminant health. The average number of OTUs for *Ruminiclostridium* was 3.53% higher in the AS group compared with the CON group. This may explain the reason for the significant positive correlation between *Ruminiclostridium* and milk yield in the AS group.

The primary limitation of this study is its small sample size. Therefore, this study should be considered a preliminary exploratory investigation. The core value of its findings lies in the proposal of hypotheses rather than their verification; these hypotheses can provide research directions and candidate targets for subsequent confirmatory studies conducted in larger-scale cohorts.

5. Conclusions

The partial replacement of alfalfa hay with alfalfa silage tends to increase the milk yield of Sanhe dairy cows and alter rumen microbial diversity. This observation may be attributed to enhanced microbial capacity for fiber degradation and utilization, thereby increasing digestible energy. However, the replacement of part of alfalfa hay with alfalfa silage may pose risks of immunosuppression or inflammation to dairy cows, and this replacement strategy also leads to a reduction in microbial crude protein. These potential issues warrant further investigation.

Author Contributions: Conceptualization: C.Z. and Z.T.; Investigation: T.X., Z.L., Z.Y. and A.J.; Analysis of data: T.X. and Z.Y.; Contributed to the writing of the manuscript: T.X. and Z.L.; Funding acquisition: C.Z. All authors have read and agreed to the published version of the manuscript.

Funding: This research was jointly supported by the Projects of Hunan Science and Technology Innovation Program (2023WK2003), the Projects of International Cooperation and Exchanges NSFC (32261143467), and the National Key Research and Development Program (2022YFD1300805). The authors confirm that there are no known conflicts of interest associated with this publication.

Institutional Review Board Statement: All procedures for animal experimentation were carried out by following guidelines approved by the Animal Care Committee, Institute of Subtropical Agriculture, Chinese Academy of Sciences (ISA-2022-0083).

Informed Consent Statement: Not applicable.

Data Availability Statement: The datasets generated and/or analyzed during the current study are available in the NCBI repository, https://dataview.ncbi.nlm.nih.gov/object/PRJNA1277443?reviewer=r3pmcq6cku1cjn94fb23meq3q1 (accessed on 18 June 2025), accession number: PRJNA1277443.

Acknowledgments: We thank all co-authors for their contributions.

Conflicts of Interest: The authors declare no conflicts of interest.

Abbreviations

The following abbreviations are used in this manuscript:

CP	crude protein
EE	ether extract

NDF	neutral detergent fiber
ADF	acid detergent fiber
DMI	dry matter intake
VFAs	volatile fatty acids
TMR	total mixed ration
TP	total protein
ALB	albumin
GLB	globulin
ALT	alanine aminotransferase
AST	aspartate aminotransferase
ALP	alkaline phosphatase
BUN	urea nitrogen
GLU	glucose
TG	triglycerides
T-CHO	total cholesterol
IgG	immunoglobulin G
IgA	immunoglobulin A
IgM	immunoglobulin M
IFN-γ	interferon-γ
IL1β	interleukin-1β
IL-2	interleukin-2
IL-6	interleukin-6
IL-8	interleukin-8
TNF-α	tumor necrosis factor-alpha
FCM	Fat-Corrected Milk
ECM	Energy-Corrected Milk
MCP	microbial crude protein
NH_3-N	ammonia nitrogen

References

1. Sikora, M.C.; Hatfield, R.D.; Kalscheur, K.F. Impact of Long-Term Storage on Alfalfa Leaf and Stem Silage Characteristics. *Agronomy* **2021**, *11*, 2505. [CrossRef]
2. Zhang, Q.; Yu, Z.; Wang, X.; Na, R. Effects of chlorpyrifos and chlorantraniliprole on fermentation quality of alfalfa (*Medicago sativa* L.) silage inoculated with or without *Lactobacillus plantarum* LP. *Anim. Sci. J.* **2017**, *88*, 456–462. [CrossRef] [PubMed]
3. Yang, J.; Zhao, S.; Zhi, W.; Lu, T.; Qiao, H.; Liu, W.; Dou, Y.; Tan, H.; Zhou, H. Improvement of silage characteristics of Lactobacillus salivarius HMC4 and improvement of silage quality of king grass. *Front. Microbiol.* **2024**, *15*, 1468577. [CrossRef]
4. Peng, W.; Zhang, L.; Wei, M.; Wu, B.; Xiao, M.; Zhang, R.; Ju, J.; Dong, C.; Du, L.; Zheng, Y.; et al. Effects of *Lactobacillus plantarum* (L) and molasses (M) on nutrient composition, aerobic stability, and microflora of alfalfa silage in sandy grasslands. *Front. Microbiol.* **2024**, *15*, 1358085. [CrossRef]
5. Vagnoni, D.B.; Broderick, G.A. Effects of supplementation of energy or ruminally undegraded protein to lactating cows fed alfalfa hay or silage. *J. Dairy Sci.* **1997**, *80*, 1703–1712. [CrossRef]
6. Luo, R.; Zhang, Y.; Wang, F.; Liu, K.; Huang, G.; Zheng, N.; Wang, J. Effects of Sugar Cane Molasses Addition on the Fermentation Quality, Microbial Community, and Tastes of Alfalfa Silage. *Animals* **2021**, *11*, 355. [CrossRef]
7. John, M.O.; Rufai, M.A.; Sunday, A.J.; Fernando, E.; Richard, K.; Eva, I.; Maidala, A.; Amos, M.; Chana, M.; Hannatu, C.; et al. Cattail (*Typha domingensis*) silage improves feed intake, blood profile, economics of production, and growth performance of beef cattle. *Trop. Anim. Health Prod.* **2022**, *54*, 48. [CrossRef]
8. Godoi, P.F.A.; Magalhães, A.L.R.; de Araújo, G.G.L.; de Melo, A.A.S.; Silva, T.S.; Gois, G.C.; Dos Santos, K.C.; Do Nascimento, D.B.; Da Silva, P.B.; de Oliveira, J.S.; et al. Chemical Properties, Ruminal Fermentation, Gas Production and Digestibility of Silages Composed of Spineless Cactus and Tropical Forage Plants for Sheep Feeding. *Animals* **2024**, *14*, 552. [CrossRef] [PubMed]
9. An, J.; Sun, L.; Liu, M.; Dai, R.; Ge, G.; Wang, Z.; Jia, Y. Influences of Growth Stage and Ensiling Time on Fermentation Characteristics, Nitrite, and Bacterial Communities during Ensiling of Alfalfa. *Plants* **2023**, *13*, 84. [CrossRef]
10. Glatter, M.; Bochnia, M.; Wensch-Dorendorf, M.; Greef, J.M.; Zeyner, A. Feed Intake Parameters of Horses Fed Soaked or Steamed Hay and Hygienic Quality of Hay Stored following Treatment. *Animals* **2021**, *11*, 2729. [CrossRef] [PubMed]

1. Sha, Y.; Hu, J.; Shi, B.; Dingkao, R.; Wang, J.; Li, S.; Zhang, W.; Luo, Y.; Liu, X. Characteristics and Functions of the Rumen Microbial Community of Cattle-Yak at Different Ages. *Biomed. Res. Int.* **2020**, *2020*, 3482692. [CrossRef] [PubMed]
2. Ponnampalam, E.N.; Priyashantha, H.; Vidanarachchi, J.K.; Kiani, A.; Holman, B. Effects of Nutritional Factors on Fat Content, Fatty Acid Composition, and Sensorial Properties of Meat and Milk from Domesticated Ruminants: An Overview. *Animals* **2024**, *14*, 840. [CrossRef] [PubMed]
3. Zhu, X.; Chen, W.; Li, M.; Liu, B.; Zhao, S.; Hu, M.; Li, J.; Li, D.; Shi, Y.; Sun, H.; et al. Comprehensive evaluation of the nutritional value and contaminants of alfalfa (*Medicago sativa* L.) in China. *Front. Nutr.* **2025**, *12*, 1539462. [CrossRef] [PubMed]
4. Sha, Y.; He, Y.; Liu, X.; Zhao, S.; Hu, J.; Wang, J.; Li, S.; Li, W.; Shi, B.; Hao, Z. Rumen Epithelial Development- and Metabolism-Related Genes Regulate Their Micromorphology and VFAs Mediating Plateau Adaptability at Different Ages in Tibetan Sheep. *Int. J. Mol. Sci.* **2022**, *23*, 16078. [CrossRef]
5. Chai, J.; Lv, X.; Zhuang, Y.; Diao, Q.; Cui, K.; Deng, F.; Li, Y.; Zhang, N. Dataset of the rumen microbiota and epithelial transcriptomics and proteomics in goat affected by solid diets. *Sci. Data* **2024**, *11*, 749. [CrossRef]
6. Kung, L.J.; Shaver, R.D.; Grant, R.J.; Schmidt, R.J. Silage review: Interpretation of chemical, microbial, and organoleptic components of silages. *J. Dairy Sci.* **2018**, *101*, 4020–4033. [CrossRef]
7. Bolsen, K.K.; Ashbell, G.; Weinberg, Z.G. Silage fermentation and silage additives—Review. *Asian Australas. J. Anim. Sci.* **1996**, *9*, 483–493. [CrossRef]
8. Ling, H.; Xiao, H.; Zhang, Z.; He, Y.; Zhang, P. Effects of Macleaya Cordata Extract on Performance, Nutrient Apparent Digestibilities, Milk Composition, and Plasma Metabolites of Dairy Goats. *Animals* **2023**, *13*, 566. [CrossRef]
9. Beauchemin, K.A.; Yang, W.Z.; Rode, L.M. Effects of particle size of alfalfa-based dairy cow diets on chewing activity, ruminal fermentation, and milk production. *J. Dairy Sci.* **2003**, *86*, 630–643. [CrossRef]
10. Liu, K.; Chen, M.; Huang, G.; Su, C.; Tang, W.; Li, N.; Yang, J.; Wu, X.; Si, B.; Zhao, S.; et al. Variations in the milk lipidomic profile of lactating dairy cows fed the diets containing alfalfa hay versus alfalfa silage. *Anim. Nutr.* **2024**, *19*, 261–271. [CrossRef]
11. Zhang, X.; Wang, H.; Guo, X. Effects of total mixed ration with various silage on growth performance, serum parameters, ruminal fermentation, and bacteria community profile in beef cattle. *Food Sci. Nutr.* **2021**, *9*, 5959–5970. [CrossRef]
12. Li, F.; Usman, S.; Huang, W.; Jia, M.; Kharazian, Z.A.; Ran, T.; Li, F.; Ding, Z.; Guo, X. Correction: Effects of inoculating feruloyl esterase-producing Lactiplantibacillus plantarum A1 on ensiling characteristics, in vitro ruminal fermentation and microbiota of alfalfa silage. *J. Anim. Sci. Biotechnol.* **2023**, *14*, 53. [CrossRef]
13. Luo, Z.; Evans, A.; Bu, D. Effects of corn silage, alfalfa hay, and oat-vetch mixed silage combinations on milk production, rumen fermentation, and nutrient digestibility in high-producing dairy cows. *J. Dairy Sci.* **2025**, *108*, 6990–7001. [CrossRef] [PubMed]
14. Cui, Z.Y.; Li, W.J.; Wang, W.K.; Wu, Q.C.; Jiang, Y.W.; Aisikaer, A.; Zhang, F.; Chen, H.W.; Yang, H.J. Wheat silage partially replacing oaten hay exhibited greater feed efficiency and fibre digestion despite low feed intake by feedlot lambs. *Anim. Nutr.* **2023**, *15*, 332–340. [CrossRef]
15. Ran, T.; Tang, S.X.; Yu, X.; Hou, Z.P.; Hou, F.J.; Beauchemin, K.A.; Yang, W.Z.; Wu, D.Q. Diets varying in ratio of sweet sorghum silage to corn silage for lactating dairy cows: Feed intake, milk production, blood biochemistry, ruminal fermentation, and ruminal microbial community. *J. Dairy Sci.* **2021**, *104*, 12600. [CrossRef]
16. Liu, Z.; Jiang, A.; Lv, X.; Fan, D.; Chen, Q.; Wu, Y.; Zhou, C.; Tan, Z. Combined Metabolomics and Biochemical Analyses of Serum and Milk Revealed Parity-Related Metabolic Differences in Sanhe Dairy Cattle. *Metabolites* **2024**, *14*, 227. [CrossRef]
17. Keulen, J.V.; Young, B.A. Evaluation of Acid-Insoluble Ash as a Natural Marker in Ruminant Digestibility Studies. *J. Anim.* **1977**, *44*, 282–287. [CrossRef]
18. Liu, Z.; Jiang, A.; Lv, X.; Zhou, C.; Tan, Z. Metabolic Changes in Serum and Milk of Holstein Cows in Their First to Fourth Parity Revealed by Biochemical Analysis and Untargeted Metabolomics. *Animals* **2024**, *14*, 407. [CrossRef]
19. Moallem, U. Future consequences of decreasing marginal production efficiency in the high-yielding dairy cow. *J. Dairy Sci.* **2016**, *99*, 2986–2995. [CrossRef] [PubMed]
20. Wang, H.R.; Chen, Q.; Chen, L.M.; Ge, R.F.; Wang, M.Z.; Yu, L.H.; Zhang, J. Effects of dietary physically effective neutral detergent fiber content on the feeding behavior, digestibility, and growth of 8- to 10-month-old Holstein replacement heifers. *J. Dairy Sci.* **2017**, *100*, 1161–1169. [CrossRef] [PubMed]
21. Wang, M.; Wang, R.; Yang, S.; Deng, J.P.; Tang, S.X.; Tan, Z.L. Effects of three methane mitigation agents on parameters of kinetics of total and hydrogen gas production, ruminal fermentation and hydrogen balance using in vitro technique. *Anim. Sci. J.* **2016**, *87*, 224–232. [CrossRef]
22. Weatherburn, M.W. Phenol-hypochlorite reaction for determination of ammonia. *Anal. Chem.* **1967**, *39*, 971–974. [CrossRef]
23. Zinn, R.A.; Owens, F.N. A rapid procedure for purine measurement and its use for estimating net ruminal protein synthesis. *Can. Vet. J. Rev. Vet. Can.* **1986**, *66*, 157–166. [CrossRef]
24. Jiao, J.; Li, X.; Beauchemin, K.A.; Tan, Z.; Tang, S.; Zhou, C. Rumen development process in goats as affected by supplemental feeding v. grazing: Age-related anatomic development, functional achievement and microbial colonisation. *Br. J. Nutr.* **2015**, *113*, 888–900. [CrossRef] [PubMed]

35. Munyaka, P.M.; Eissa, N.; Bernstein, C.N.; Khafipour, E.; Ghia, J.E. Antepartum Antibiotic Treatment Increases Offspring Susceptibility to Experimental Colitis: A Role of the Gut Microbiota. *PLoS ONE* **2015**, *10*, e142536. [CrossRef]
36. Jiao, J.; Lu, Q.; Forster, R.J.; Zhou, C.; Wang, M.; Kang, J.; Tan, Z. Age and feeding system (supplemental feeding versus grazing) modulates colonic bacterial succession and host mucosal immune maturation in goats. *J. Anim. Sci.* **2016**, *94*, 2506–2518. [CrossRef]
37. Amato, K.R.; Yeoman, C.J.; Kent, A.; Righini, N.; Carbonero, F.; Estrada, A.; Gaskins, H.R.; Stumpf, R.M.; Yildirim, S.; Torralba, M.; et al. Habitat degradation impacts black howler monkey (Alouatta pigra) gastrointestinal microbiomes. *ISME J.* **2013**, *7*, 1344–1353. [CrossRef] [PubMed]
38. Segata, N.; Izard, J.; Waldron, L.; Gevers, D.; Miropolsky, L.; Garrett, W.S.; Huttenhower, C. Metagenomic biomarker discovery and explanation. *Genome Biol.* **2011**, *12*, R60. [CrossRef]
39. Zhu, J.; Wu, Y.; Jiang, A.; Li, B.; Ran, T.; Zhou, C.; Xiao, D.; Tan, Z. Effects of dietary N-carbamylglutamate on rumen fermentation parameters, and bacterial community diversity of Holstein dairy cows in Tibet. *Front. Microbiol.* **2023**, *14*, 1101620. [CrossRef]
40. Jiang, A.; Liu, Z.; Yang, Z.; Zhang, S.; Wu, J.; Zhou, C.; Tan, Z. Direct-fed microbials optimize ruminal fermentation, microbial ecosystem and milk quality to enhance the lactation performance of Sanhe dairy cows. *Anim. Microbiome* **2025**, *7*, 75. [CrossRef]
41. Sun, M.; Qian, X.; Chen, C.; Cheng, S.; Jia, B.; Zhu, Y.; Sun, X. Ectopic Expression of GsSRK in *Medicago sativa* Reveals Its Involvement in Plant Architecture and Salt Stress Responses. *Front. Plant Sci.* **2018**, *9*, 226. [CrossRef] [PubMed]
42. Tian, Y.; Zhang, X.; Li, S.; Liu, K.; Guo, P. Effect of Harvest Time and Microbial Anaerobic Fermentation at Ruminal Degradability, In Vitro Digestibility to Milk Production and Milk Quality for Whole Plant Zhang Hybrid Millet in Dairy Cows. *Animals* **2019**, *9*, 749. [CrossRef]
43. Wang, D.; Liang, G.; Wang, B.; Sun, H.; Liu, J.; Guan, L.L. Systematic microRNAome profiling reveals the roles of microRNAs in milk protein metabolism and quality: Insights on low-quality forage utilization. *Sci. Rep.* **2016**, *6*, 21194. [CrossRef] [PubMed]
44. Wang, Y.; Yu, Q.; Wang, X.; Song, J.; Lambo, M.T.; Huang, J.; He, P.; Li, Y.; Zhang, Y. Replacing alfalfa hay with industrial hemp ethanol extraction byproduct and Chinese wildrye hay: Effects on lactation performance, plasma metabolites, and bacterial communities in Holstein cows. *Front. Vet. Sci.* **2023**, *10*, 1061219. [CrossRef] [PubMed]
45. Kang, J.; Tang, S.; Zhong, R.; Tan, Z.; Wu, D. Alfalfa Silage Treated with Sucrose Has an Improved Feed Quality and More Beneficial Bacterial Communities. *Front. Microbiol.* **2021**, *12*, 670165. [CrossRef]
46. Joch, M.; Kudrna, V.; Výborná, A.; Tyrolová, Y.; Jančík, F.; Kubelková, P.; Vadroňová, M.; Tichá, D. Effect of corn shredlage on feed intake, rumen fermentation, and lactation performance of dairy cows fed a low-fibre diet. *Ital. J. Anim. Sci.* **2023**, *22*, 116–124. [CrossRef]
47. Wang, Y.; Ying, G.; Zhang, Z.; Tang, Y.; Zhang, Y.; Chen, L. Bacillus velezensis promotes the proliferation of lactic acid bacteria and influences the fermentation quality of whole-plant corn silage. *Front. Plant Sci.* **2024**, *15*, 1285582. [CrossRef]
48. Calberry, J.M.; Plaizier, J.C.; Einarson, M.S.; McBride, B.W. Effects of Replacing Chopped Alfalfa Hay with Alfalfa Silage in a Total Mixed Ration on Production and Rumen Conditions of Lactating Dairy Cows. *J. Dairy Sci.* **2003**, *86*, 3611–3619. [CrossRef]
49. Thomas, J.W.; Brown, L.D.; Emery, R.S.; Benne, E.J.; Huber, J.T. Comparisons between alfalfa silage and hay. *J. Dairy Sci.* **1969**, *52*, 195. [CrossRef]
50. Niu, J.; Liu, X.; Xu, J.; Li, F.; Wang, J.; Zhang, X.; Yang, X.; Wang, L.; Ma, S.; Li, D.; et al. Effects of Silage Diet on Meat Quality through Shaping Gut Microbiota in Finishing Pigs. *Microbiol. Spectr.* **2023**, *11*, e241622. [CrossRef]
51. Caparra, P.; Chies, L.; Scerra, M.; Foti, F.; Bognanno, M.; Cilione, C.; De Caria, P.; Claps, S.; Cifuni, G.F. Effect of Dietary Ensiled Olive Cake Supplementation on Performance and Meat Quality of Apulo-Calabrese Pigs. *Animals* **2023**, *13*, 2022. [CrossRef] [PubMed]
52. Jawaid, M.Z.; Ashfaq, M.Y.; Al-Ghouti, M.; Zouari, N. Insights into population adaptation and biodiversity of lactic acid bacteria in challenged date palm leaves silaging, using MALDI-TOF MS. *Curr. Res. Microb Sci.* **2024**, *6*, 100235. [CrossRef]
53. Khan, M.R.; Azam, S.; Ahmad, S.; Ali, Q.; Liaqat, Z.; Rehman, N.; Khan, I.; Alharbi, M.; Alshammari, A. Molecular Characterization and Epidemiology of Antibiotic Resistance Genes of beta-Lactamase Producing Bacterial Pathogens Causing Septicemia from Tertiary Care Hospitals. *Antibiotics* **2023**, *12*, 617. [CrossRef]
54. Kargar, S.; Kanani, M.; Albenzio, M.; Caroprese, M. Substituting corn silage with reconstituted forage or nonforage fiber sources in the starter diets of Holstein calves: Effects on performance, ruminal fermentation, and blood metabolites. *J. Anim. Sci.* **2019**, *97*, 3046–3055. [CrossRef]
55. Shi, J.; Zhang, G.; Ke, W.; Pan, Y.; Hou, M.; Chang, C.; Sa, D.; Lv, M.; Liu, Y.; Lu, Q. Effect of endogenous sodium and potassium ions in plants on the quality of alfalfa silage and bacterial community stability during fermentation. *Front. Plant Sci.* **2023**, *14*, 1295114. [CrossRef]
56. de Souza, D.A.P.; Mendonca, C.; Moreno, A.; Bueno, A.; de Almeida, S.; Seibert, L.; Converti, A.; Watanabe, I.S.; Gierus, M.; de Souza, O.R. Antibacterial and antifungal activity of crude and freeze-dried bacteriocin-like inhibitory substance produced by *Pediococcus pentosaceus*. *Sci. Rep.* **2020**, *10*, 12291. [CrossRef] [PubMed]

57. Muhmood, A.; Liu, J.; Liu, D.; Liu, S.; Azzam, M.M.; Junaid, M.B.; Hou, L.; Le, G.; Huang, K. Mitigation of Deoxynivalenol (DON)- and Aflatoxin B1 (AFB1)-Induced Immune Dysfunction and Apoptosis in Mouse Spleen by Curcumin. *Toxins* **2024**, *16*, 356. [CrossRef] [PubMed]
58. Ma, J.; Wang, C.; Wang, Z.; Cao, G.; Hu, R.; Wang, X.; Zou, H.; Kang, K.; Peng, Q.; Xue, B.; et al. Active dry yeast supplementation improves the growth performance, rumen fermentation, and immune response of weaned beef calves. *Anim. Nutr.* **2021**, *7*, 1352–1359. [CrossRef]
59. Cheng, Z.; Meng, Z.; Tan, D.; Datsomor, O.; Zhan, K.; Lin, M.; Zhao, G. Effects of supplementation of sodium acetate on rumen fermentation and microbiota in postpartum dairy cows. *Front. Microbiol.* **2022**, *13*, 1053503. [CrossRef]
60. Zou, Y.; Zou, X.P.; Cao, X.L.G.J. Substituting oat hay or maize silage for portion of alfalfa hay affects growth performance, ruminal fermentation, and nutrient digestibility of weaned calves. *Anim. Biosci.* **2018**, *31*, 369–378. [CrossRef]
61. Barbato, O.; Holmes, B.; Filipescu, I.E.; Celi, P. Dietary Supplementation of Yerba Mate (Ilex paraguariensis) during the Dry Period Improves Redox Balance in Lactating Dairy Cows. *Antioxidants* **2019**, *8*, 38. [CrossRef]
62. Kong, F.; Wang, S.; Dai, D.; Cao, Z.; Wang, Y.; Li, S.; Wang, W. Preliminary Investigation of the Effects of Rosemary Extract Supplementation on Milk Production and Rumen Fermentation in High-Producing Dairy Cows. *Antioxidants* **2022**, *11*, 1715. [CrossRef]
63. Qu, X.; Raza, S.; Zhao, Y.; Deng, J.; Ma, J.; Wang, J.; Alkhorayef, N.; Alkhalil, S.S.; Pant, S.D.; Lei, H.; et al. Effect of Tea Saponins on Rumen Microbiota and Rumen Function in Qinchuan Beef Cattle. *Microorganisms* **2023**, *11*, 374. [CrossRef] [PubMed]
64. Jing, X.; Ding, L.; Zhou, J.; Huang, X.; Degen, A.; Long, R. The adaptive strategies of yaks to live in the Asian highlands. *Anim. Nutr.* **2022**, *9*, 249–258. [CrossRef]
65. Dai, R.; Ma, X.; Dingkao, R.; Huang, C.; La, Y.; Li, X.; Ma, X.; Wu, X.; Chu, M.; Guo, X.; et al. Effects of dietary crude protein levels in the concentrate supplement after grazing on rumen microbiota and metabolites by using metagenomics and metabolomics in Jersey-yak. *Front. Microbiol.* **2023**, *14*, 1124917. [CrossRef]
66. Beaver, A.; Petersen, C.; Weary, D.M.; Finlay, B.B.; von Keyserlingk, M.A.G. Differences in the fecal microbiota of dairy calves reared with differing sources of milk and levels of maternal contact. *JDS Commun.* **2021**, *2*, 200–206. [CrossRef]
67. Li, M.; Zi, X.; Lv, R.; Zhang, L.; Ou, W.; Chen, S.; Hou, G.; Zhou, H. Cassava Foliage Effects on Antioxidant Capacity, Growth, Immunity, and Ruminal Microbial Metabolism in Hainan Black Goats. *Microorganisms* **2023**, *11*, 2320. [CrossRef]
68. Tan, R.; Jin, M.; Shao, Y.; Yin, J.; Li, H.; Chen, T.; Shi, D.; Zhou, S.; Li, J.; Yang, D. High-sugar, high-fat, and high-protein diets promote antibiotic resistance gene spreading in the mouse intestinal microbiota. *Gut Microbes.* **2022**, *14*, 2022442. [CrossRef] [PubMed]
69. Wang, Y.; Takano, T.; Zhou, Y.; Wang, R.; Toshimitsu, T.; Sashihara, T.; Tanokura, M.; Miyakawa, T.; Nakajima-Adachi, H.; Hachimura, S. Orally administered Lactiplantibacillus plantarum OLL2712 decreased intestinal permeability, especially in the ileum: Ingested lactic acid bacteria alleviated obesity-induced inflammation by collaborating with gut microbiota. *Front. Immunol.* **2023**, *14*, 1123052. [CrossRef] [PubMed]
70. Antoine, D.; Chupikova, I.; Jalodia, R.; Singh, P.K.; Roy, S. Chronic Morphine Treatment and Antiretroviral Therapy Exacerbate HIV-Distal Sensory Peripheral Neuropathy and Induce Distinct Microbial Alterations in the HIV Tg26 Mouse Model. *Int. J. Mol. Sci.* **2024**, *25*, 1569. [CrossRef]
71. Huang, F.; Pan, H.; Tan, Z.; Chen, L.; Li, T.; Liu, Y. Prevotella histicola Prevented Particle-Induced Osteolysis via Gut Microbiota-Dependent Modulation of Inflammation in Ti-Treated Mice. Probiotics Antimicrob. *Proteins* **2024**, *16*, 383–393.
72. Hartinger, T.; Pacifico, C.; Sener-Aydemir, A.; Poier, G.; Kreuzer-Redmer, S.; Terler, G.; Klevenhusen, F.; Zebeli, Q. Dietary carbohydrate sources differently prime the microbial ecosystem but not the epithelial gene expression profile along the complete gut of young calves. *Anim. Microbiome* **2024**, *6*, 12. [CrossRef] [PubMed]
73. Yin, X.; Wang, P.; Yan, Z.; Yang, Q.; Huang, X.; Gun, S. Effects of Whole-Plant Corn Silage on Growth Performance, Serum Biochemical Indices, and Fecal Microorganisms in Hezuo Pigs. *Animals* **2024**, *14*, 662. [CrossRef] [PubMed]
74. Nathani, N.M.; Patel, A.K.; Mootapally, C.S.; Reddy, B.; Shah, S.V.; Lunagaria, P.M.; Kothari, R.K.; Joshi, C.G. Effect of roughage on rumen microbiota composition in the efficient feed converter and sturdy Indian Jaffrabadi buffalo (*Bubalus bubalis*). *BMC Genom.* **2015**, *16*, 1116. [CrossRef]
75. Ransom-Jones, E.; Jones, D.L.; McCarthy, A.J.; McDonald, J.E. The Fibrobacteres: An important phylum of cellulose-degrading bacteria. *Microb. Ecol.* **2012**, *63*, 267–281. [CrossRef] [PubMed]
76. Xue, M.Y.; Wu, J.J.; Xie, Y.Y.; Zhu, S.L.; Zhong, Y.F.; Liu, J.X.; Sun, H.Z. Investigation of fiber utilization in the rumen of dairy cows based on metagenome-assembled genomes and single-cell RNA sequencing. *Microbiome* **2022**, *10*, 11. [CrossRef]
77. Duniere, L.; Ruiz, P.; Lebbaoui, Y.; Guillot, L.; Bernard, M.; Forano, E.; Chaucheyras-Durand, F. Effects of rearing mode on gastro-intestinal microbiota and development, immunocompetence, sanitary status and growth performance of lambs from birth to two months of age. Anim. *Microbiome* **2023**, *5*, 34. [CrossRef]
78. Ianni, A.; Martino, G. Dietary Grape Pomace Supplementation in Dairy Cows: Effect on Nutritional Quality of Milk and Its Derived Dairy Products. *Foods* **2020**, *9*, 168. [CrossRef]

79. de Araújo, A.B.; Azul, F.V.C.S.; Silva, F.R.M.; de Almeida, T.S.; Oliveira, J.V.N.; Pimenta, A.T.Á.; Bezerra, A.M.E.; Machado, N.J.; Leal, L.K.A.M.; Rauf, A. Antineuroinflammatory Effect of Amburana cearensis and Its Molecules Coumarin and Amburoside A by Inhibiting the MAPK Signaling Pathway in LPS-Activated BV-2 Microglial Cells. *Oxidative Med. Cell. Longev.* **2022**, *2022*, 6304014–6304087. [CrossRef]
80. Lam, M.Q.; Oates, N.C.; Leadbeater, D.R.; Goh, K.M.; Yahya, A.; Md, S.M.; Ibrahim, Z.; Bruce, N.C.; Chong, C.S. Genomic Analysis to Elucidate the Lignocellulose Degrading Capability of a New Halophile Robertkochia solimangrovi. *Genes* **2022**, *13*, 2135. [CrossRef]

animals

MDPI

Article

Mixed Ensiling Increases Degradation Without Altering Attached Microbiota Through *In Situ* Ruminal Incubation Technique

Xuanxuan Pu [1,2,3,†], Min Zhang [4,†], Jianjun Zhang [4], Xiumin Zhang [4,*], Shizhe Zhang [4], Bo Lin [5], Tianwei Wang [6], Zhiliang Tan [4] and Min Wang [4,*]

1 College of Animal Science and Technology, Tarim University, Alar 843300, China; puxuanxuan@taru.edu.cn
2 Key Laboratory of Tarim Animal Husbandry Science and Technology of Xinjiang Production and Construction Group, Alar 843300, China
3 Key Laboratory of Livestock and Forage Resources Utilization around Tarim, Ministry of Agriculture and Rural Affairs, Alar 843300, China
4 CAS Key Laboratory for Agro-Ecological Processes in Subtropical Region, Hunan Provincial Engineering Research Center for Healthy Livestock and Poultry Production, Institute of Subtropical Agriculture, Chinese Academy of Sciences, Changsha 410125, China; zmlalala123@outlook.com (M.Z.); zhangjianjun0522@163.com (J.Z.); zszdyxhm@126.com (S.Z.); zltan@isa.ac.cn (Z.T.)
5 College of Animal Science and Technology, Guangxi University, Nanning 530004, China; linbo@gxu.edu.cn
6 State Key Laboratory of Microbial Resources, Institute of Microbiology, Chinese Academy of Sciences, Beijing 100101, China; wangtw@im.ac.cn
* Correspondence: xmzhang@isa.ac.cn (X.Z.); mwang@isa.ac.cn (M.W.)
† These authors contributed equally to this work.

Simple Summary

Better utilization of rape straw can provide alternative strategies for the sustainability of ruminants and food production. In the present study, rape straw was mixed ensiled with whole crop corn, and obtained good fermentation quality by enriching lactic acid bacteria. For a better understanding of mixed silage effects on the nutrition and utilization value of rape straw, rape straw was mixed with whole crop corn silage and used as a control group. Results showed that mixed silage destroyed the fiber structure of rape straw, and enhanced *in situ* nutrient degradation by increasing both rapidly and potentially degradable nutrients. In summary, mixed silage provides a potential strategy to improve nutrient and utilization value of rape straw.

Abstract

Mixed silage can disrupt the girder structure of rape straw, and thus facilitate ruminal degradation. Further investigation is warranted to validate this observation in vivo. The objective of this study was to investigate the degradation kinetics and bacterial colonization of mixed silage during digestion using an *in situ* ruminal incubation technique. The experiment comprised two treatments: a mixture of rape straw and corn silage (control), and a mixed silage treatment of rape straw and whole crop corn (mixed silage). Three ruminally cannulated Holstein bulls were employed. Substrates were incubated for varying durations (4, 12, 24, 48, 72, 96, 120 and 216 h) to assess substrate degradation kinetics. Bacterial colonization were analyzed after 4- and 48-h incubation time. Mixed ensiling disrupted the fiber structure of rape straw, and thus had lower fiber content compared to the control, as NDF and ADF content decreased by 55 g/kg (678 vs. 623 g/kg) and 27 g/kg (440 vs. 413 g/kg), respectively. Compared to the control group, ruminal DM disappearance of mixed silage significantly ($p \leq 0.05$) increased from 315 to 366 g/kg (+16.2%) at an incubation time of 4 h, 552 to 638 g/kg (+15.6%) at 120 h, and 563 to 651 g/kg (+15.6%) at 216 h. Similarly, compared to the control group, NDF disappearance of mixed silage significantly ($p \leq 0.05$) rose from

112 to 201 g/kg (+79.5%) at 4 h, 405 to 517 g/kg (+27.7%) at 120 h, and 429 to 532 g/kg (+24.0%) at 216 h. Compared to the control group, soluble and washout nutrient fractions (*a*) of DM or NDF fraction in mixed silage significantly ($p \leq 0.05$) rose from 289 to 340 g/kg (+17.6%), potentially degradable fractions (*b*) of NDF increased from 310 to 370 g/kg (+19.4%), and the undegraded fraction of NDF (μNDF) decreased from 582 to 471 g/kg (−19.1%). Incubation time, apart from in the mixed ensiling treatment, altered the bacterial community. The study highlights that higher total potentially degradable fractions account for enhanced ruminal substrate degradation of mixed silage.

Keywords: mixed silage; rape straw; ruminal fiber degradation; micro-structure; ruminal bacteria

1. Introduction

About 4.05 million tons of rape straw are produced annually in China, and its disposal has become a concern [1]. Burning, as a traditional way to treat rape straw, causes both environmental pollution and resource waste [1]. Rape straw can be used as a potential feedstuff for ruminants, but the high lignified fiber makes it difficult for ruminants to degrade and utilize [2]. Therefore, strategies are needed to improve the availability of rape straw as ruminant feed. Mixed silage of straw with materials that are rich in water soluble carbohydrates (WSC), such as whole crop corn, promises to exhibit good fermentation quality for long-term storage as a rapid decrease in pH [3,4]. The microorganism enriched during the ensiling process could break down the ligno-cellulosic bonds of the plant cell wall [5,6], which facilitates ruminal microbial colonization and thus enhances the feeding value of straw for ruminants. However, no study has yet been conducted to evaluate effects of mixed silage on the carbohydrate composition of rape straw, which is critical in improving potential utilization of rape straw.

Microorganisms colonizing the rumen play crucial roles in digesting plant materials [7], which can change with prolongation of incubation time [8]. Disparate forages can alter attached microbiota, owing to their chemical composition and ruminal availability differences [9]. For instance, rice straw treated with sodium hydroxide can enrich fibrolytic species compared to untreated rice straw [10]. The hydrolyzed plant cell walls may enhance bacterial colonization by generating substrates that favor specific microorganisms [11]. However, it remains unclear whether changes in the micro-structure of rape straw during the ensiling process could impact ruminal bacterial attachment, and consequently influence ruminal substrate degradation.

We hypothesized that mixed ensiling rape straw with whole crop corn could produce good fermentation quality, and disrupt the fiber structure of rape straw, and thus result in better nutrient and ruminal utilization value of rape straw. This study was designed to investigate fermentation quality in a mixed silage treatment, and observed the fiber structure changes of rape straw during the mixed ensiling process. An *in situ* ruminal incubation technique was employed to investigate the effects of the mixed silage treatment on ruminal degradation kinetics and the attached bacterial community.

2. Materials and Methods

2.1. Silage Preparation and Sample Collection

Whole crop corn was harvested with a stem cut about 15 cm above the ground level when 50% to 60% of the grain had the characteristics of the dough stage. Rape was harvested 30 d after the end of flowering, and rape straw was air-dried after it was obtained. Both rape straw and whole crop corn were harvested in Hulunbuir City (47°05′ N to 53°20′ N), Inner Mongolia Autonomous Region, China. Prior to ensiling, both rape straw and whole crop corn were chopped into 2–3 cm pieces. For the control treatment, a portion of whole crop corn was packed and sealed into plastic bags (20 cm × 30 cm), and anaerobically fermented for 60 d at room temperature to obtain corn silage, which was then mixed with rape straw at the dry matter (DM) ratio of 1:2 to obtain the mixture of rape straw and corn silage (control). In the preparation of mixed silage, rape straw was combined with whole crop corn at the DM ratio of 2:1. The mixture was sprayed with distilled water to adjust moisture content to around 60%, as described in our previous study [12]. Subsequently, the mixture was packed and sealed into plastic bags (20 cm × 30 cm), and anaerobically fermented for 60 d at room temperature to obtain mixed silage. Both the control and mixed silage were prepared with four replicates. Samples collected for the analysis of fermentation parameters, nutrient compositions and bacterial compositions were according to procedures outlined in our previous studies [12].

2.2. Animals

The experiment was conducted at the beef cattle farm of Guangxi Taiminxing Animal Husbandry development Co., Ltd., Nanning, China. Three ruminally cannulated Holstein bulls (12 mo, 280 ± 7.5 kg) were employed for the *in situ* incubation. Cattle were offered a high-forage diet consisting of wheat straw and concentrates (Table 1), and fed twice daily at 0800 and 1700 in individual stalls.

Table 1. The ingredients and nutrient composition of the cattle diet (g/kg DM).

Item *	Value
Ingredients, g/kg	
Wheat straw	800
Corn	115.9
Soybean meal	13.3
Cottonseed meal	15.2
Wheat bran	15.2
Soybean hull	11.4
DDGS	9.5
$NaHCO_3$	1.9
NaCl	1.9
Premix	5.7
Urea	10.0
Chemical composition, g/kg	
DM	921
OM	902
CP	82
NDF	634
ADF	377
NEm, MJ/kg	4.19
NEf, MJ/kg	2.18

* DM, dry matter; OM, organic matter; CP, crude protein; NDF, neutral detergent fiber; ADF, acid detergent fiber; NEm, net energy for maintenance; NEf, net energy for fattening. The premix provided vitamins and microelements per kilogram as follows: 1,000,000 IU of vitamin A; 200,000 IU of vitamin E; 8000 mg of Zn; 80 mg of Se; 120 mg of I; 2000 mg of Fe; 40 mg of Co; 2500 mg of Mn; 2000 mg of Cu. The NEm and NEf were calculated according to Feeding Standard of Beef Cattle (NY/T 815-2004, China [13]).

2.3. *In Situ* Ruminal Incubation

Holstein bulls were given 30 d to adapt to the diet and environment before the *in situ* ruminal incubation. The procedure of incubation with nylon bags in the rumen of animal was slightly modified according to the procedure described by Zhang et al. [14]. Briefly, approximately 6.5 g of substrate (either control or mixed silage) was weighed into polyester bags. For each Holstein bull, a total of 20 bags were prepared for each forage, with 16 bags allocated for degradation kinetics analysis and 4 bags for assessing microbial community changes. Incubation was staggered in time to ensure no more than 12 bags were present in the rumen for each Holstein bull at each time point, ensuring adequate contact between the nylon bags and rumen microbes. For the degradation kinetics analysis, incubation times were set at 4, 12, 24, 48, 72, 96, 120 and 216 h, with duplicate bags collected at each sampling time. For the microbial community analysis, incubation times were set at 4 and 48 h, with duplicate bags collected at each time point. Upon removal from the rumen, bags for degradation kinetics analyses were put into ice water to arrest microbial activity and then dried at 65 °C for the subsequent chemical analysis. Microbial bags were cleaned with phosphate buffer saline (pH = 7.4), and the liquid was gently wrung out. Then, the inner content of the bags was put into a 15-mL tube and stored at −80 °C for microbial DNA extraction.

2.4. Analytical Methods

The DM and crude protein (CP) were determined as outlined in AOAC methods [15]. NDF and ADF were measured as outlined in Van Soest et al. [16], and hemicellulose content was calculated as NDF minus ADF. Scanning electron microscopy (model SU8010, Hitachi, Tokyo, Japan) was employed to obtain images of rape straw according to the manufacturer's instructions. Lactate was measured according to the method described by Taylor [17]. Individual VFA concentrations were measured using a gas chromatograph (Agilent 7890 A, Agilent Inc., Palo Alto, CA, USA) [18]. The ammonia-N concentration was measured using the phenol–hypochlorite colorimetric method [19].

2.5. DNA Extraction and High-Throughput Sequencing

DNA extraction was performed using methodology modified from that of Ma et al. [20]. Illumina sequencing was performed using the MiSeq platform (Illumina, San Diego, CA, USA) at the Shanghai Biozeron Biological Technology Co., Ltd. (Shanghai, China). After PCR amplification, all amplicon libraries were sequenced and the barcodes and sequencing primers were removed before data processing. Raw reads were processed using SMRT Link Analysis software version 11.0 to obtain demultiplexed circular consensus sequence (CCS) reads with the following settings: minimum number of passes = 3, minimum predicted accuracy = 0.99. Raw reads were processed through SMRT Portal to filter sequences for length (>1000 or <1800 bp) and quality. Sequences were further filtered by removing barcode and primer sequences with Lima Pipeline (Pacific Biosciences demultiplexing barcoded software, https://lima.how/). The circular consensus sequencing reads were clustered into amplicon sequences variants using QIIME 2-data2 [21]. The phylogenetic affiliation of each 16S rRNA gene sequence was analyzed by RDP Classifier (https://mothur.s3.us-east-2.amazonaws.com/wiki/trainset19_072023.rdp.tgz, accessed on 15 July 2025) against the SILVA (SSU138) 16S rRNA database using a confidence threshold of 70% [22].

2.6. Calculations

The exponential model of Ørskov and McDonald [23] was used to analyze degradation kinetics parameters. The effective rumen degradation (*ED*, g/kg) was calculated according to the method described by McDonald [24]. The undegradable fraction of NDF (μNDF, g/kg) was estimated by 1000 minus NDF disappearance measured at 216-h incubation time.

2.7. Statistical Analysis

Data was analyzed using SPSS 26.0 software (Chicago, IL, USA). Exponential parameters were tested by paired *T*-tests based on each animal. Nutrient disappearances at each incubation time were analyzed using a linear mixed model with treatment ($n = 2$) as fixed effect and animal ($n = 3$) as random effect. A Wilcoxon rank-sum test was employed for analyzing attached bacteria compositions. Principal coordinate analysis (PCoA) using amplicon sequences variants with adonis in vegan of R 4.0.3 based on the Bray–Curtis distance was used to test effects of treatment and incubation time on the attached bacterial community in the substrate. $p \leq 0.05$ indicates that a finding was significant, and $0.05 < p \leq 0.10$ indicates a tendency to be significant.

3. Results

Mixed silage exhibited good silage quality, as indicated by the low pH (4.22), ammonia-N (2.66 g/kg), propionate (2.29 g/kg) and butyrate (0.01 g/kg), and good lactate (32.3 g/kg) (Table 2). The mixed silage was dominated by the genuses of *Lentilactobacillus* (52.6%) and *Pediococcus* (23.6%) (Table 2).

Table 2. Fermentation parameters and bacterial compositions in the control and mixed silage.

Items *	Control	Mixed Silage
Fermentation end products, g/kg DM		
pH	-	4.22 ± 0.15
Ammonia-N, g/kg DM	3.52 ± 0.32	2.66 ± 0.18
Lactate, g/kg DM	20.5 ± 1.22	32.3 ± 1.36
Acetate, g/kg DM	9.23 ± 1.23	24.2 ± 1.56
Propionate, g/kg DM	0.26 ± 0.19	2.29 ± 0.22
Butyrate, g/kg DM	0.05 ± 0.02	0.01 ± 0.01
Nutrient composition, g/kg DM		
DM	408 ± 1.25	405 ± 2.00
CP	60.5 ± 0.78	61.3 ± 0.89
NDS	322 ± 2.12	377 ± 1.56
NDF	678 ± 2.12	623 ± 1.56
ADF	440 ± 1.56	413 ± 0.98
Hemicellulose	238 ± 1.76	210 ± 1.35
Bacterial compositions (genus), %		
Lentilactobacillus	-	52.6 ± 2.35
Pediococcus	-	23.6 ± 2.89
Pantoea	-	5.29 ± 1.23
Stenotrophomonas	-	3.23 ± 0.98
Levilactobacillus	-	2.34 ± 0.78
Erwinia	-	2.17 ± 1.35
Secundilactobacillus	-	1.23 ± 0.46
Escherichia	-	0.82 ± 0.50
Salmonella	-	0.61 ± 0.23
Sphingobacterium	-	0.61 ± 0.12
Sarcina	-	0.53 ± 0.13
others	-	6.97 ± 2.67

* DM, dry matter; CP, crude protein; NDS, neutral detergent solubles; NDF, neutral detergent fiber; ADF, acid detergent fiber. Control, a mixture of rape straw and whole crop corn silage; Mixed silage, ensiling treatment of rape straw and whole crop corn mixture; -, not measured. The same as below.

Compared with the control group, mixed silage had lower NDF (623 vs. 678 g/kg), ADF (413 vs. 440 g/kg) and hemicellulose (210 vs. 238 g/kg) contents, and higher NDS content (377 vs. 322 g/kg) (Table 2). Mixed silage altered the physical structure of rape straw, as indicated by the scanning electron microscopy images (Figure 1). The surface of rape straw was smooth and complete in the control group, while the fiber surface connection in mixed silage was broken, leading to partially regular or irregular pore structures with large numbers of microbes colonized in.

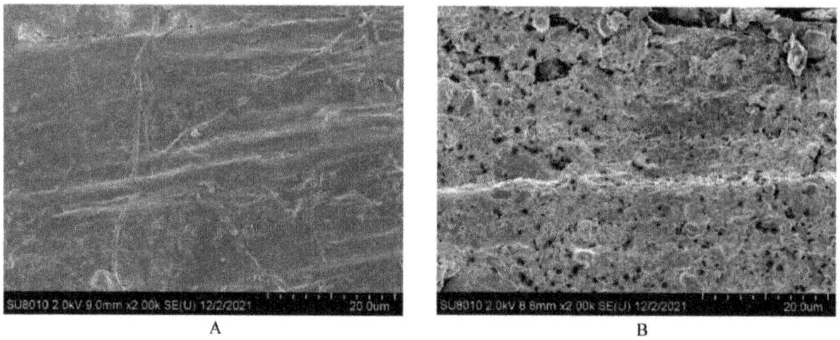

Figure 1. Scanning electron microscopy images (scale bar = 20 μm) of rape straw in treatments without (i.e., control, (**A**)) or with mixed ensiling (i.e., mixed silage, (**B**)). Control, a mixture of rape straw and whole crop corn silage; Mixed silage, mixed silage of rape straw and whole crop corn.

In the present study, mixed silage had distinct kinetics of DM or NDF disappearance compared to those of the control, and exhibited greater ($p \leq 0.05$) DM or NDF disappearance at the incubation times of 4, 120 and 216 h than the control (Figure 2).

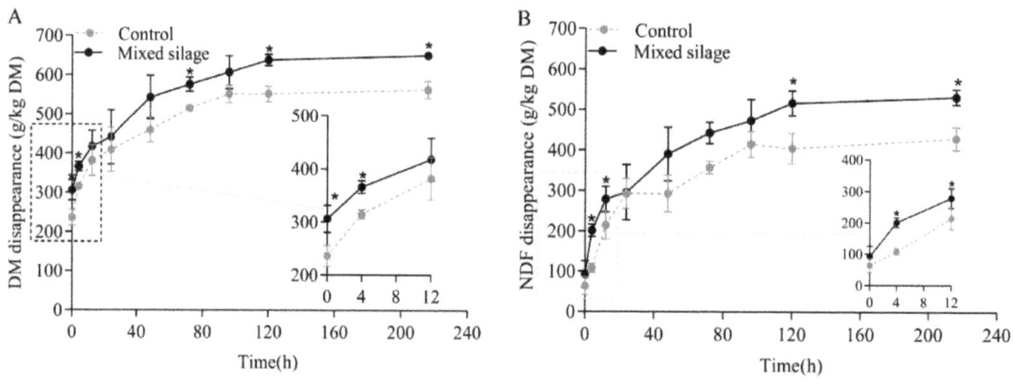

Figure 2. Effects of mixed silage on the kinetics of *in situ* ruminal disappearance of DM (**A**) and NDF (**B**). Control, a mixture of rape straw and whole crop corn silage; Mixed silage, mixed silage of rape straw and whole crop corn. * indicates significant difference ($p \leq 0.05$).

The control and mixed silage showed distinct *in situ* kinetics of ruminal disappearance in DM and NDF (Table 3). Mixed silage had greater ($p \leq 0.05$) a, $a + b$, ED_2 and ED_6 for both *in situ* DM and NDF disappearance. Mixed silage had greater ($p \leq 0.05$) b for *in situ* NDF disappearance, and tended to have greater b and lower lag for *in situ* DM disappearance ($0.05 < p \leq 0.10$). Mixed silage exhibited greater ($p \leq 0.05$) ED and lower ($p \leq 0.05$) μNDF than the control group.

Table 3. Effects of mixed ensiling on the kinetics of *in situ* ruminal disappearance of DM and NDF.

Items [a]	Treatments [b]		SEM	*p*-Value
	Control	Mixed Silage		
		DM		
a, g/kg	289	340	12.6	0.01
b, g/kg	272	318	13.0	0.08
$a + b$, g/kg	561	659	22.3	0.01
c,% h^{-1}	3.18	2.37	0.49	0.29
Lag, h	2.82	2.46	0.232	0.08
ED_2, g/kg	449	507	17.5	0.02
ED_6, g/kg	379	427	16.2	0.03
		NDF		
a, g/kg	108	159	12.1	0.01
b, g/kg	310	370	14.5	0.04
$a + b$, g/kg	418	529	26.0	0.01
c,% h^{-1}	3.16	2.51	0.42	0.31
lag, h	3.13	2.36	0.292	0.31
ED_2, g/kg	293	361	19.4	0.02
ED_6, g/kg	212	266	16.4	0.04
μNDF, g/kg	582	471	26.0	0.01

[a] *a* soluble and washout nutrient fractions and represents the rapidly degraded component; *b*, potentially degradable nutrient fractions and represents the slowly degraded component; *a* + *b*, total potentially degradable fraction; *c*, disappearance rate of fraction *b*; *lag*, lag time before the initial degradation of fraction *b*; ED_2 and ED_6, effective ruminal degradation calculated with passage rate being 0.02/h and 0.06/h respectively; μNDF, undegraded fraction of NDF at 216 h of incubation time. [b] Control, a mixture of rape straw and whole crop corn silage; Mixed silage, mixed ensiling treatment of rape straw and whole crop corn mixture.

The bacterial communities attached to the incubated substrate were greatly affected by the incubation time, independent of the treatment. Mixed silage had a similar relative abundance of bacteria at phylum and genus levels ($p > 0.10$), when compared with the control group (Table 4). Extending incubation time from 4 to 48 h enriched the phyla Proteobacteria, Fibrobacteres and Actinobacteria, and genuses *Kineothrix*, *Lacrimispora*, *Negativibacillus*, *Stenotrophomonas* and *Fibrobacter* ($p \leq 0.05$), and decreased the phylum Bacterodietes, and genuses *Prevotella* and *Paraprevotella* ($p \leq 0.05$) (Table 4). The PCoA analyses showed that incubation time, rather than treatment, led to the distinct clustering, with PC 1 and PC 2 explaining 30% and 23% of variation in bacterial amplicon sequences variants, respectively (Figure 3, $P_{\text{Incubation time}} \leq 0.05$).

Table 4. Bacterial compositions with relative abundance larger than 0.5% attached to incubated substrate in control or mixed silage.

Items *	Treatment		Incubation Time (h)		SEM	*p*-Value		
	Control	Mixed Silage	4	48		Treatment	Incubation Time	Treatment × Incubation Time
Phylum Firmicutes	43.0	45.5	43.5	45.0	1.12	0.33	0.55	0.48
Papillibacter	2.29	2.31	0.98	3.62	0.414	0.92	<0.001	0.47
Kineothrix	1.46	1.69	1.29	1.86	0.151	0.29	0.03	0.06
Lacrimispora	1.58	2.11	1.06	2.63	0.351	0.26	0.01	0.27
Negativibacillus	0.59	0.54	0.29	0.83	0.120	0.70	0.01	0.66
Phylum Bacteroidetes	32.2	31.2	38.7	24.6	2.27	0.59	<0.001	0.66
Prevotella	16.3	16.8	23.0	10.2	2.36	0.88	0.01	0.93
Paraprevotella	5.89	4.73	7.12	3.50	0.752	0.21	0.01	0.55
Phylum Proteobacteria	16.4	14.9	11.3	20.0	1.50	0.23	<0.001	0.73
Stenotrophomonas	9.98	8.54	4.54	14.0	1.51	0.14	<0.001	0.13
Phylum Fibrobacteres	1.79	2.00	0.73	3.06	0.459	0.63	0.001	0.15
Fibrobacter	1.75	1.94	0.68	3.01	0.458	0.66	0.001	0.14
Phylum Actinobacteria	0.61	0.49	0.30	0.79	0.094	0.37	0.01	0.93

* Control, a mixture of rape straw and whole crop corn silage; Mixed silage, mixed silage of rape straw and whole crop corn.

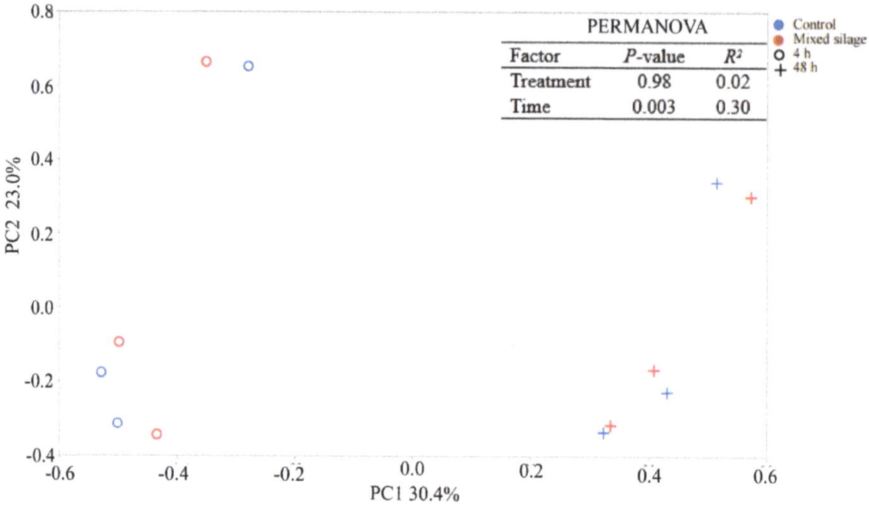

Figure 3. Principal coordinate analysis (PCoA) of bacterial community attached to control and mixed silage at 4-h and 48-h incubation. Control, a mixture of rape straw and whole crop corn silage; Mixed silage, mixed silage of rape straw and whole crop corn.

4. Discussion

Farmers quickly evaluate the quality of silage by its appearance, color, smell, water content and pH [25,26]. It has been reported that a suitable water content of silage can reduce the likelihood of bad fermentation, and that silage ferments well when the pH is below 4.2 [4,27]. Lactic acid bacteria in silage can use carbohydrates to produce lactate, which reduces pH value and inhibits the multiplication of other microorganisms [28,29]. Butyrate is a product of poorly fermented silage and carries the risk of causing clinical ketosis in livestock, and was present at a low level as 0.01 g/kg in the product of mixed silage. The pH, desirable lactate content and low butyrate content indicated that the mixed silage we produced was of good quality.

Enrichment of homofermentative *Pediococcus* [30] and heterofermentative *Lentilactobacillus* [31,32] contributed to the favorable fermentative quality in the mixed silage. *Erwinia, Pantoea* and *Stenotrophomonas* are widely distributed in soil and plants [33–35], and their appearance in the mixed silage might be due to the production process of rape straw. In summary, mixed silage exhibited good fermentation quality by enriching *Lentilactobacillus* and *Pediococcus*.

Silage could effectively reduce the fiber content in crop straw. Yang et al. [36] have reported that the NDF and ADF content in *Leymus chinensis* is decreased after ensiling. A previous study showed that silage treatment decomposed the platelet structure of epicuticular wax crystals on the leaves of fresh alfalfa [37], so the fermentation of microorganisms in the mixed silage could impact on the micro-structure of forage. Similarly, the mixed silage treatment reduced the NDF and ADF content of crop straw in the present study, as the surface of rape straw was destroyed. This change comes from the microorganisms that decompose and ferment substances in straw to obtain energy and nutrients during the ensiling process [38].

Disruption of fiber structure helps to partially hydrolyze hemicellulose and solubilize lignin, leading to an increased surface for the colonization of ruminal microorganisms [39], and thus facilitates rumen microbial degradation, especially for complicated plant cell carbohydrates [40,41], which is important for the utilization of straw [38]. That is also the reason for the greater ruminal degradation of mixed silage in the present study. Such enhanced ruminal microbial degradation can be attributed to the destruction of rape straw during the ensiling process. In future, an in vivo study will be conducted to further analyze the effects of silage feeding on production parameters.

Mixed silage had greater *a* for both DM and NDF disappearance in our study. The greater *a* in the mixed silage could be attributed to the greater organic acids and monosaccharides contents. Similar findings were reported by Thomas et al. [42], who observed that the greater washout fractions in sorghum silage are associated with higher water-soluble carbohydrate concentrations. Furthermore, the damage to the surface of rape straw during the ensiling process may lead to the detachment of minute fragments, allowing them to pass through the aperture of the polyester bags, leading to greater rapidly soluble and washout nutrient fractions in the mixed silage. Mixed silage enhanced ruminal NDF availability in the present study, which was consistent with findings observed by Beauchemin et al. [43]. However, some studies have suggested that the reduction in structure carbohydrate content of straw may lead to decreased fiber disappearance [44]. Although straw treatments can increase the fraction of the more readily available portion of the NDF [45], they may also result in a greater proportion of undegradable fiber in NDF residue [46]. In the current study, mixed ensiling demonstrated the ability to reduce undegradable fiber in rape straw, as evidenced by the increased $a + b$ and *ED*.

Bacteria play important roles in the carbohydrate degradation and utilization of straw by attaching to the surface [41,47]. The change of remaining biodegradable nutrients in the nylon bags due to pasture degradation will select the microorganism attached to it [9]. In our study, the mixed silage treatment scarcely altered the bacteria composition attached to the incubated substrate. Pu et al. [48] also reported that the bacterial community attached was not affected by the type of straw (rice or wheat straw). However, Liu et al. [9] reported that the microbial community attached was strongly affected by the straw type (rice straw or alfalfa hay). These different results may be partly explained by differences in the content of chemical ingredients and availability between the straws. The enhanced rumen degradation observed for mixed silage may be attributed to easier access of carbohydrate enzymes resulting from the disruption of rape straw during the ensiling process, rather than relying on the bacterial attachment. Incubation time was another factor affecting ruminal microbiota. The phylum Fibrobacteres is adept at degrading plant cell walls [49], facilitating the degradation of intricate crystalline regions of straw in later incubation stages [41,47]. Both the phyla Bacteroidetes and the genus *Prevotella* are adept at colonizing and utilizing soluble carbohydrates in the early incubation period [8]. That is the reason for the enrichment of the phyla Bacteroidetes and genus *Prevotella* in the early incubation period, and the phyla Fibrobacteres in the later incubation time in the present study. Furthermore, 48 h of incubation time also enriched the genuses *Stenotrophomonas, Papillibacter, Kineothrix, Lacrimispora* and *Negativibacillus*, which might also contribute to the degradation of structural carbohydrate and warrant further investigation.

5. Conclusions

Mixed silage disrupts the fiber structure of rape straw, resulting in higher NDS content and lower fiber content in the mixed silage. This treatment enhances *in situ* ruminal DM and NDF degradation by a creating higher total potentially degradable fractions in the mixed silage. Such enhanced rumen degradation is not associated with changes in attached bacterial communities. Instead, incubation time had a great influence on the bacterial communities, enriching the phyla Fibrobacteres and the genuses *Stenotrophomonas, Papillibacter, Kineothrix, Lacrimispora* and *Negativibacillus* at 48 h of incubation. In summary, the mixed silage treatment improve the feed utilization of rape straw by creating easier physical accessibility.

Author Contributions: Conceptualization, X.Z. and M.W.; methodology, X.P., M.Z., X.Z. and M.W.; validation, J.Z., S.Z., B.L., T.W. and Z.T.; formal analysis, X.P. and M.Z.; investigation, X.P. and M.Z.; resources, B.L. and T.W.; data curation, M.Z., X.P., J.Z. and S.Z.; writing—original draft, X.P. and M.Z.; writing—review and editing, X.Z. and M.W.; supervision, X.Z., M.W. and Z.T.; project administration, M.W.; funding acquisition, X.P., M.W. and X.Z. All authors have read and agreed to the published version of the manuscript.

Funding: This work was supported by the Hunan Province Science and Technology Plan (2022NK2021), The Science and Technology Innovation Program of Hunan Province (2023RC3206), the Strategic Priority Research Program of the Chinese Academy of Sciences (Grant No. XDA26040203), the National Key R&D Program of China (2021YFD1301003, 2023YFD1300900), and the President's Fund of Tarim University (TDZKBS202518).

Institutional Review Board Statement: Animal related procedures were approved by the animal care committee (approval number PA20250326019), College of Animal Science and Technology, Tarim University, Alar, Xinjiang, China.

Informed Consent Statement: Not applicable.

Data Availability Statement: The data presented in this study are available upon request from the corresponding author.

Acknowledgments: We gratefully acknowledge the provision of materials needed in our experiment by Yiming Cheng.

Conflicts of Interest: The authors declare no conflicts of interest.

References

1. Gaballah, E.S.; Abomohra, A.E.; Xu, C.; Elsayed, M.; Abdelkader, T.K.; Lin, J.; Yuan, Q. Enhancement of biogas production from rape straw using different co-pretreatment techniques and anaerobic co-digestion with cattle manure. *Bioresour. Technol.* **2020**, *309*, 123311. [CrossRef] [PubMed]
2. Griffith, C.; Ribeiro, G.O.; Oba, M.; McAllister, T.A.; Beauchemin, K.A. Potential for improving fiber digestion in the rumen of cattle (*Bos taurus*) through microbial inoculation from bison (*Bison bison*): *In situ* fiber degradation. *J. Anim. Sci.* **2017**, *95*, 2156–2167. [CrossRef] [PubMed]
3. Zhang, Q.; Zhao, M.; Wang, X.; Yu, Z.; Na, R. Ensiling alfalfa with whole crop corn improves the silage quality and in vitro digestibility of the silage mixtures. *Grassl. Sci.* **2017**, *63*, 211–217. [CrossRef]
4. Zhao, C.; Wang, L.; Ma, G.; Jiang, X.; Yang, J.; Lv, J.; Zhang, Y. Cellulase Interacts with Lactate Bacteria to Affect Fermentation Quality, Microbial Community, and Ruminal Degradability in Mixed Silage of Soybean Residue and Corn Stover. *Animals* **2021**, *11*, 334. [CrossRef] [PubMed]
5. Xing, L. Study on the Effects of Lactobacillus and Cellulase Additives on the Quality of Different Silage. Master's Thesis, China Agricultural University, Beijing, China, 2004.
6. Jiang, B.W. Effects of Enzyme Bacteria Mixed Treatment of Roughage on Fiber Structure, Growth Performance and Rumen Microflora of Tan Sheep. Ph.D. Thesis, Ningxia University, Ningxia, China, 2021.
7. Mizrahi, I.; Wallace, R.J.; Moraïs, S. The rumen microbiome: Balancing food security and environmental impacts. *Nat. Rev. Microbiol.* **2021**, *19*, 553–566. [CrossRef] [PubMed]
8. Cheng, Y.; Wang, Y.; Li, Y.; Zhang, Y.; Liu, T.; Wang, Y.; Sharpton, T.J.; Zhu, W. Progressive colonization of bacteria and disappearance of rice straw in the rumen by Illumina Sequencing. *Front. Microbiol.* **2017**, *8*, 2165. [CrossRef] [PubMed]
9. Liu, J.; Zhang, M.; Xue, C.; Zhu, W.; Mao, S. Characterization and comparison of the temporal dynamics of ruminal bacterial microbiota colonizing rice straw and alfalfa hay within ruminants. *J. Dairy Sci.* **2016**, *99*, 9668–9681. [CrossRef] [PubMed]
10. Koike, S.; Yabuki, H.; Kobayashi, Y. Interaction of rumen bacteria as assumed by colonization patterns on untreated and alkali-treated rice straw. *J. Anim. Sci.* **2014**, *85*, 524–531. [CrossRef]
11. Meale, S.J.; Beauchemin, K.A.; Hristov, A.N.; Chaves, A.V.; McAllister, T.A. Board-invited review: Opportunities and challenges in using exogenous enzymes to improve ruminant production. *J. Anim. Sci.* **2014**, *92*, 427–442. [CrossRef] [PubMed]

12. Pu, X.X.; Zhang, X.M.; Yi, S.Y.; Wang, R.; Li, Q.S.; Zhang, W.Q.; Qu, J.J.; Huo, J.B.; Lin, B.; Tan, B.; et al. Mixed ensiling plus nitrate destroy fiber structure of rape straw, increase degradation and reduce methanogenesis through in vitro ruminal fermentation. *J. Sci. Food Agric.* **2024**, *104*, 3428–3436. [CrossRef] [PubMed]
13. *NY/T 815-2004*; Beef Cattle Feeding Standards. The Standardization Administration of the People's Republic of China: Beijing, China, 2004.
14. Zhang, X.M.; Gruninger, R.J.; Alemu, A.W.; Wang, M.; Tan, Z.L.; Kindermann, M.; Beauchemin, K.A. 3-Nitrooxypropanol supplementation had little effect on fiber degradation and microbial colonization of forage particles when evaluated using the *in situ* ruminal incubation technique. *J. Dairy Sci.* **2020**, *103*, 8986–8997. [CrossRef] [PubMed]
15. Association of Official Analytical Chemists (AOAC). *Official Methods of Analysis*, 18th ed.; AOAC International: Gaithersburg, MD, USA, 2005.
16. Van Soest, P.J.; Robertson, J.B.; Lewis, B.A. Methods for dietary fiber, neutral detergent fiber, and nonstarch polysaccharides in relation to animal nutrition. *J. Dairy Sci.* **1991**, *74*, 3583–3597. [CrossRef] [PubMed]
17. Taylor, K.A.C.C. A simple colorimetric assay for muramic acid and lactic acid. *Appl. Biochem. Biotech.* **1996**, *56*, 49–58. [CrossRef]
18. Wang, M.; Sun, X.Z.; Janssen, P.H.; Tang, S.X.; Tan, Z.L. Responses of methane production and fermentation pathways to the increased dissolved hydrogen concentration generated by eight substrates in in vitro ruminal cultures. *Anim. Feed. Sci. Technol.* **2014**, *194*, 1–11. [CrossRef]
19. Weatherburn, M. Phenol-hypochlorite reaction for determination of ammonia. *Anal. Chem.* **1967**, *39*, 971–974. [CrossRef]
20. Ma, Z.; Wang, R.; Wang, M.; Zhang, X.; Mao, H.; Tan, Z. Short communication: Variability in fermentation end-products and methanogen communities in different rumen sites of dairy cows. *J. Dairy Sci.* **2018**, *101*, 5153–5158. [CrossRef] [PubMed]
21. Callahan, B.J.; McMurdie, P.J.; Rosen, M.J.; Han, A.W.; Johnson, A.J.; Holmes, S.P. DADA2: High-resolution sample inference from Illumina amplicon data. *Nat. Methods* **2016**, *13*, 581–583. [CrossRef] [PubMed]
22. Amato, K.R.; Yeoman, C.J.; Kent, A.; Righini, N.; Carbonero, F.; Estrada, A.; Gaskins, H.R.; Stumpf, R.M.; Yildirim, S.; Torralba, M.; et al. Habitat degradation impacts black howler monkey (*Alouatta pigra*) gastrointestinal microbiomes. *J. ISME* **2013**, *7*, 1344–1353. [CrossRef] [PubMed]
23. Ørskov, E.; McDonald, I. The estimation of protein degradability in the rumen from incubation measurements weighted according to rate of passage. *J. Agric. Sci.* **1979**, *92*, 499–503. [CrossRef]
24. McDonald, I. A revised model for the estimation of protein degradability in the rumen. *J. Agric. Sci.* **1981**, *96*, 251–252. [CrossRef]
25. Borreani, G.; Tabacco, E.; Schmidt, R.J.; Holmes, B.J.; Muck, R.E. Silage review: Factors affecting dry matter and quality losses in silages. *J. Dairy Sci.* **2018**, *101*, 3952–3979. [CrossRef] [PubMed]
26. Carvalho, B.F.; Sales, G.F.C.; Schwan, R.F.; Ávila, C.L.S. Criteria for lactic acid bacteria screening to enhance silage quality. *J. Appl. Microbiol.* **2021**, *130*, 341–355. [CrossRef] [PubMed]
27. Ni, K.; Wang, F.; Zhu, B.; Yang, J.; Zhou, G.; Pan, Y.; Tao, Y.; Zhong, J. Effects of Lactate bacteria and molasses additives on the microbial community and fermentation quality of soybean silage. *Bioresour. Technol.* **2017**, *238*, 706–715. [CrossRef] [PubMed]
28. Bonaldi, D.S.; Carvalho, B.F.; Ávila, C.L.D.S.; Silva, C.F. Effects of Bacillus subtilis and its metabolites on corn silage quality. *Lett. Appl. Microbiol.* **2021**, *73*, 46–53. [CrossRef] [PubMed]
29. Su, R.; Ni, K.; Wang, T.; Yang, X.; Zhang, J.; Liu, Y.; Shi, W.; Yan, L.; Jie, C.; Zhong, J. Effects of ferulic acid esterase-producing Lactobacillus fermentum and cellulase additives on the fermentation quality and microbial community of alfalfa silage. *PeerJ* **2019**, *7*, e7712. [CrossRef] [PubMed]
30. Dong, Z.; Shao, T.; Li, J.; Yang, L.; Yuan, X. Effect of alfalfa microbiota on fermentation quality and bacterial community succession in fresh or sterile Napier grass silages. *J. Dairy Sci.* **2020**, *103*, 4288–4301. [CrossRef] [PubMed]
31. Muck, R.E.; Nadeau, E.M.G.; McAllister, T.A.; Contreras-Govea, F.E.; Santos, M.C.; Kung, L., Jr. Silage review: Recent advances and future uses of silage additives. *J. Dairy Sci.* **2018**, *101*, 3980–4000. [CrossRef] [PubMed]
32. Wu, B.; Hu, Z.; Wei, M.; Yong, M.; Niu, H. Effects of inoculation of *Lactiplantibacillus plantarum* and *Lentilactobacillus buchneri* on fermentation quality, aerobic stability, and microbial community dynamics of wilted *Leymus chinensis* silage. *Front. Microbiol.* **2022**, *13*, 928731. [CrossRef] [PubMed]
33. Dutkiewicz, J.; Mackiewicz, B.; Lemieszek, M.K.; Golec, M.; Milanowski, J. *Pantoea agglomerans*: A mysterious bacterium of evil and good. Part IV beneficial effects. *Ann. Agric. Environ. Med.* **2016**, *23*, 206–222. [CrossRef] [PubMed]
34. Yao, B.; Huang, R.; Zhang, Z.; Shi, S. Seed-Borne *Erwinia persicina* Affects the Growth and Physiology of Alfalfa (*Medicago sativa* L.). *Front. Microbiol.* **2022**, *13*, 891188. [CrossRef] [PubMed]
35. Kaparullina, E.; Doronina, N.; Chistyakova, T.; Trotsenko, Y. *Stenotrophomonas chelatiphaga* sp. nov., a new aerobic EDTA-degrading bacterium. *Syst. Appl. Microbiol.* **2009**, *32*, 157–162. [CrossRef] [PubMed]
36. Yang, H.; Zhang, Q.; Hou, J.; Yu, Z. Effects of biological additives on ultrastructure and fiber content of *Leymus chinensis* silage. *Acta Pratacult. Sin.* **2016**, *25*, 94–101.

37. Hu, Z.; Niu, H.; Tong, Q.; Chang, J.; Yu, J.; Li, S.; Zhang, S.; Ma, D. The Microbiota Dynamics of Alfalfa Silage During Ensiling and After Air Exposure, and the Metabolomics After Air Exposure Are Affected by Lactobacillus casei and Cellulase Addition. *Front. Microbiol.* **2020**, *11*, 519121. [CrossRef] [PubMed]
38. Ávila, C.L.S.; Carvalho, B.F. Silage fermentation-updates focusing on the performance of micro-organisms. *J. Appl. Microbiol.* **2020**, *28*, 966–984. [CrossRef] [PubMed]
39. Balan, V.; Bals, B.; Chundawat, S.P.; Marshall, D.; Dale, B.E. Lignocellulosic biomass pretreatment using AFEX. *Methods Mol. Biol.* **2009**, *581*, 61–77. [CrossRef] [PubMed]
40. Adesogan, A.T.; Arriola, K.G.; Jiang, Y.; Oyebade, A.; Paula, E.M.; Pech-Cervantes, A.A.; Romero, J.J.; Ferraretto, L.F.; Vyas, D. Symposium review: Technologies for improving fiber utilization. *J. Dairy Sci.* **2019**, *102*, 5726–5755. [CrossRef] [PubMed]
41. Moraïs, S.; Mizrahi, I. Islands in the stream: From individual to communal fiber degradation in the rumen ecosystem. *FEMS Microbiol. Rev.* **2019**, *43*, 362–379. [CrossRef] [PubMed]
42. Thomas, M.E.; Foster, J.L.; McCuistion, K.C.; Redmon, L.A.; Jessup, R.W. Nutritive value, fermentation characteristics, and *in situ* disappearance kinetics of sorghum silage treated with inoculants. *J. Dairy Sci.* **2013**, *96*, 7120–7131. [CrossRef] [PubMed]
43. Beauchemin, K.A.; Ribeiro, G.O.; Ran, T.; Marami, M.R.; Yang, W.Z.; Khanaki, H.; Gruninger, R.; Tsang, A.; McAllister, T.A. Recombinant fibrolytic feed enzymes and ammonia fibre expansion (AFEX) pretreatment of crop residues to improve fibre degradability in cattle. *Anim. Feed. Sci. Technol.* **2019**, *256*, 114260. [CrossRef]
44. Zhang, X.M.; Wang, M.; Yu, Q.; Ma, Z.Y.; Beauchemin, K.A.; Wang, R.; Wen, J.N.; Lukuyu, B.A.; Tan, Z.L. Liquid hot water treatment of rice straw enhances anaerobic degradation and inhibits methane production during in vitro ruminal fermentation. *J. Dairy Sci.* **2020**, *103*, 4252–4261. [CrossRef] [PubMed]
45. Hall, M.B.; Pell, A.N.; Chase, L.E. Characteristics of neutral detergent-soluble fiber fermentation by mixed ruminal microbes. *Anim. Feed. Sci. Technol.* **1988**, *70*, 23–39. [CrossRef]
46. Robinson, P.H.; Brice, R.E.; Morrison, I.M. Effect of straw treatment and laboratory of incubation on degradation of neutral detergent residue in the rumen. *Anim. Feed. Sci. Technol.* **1987**, *18*, 75–82. [CrossRef]
47. Huws, S.A.; Creevey, C.J.; Oyama, L.B.; Mizrahi, I.; Denman, S.E.; Popova, M.; Muñoz-Tamayo, R.; Forano, E.; Waters, S.M.; Hess, M.; et al. Addressing Global Ruminant Agricultural Challenges Through Understanding the Rumen Microbiome: Past, Present, and Future. *Front. Microbiol.* **2018**, *9*, 2161. [CrossRef] [PubMed]
48. Pu, X.X.; Zhang, X.M.; Li, Q.S.; Wang, R.; Zhang, M.; Zhang, S.Z.; Lin, B.; Tan, B.; Tan, Z.L.; Wang, M. Comparison of *in situ* ruminal straw fiber degradation and bacterial community between buffalo and Holstein fed with high-roughage diet. *Front. Microbiol.* **2023**, *13*, 1079056. [CrossRef] [PubMed]
49. Sha, Y.; Hu, J.; Shi, B.; Dingkao, R.; Wang, J.; Li, S.; Zhang, W.; Luo, Y.; Liu, X. Characteristics and Functions of the Rumen Microbial Community of Cattle-Yak at Different Ages. *Biomed. Res. Int.* **2020**, *2020*, 3482692. [CrossRef] [PubMed]

animals

MDPI

Article

Effects of Yeast Cultures on Growth Performance, Fiber Digestibility, Ruminal Dissolved Gases, Antioxidant Capacity and Immune Activity of Beef Cattle

Siyu Yi [1,†], Xu Tian [2,†], Xianwu Qin [3], Yan Zhang [3], Shuang Guan [2], Zhongping Chen [3], Daliang Cai [3], Duanqin Wu [2], Rong Wang [1], Zhiyuan Ma [4,*], Min Wang [1] and Xiumin Zhang [1,*]

1 State Key Laboratory of Forage Breeding-by-Design and Utilization, National Engineering Laboratory for Pollution Control and Waste Utilization in Livestock and Poultry Production, and Hunan Provincial Key Laboratory of Animal Nutritional Physiology and Metabolic Process, Institute of Subtropical Agriculture, Chinese Academy of Sciences, Changsha 410125, China; siyuyi1996@163.com (S.Y.); rongwang@isa.ac.cn (R.W.); mwang@isa.ac.cn (M.W.)
2 Institute of Bast Fiber Crops, Chinese Academy of Agricultural Sciences, Changsha 410205, China; tx17723618850@163.com (X.T.); 15670246826@163.com (S.G.); wuduanqin@caas.cn (D.W.)
3 National Key Laboratory of Agricultural Microbiology, The Hubei Provincial Engineering Research Center of Yeast, Yichang 443000, China; qxw@angelyeast.com (X.Q.); zhangyan@angelyeast.com (Y.Z.); chenzp@angelyeast.com (Z.C.); caidl@angelyeast.com (D.C.)
4 College of Pastoral Agriculture Science and Technology, Lanzhou University, Lanzhou 730020, China
* Correspondence: mzy@lzu.edu.cn (Z.M.); xmzhang@isa.ac.cn (X.Z.)
† These authors contributed equally to this work.

Simple Summary: Yeast culture (YC) has been widely used as an additive in ruminants, and has various functions of improving growth performance, nutrient digestibility and antioxidant and immune capacities. However, the mechanism of metabolic components in YC and their inhibitory potential on methane (CH_4) emissions from ruminants are still unclear. This study explored supplementation with Diamon V (XP) and Keliben (KLB) to the diet of finishing Simmental beef cattle and found that XP was more effective in improving antioxidant capacity and reducing CH_4 production, and that KLB was more effective in improving the immune capacity. Thus, YC is an effective additive for improving production performance and health, with potential for the suppression of CH_4 production in beef cattle.

Abstract: This study aimed to evaluate the effects of yeast culture (YC) supplementation on growth performance, dietary nutrient digestibility, ruminal fermentation characteristics, methane (CH_4) synthesis potential, ruminal bacterial composition, antioxidant and immune capacities in beef cattle. Thirty-six finishing Simmental beef cattle were employed for this experiment. The experiment included three dietary treatments: the basal diet (CON), the basal diet supplemented with Diamon V XP (XP; 50 g/day per cattle), and the basal diet supplemented with Keliben (KLB; 50 g/day per cattle). Various metabolites, such as acetophenone (12.7%), ascorbic acid (10.3%), citric acid (7.25%), D-(+)-proline (6.42%), succinic acid (5.70%), betaine (5.65%) and DL-malic acid (2.62%) were abundant in XP; and ascorbic acid (14.0%), oleamide (9.23%), citric acid (6.03%), betaine (5.88%), succinic acid (4.42%), indole-3-acrylic acid (2.85%) and DL-malic acid (1.73%) were abundant in KLB. Supplementing YC to the cattle increased the apparent total-tract digestibility of DM, OM, NDF and ADF, and tended to increase average daily gain. The supplementation of YC to the cattle had no effect on the rumen fermentation pathway as reflected by the unaltered molar percentage of acetate or propionate. The supplementation of XP decreased the concentration of rumen dissolved CH_4, although no effect on the concentration of dissolved hydrogen was observed. The supplementation of KLB to the cattle increased *Paraprevotella* relative abundance, while the supplementation of XP decreased Euryarchaeota relative abundance in the rumen. Supplementing XP to the cattle increased serum GSH-Px, catalase and T-AOC concentrations, and was accompanied by decreased MAD concentration, indicating improved antioxidant capacity. Supplementing XP to the cattle increased the concentrations of serum IgA, IgG, IL-2, IL-10, IFN-γ and C4, compared with the CON group, and the KLB group had higher concentrations of serum IgA, IgG, IgM, IL-10, IFN-γ, C3 and C4 compared with the CON and XP groups, which revealed that both XP and KLB can improve the immune function, and that KLB showed a stronger effect. Overall, the supplementation of YC is beneficial to the nutrient digestibility, growth performance and health of beef cattle. Furthermore, XP was more effective than KLB in improving antioxidant capacity and reducing CH_4 production, while KLB was more effective in improving the immune capacity of beef cattle than XP.

Keywords: yeast culture; methane; antioxidant; immune; growth performance

1. Introduction

Yeast culture (YC) is a natural fermentation product produced by *Saccharomyces cerevisiae* in anaerobic environments, and it has been widely used as an additive in ruminants [1–3]. It has been found that YC plays a beneficial role in the growth performance of beef cattle, which may be attributed to the improvement of dietary neutral detergent fiber (NDF) and crude protein (CP) digestibility [4]. Moreover, Chen et al. [5] found that YC can improve the serum total antioxidant capacity (T-AOC) and catalase and glutathione peroxidase (GSH-Px) concentrations of lambs. Chen et al. [5] and Li et al. [6] found that the supplementation of YC to ruminants increased the levels of immunoglobulin A (IgA), immunoglobulin G (IgG), immunoglobulin M (IgM) and interferon-γ (IFN-γ) in blood, which means that YC can enhance the antioxidant capacity and immunity of ruminants. However, Zhang et al. [7] found that the supplementation of YC had no effects on the immunity of cattle, which may be caused by the differences in effective components of different YC. Yeast culture includes the cells and metabolites of *Saccharomyces cerevisiae* and culture medium components, including organic acids, mannan, β-glucan, active peptides, nucleotides, vitamins, oligosaccharides, etc. [1], which may contribute to the various biological functions of YC. However, it is not clear how the specific metabolites in YC affect the growth performance and health of animals.

The enteric methane (CH_4) of ruminants is an important anthropogenic greenhouse gas emission source globally, which is accompanied by a large amount of energy loss of ruminants, which reduces the breeding efficiency [8,9]. Hydrogen (H_2) is produced during carbohydrate fermentation and is mainly utilized by methanogens in the dissolved form to produce CH_4 in the rumen [10]. The remaining H_2 can be reduced to other reduction products, such as microbial protein (MCP) and propionate, or emitted in gaseous form of [10,11]. The amount of H_2 production is usually determined by the volatile fatty acids (VFA) production pattern in the rumen, and the formation of acetate and propionate leads to its production and consumption, respectively [9]. Although previous studies have found that [12,13] active dried *Saccharomyces cerevisiae* did not reduce CH_4 emissions in ruminants, the potential and mechanism of CH_4 emission reduction of YC need to be further explored. The supplementation of YC usually unalters the fermentation characteristics of ruminal VFA in ruminants, such as the molar percentages of propionate, acetate and butyrate [2,14], which means that YC supplementation may not reduce enteric CH_4 production by reducing the H_2 production. Methanogenic archaea dominate the synthesis of CH_4 in the rumen, and inhibiting its activity is also beneficial to CH_4 emission reduction [15]. Therefore, it is necessary to further explore the CH_4 inhibition effects of YC by combining the VFA profile and the relative abundance of methanogenic archaea in the rumen.

Herein, our purpose is to explore the effects of YC on growth performance and health of beef cattle, and its potential to inhibit methanogenesis in the rumen. Thus, we analyzed the nutrient and metabolite components of two different YC, Diamond V XP (XP) and Keliben (KLB), and explored the effects of YC on the growth performance, dietary nutrients digestibility, ruminal fermentation characteristics and dissolved gases, ruminal bacterial composition and antioxidant and immune capacities of beef cattle.

2. Materials and Methods

2.1. Sources of YC

The samples of XP and KLB were purchased from Diamond V Bio-fermentation Technology Co., Ltd., Shenzhen, China and Angel yeast Co., Ltd., Yichang, China, respectively, and were stored in sealed bags at room temperature.

2.2. Animals, Experimental Design and Feeding

A total of 36 finishing male Simmental beef cattle with similar age (18.3 ± 0.70 months) and weight (460 ± 17.4 kg) were randomly divided into 3 groups: CON (the basal diet; Table 1), XP (the basal diet supplemented with 50 g/day per cattle of XP) and KLB (the basal diet supplemented with 50 g/day per cattle of KLB), with 12 cattle in each group. The experiment was conducted on a beef cattle farm of Xindalinmu Science and Technology Development Co., Ltd., Yuanjiang, Yiyang, Hunan, China (29°07′ N, 112°34′ E). Cattle were kept in separate pens, fed with total mixed rations (TMR; mixed by a horizontal mixer) at 07:00 and 16:00 every day with an independent feeder, and were free to drink water. Yeast culture was evenly top-dressed on the TMR diet of each cattle. The experiment lasted for 70 days (including 10 days of a dietary adaptation period).

Table 1. Composition of ingredients and nutrients of the basal diet, as dry matter basis.

Ingredients [1]	Content, %	Nutrients [2]	Content, g/kg
Corn stover silage	45.0	OM	909
Corn	23.1	CP	113
Soybean meal	5.50	NDF	476
Wheat bran	4.40	ADF	350
Cron gluten	4.40	Starch	165
DDGS	8.25	GE, Mcal/kg	4.25
Cotton meal	2.20	NEm, Mcal/kg	2.47
Molasses	1.10	NEg, Mcal/kg	1.49
Rice bran meal	2.75		
$NaHCO_3$	0.55		
Premix	2.75		

[1] Premix contains 250,000 IU of vitamin A, 70,000 IU of vitamin D3, vitamin E 1000 IU, 0.8 g of copper, 1.5 g of zinc, 1.3 g of manganese, 27 mg of iodine, 11.5 mg of selenium, 23 mg of cobalt, 17% of calcium and 3.5% of phosphorus per kg premix. [2] OM: organic matter; CP: crude protein; NDF: neutral detergent fiber; ADF: acid detergent fiber; GE: gross energy; NEm: net energy for maintenance; NEg: net energy for gain. The values of NEm and Neg were calculated according to NY/T 815 [16].

2.3. Apparent Total-Tract Digestibility, Nutrient Components and Growth Performance

During the 11–70th days of the experimental period, the dry matter intake (DMI) was calculated by recording the feed intake and refusals daily. During the 60–64th days of the experimental period, about 200 g fecal samples were collected from the rectum at 08:00 and 17:00 and divided into two subsamples on average. One subsample was added with

10 mL 10% (v/v) dilute sulfuric acid for nitrogen fixation to determine CP, and the other subsample was used to determine other nutrient components, and then immediately stored at −20 °C. After that, the fecal samples were dried and crushed at 65 °C, and the samples of the same cattle for 5 days were mixed in equal proportions with the DMI and then stored in a sealed bag. Feed samples were randomly collected every 7 days, mixed and dried at 65 °C, and kept in sealed bags for determination. On the 34th and 70th days of the experimental period, all cattle were weighed before the morning feeding, and the average daily gain (ADG) was calculated.

The dry matter (DM; method 934.01), organic matter (OM; method 942.05) and CP (method 990.03) contents were determined according to the AOAC [17]. The gross energy (GE) was measured by an isothermal automatic calorimeter (5E-AC8018; Changsha kaide measurement and control instruments Co., Ltd., Changsha, China). The contents of NDF and acid detergent fiber (ADF) were determined according to the method previously described by Van Soest et al. [18], and the heat-stable amylase and sodium sulfite were added when the NDF was determined. The starch content was determined following the method previously described by Kartchner et al. [19]. The acid-insoluble ash content of feed and feces samples was determined in accordance with the GB/T 23,742 method [20], and the apparent total-tract digestibility of nutrients was calculated according to the method described by Van Keulen et al. [21] with the acid-insoluble ash contents.

Mannan content in YC was analyzed according to T/QBAA 001 method [22] by high-performance liquid chromatography (HPLC; LC1290; Agilent Inc., Santa Clara, CA, USA). The contents of free amino acids, acid-soluble protein and total organic acids in YC were analyzed according to GB/T 18,246 [23], NY/T 3801 [24] and GB/T 12,456 [25] methods. The metabolites content of YC was determined by Nuohezhiyuan Technology Co., Ltd., Beijing, China. The content of metabolites was analyzed by a liquid chromatography–mass spectrometer (LC-MS; Agilent Inc., Santa Clara, CA, USA) [26,27]. The original data of metabolite determination are preprocessed by Compound Discoverer 3.1 data processing software. Then, the metabolites were identified by comparing the high-resolution secondary spectrogram databases of mzCloud and mzVault with the primary database of MassList. Metabolites with a coefficient of variance less than 30% in the samples were retained as the final result [28]. The chromatographic peaks detected in the samples are integrated, in which the peak area of each characteristic peak represents the relative quantitative value of a metabolite, and the peak area quantitative result of each characteristic peak is standardized by using the total peak area. The analysis of the nutrients and metabolites components was performed in triplicate and sextuplicate, respectively.

2.4. Collection and Determination of Ruminal Contents

During the 67–70th days of the experimental period, ruminal contents were taken from the middle of the rumen at 0 and 2.5 h after feeding in the morning through a rumen cannula. About 100 mL of ruminal contents collected at the beginning were discarded to avoid saliva pollution. The pH value of ruminal contents was measured immediately by a pH meter (Starter 300; Ohaus Instruments Co., Ltd., Shanghai, China). The dissolved hydrogen (dH_2), dissolved CH_4 (dCH_4) and VFA concentrations were analyzed as described by Wang et al. [29], by a gas chromatograph (7890A; Agilent Inc., Santa Clara, CA, USA). Briefly, 35 mL of ruminal contents were transferred to a 50 mL-plastic syringe, and the dissolved gases in ruminal contents were injected into another connected 20 mL-syringe with 10 mL of nitrogen by a shaker (Huichen Biotech Co., Ltd., Wuhan, China), and the gases were collected into a sealed bottle for determination. About 2 mL of ruminal contents were centrifuged at 12,000× g for 10 min at 4 °C, and the supernatants (1.5 mL) were acidified with 0.15 mL of 25% (w/v) metaphosphoric acid and stored at −20 °C for the determination of VFA. About 5 mL of ruminal contents were immediately frozen in liquid nitrogen and then stored at −20 °C for MCP content analysis according to the method described by Zinn et al. [30]. In addition, about 2 mL of ruminal contents were immediately frozen in liquid nitrogen and then stored at −80 C for subsequent DNA extraction.

2.5. Collection and Determination of Blood Samples

On the 65–66th days of the experimental period, blood samples were collected through the jugular vein at 2 h before feeding in the morning using a 5 mL vacuum blood collection tube. The collected blood was immediately transferred to an incubator for incubation for 30 min, then centrifuged at 3000× *g* for 15 min at room temperature to separate the serum samples and stored at −20 °C.

Serum metabolites were determined by Roche automatic biochemical analyzer (cobas c311; Roche Inc., Basel, Switzerland). The serum antioxidant index and immune index were determined by unitized biochemical and ELISA kits (Nanjing jiancheng bioengineering institute, Nanjing, China), respectively.

2.6. Analysis of Ruminal Microbiota Composition

Samples of ruminal contents collected at 0 h and 2.5 h after feeding were mixed, and the total microbial DNA was extracted according to Ma et al. [31]. The 16S rDNA sequencing was carried out by Biozeron Biotechnology Co., Ltd., Shanghai, China, on the MiSeq platform (Illumina Inc., San Diego, CA, USA), and the PCR amplification process was as described by Zakrzewski et al. [32]. The universal primer of 341F: (5′-CCTAYGGGRBGCASCAG-3′) and 806R: (5′-GGACTACNNGGGTATCTAAT-3′) was used to amplify the V3-V4 region of bacteria. The operational taxonomic units were clustered with a 98.7% similarity cutoff by UPARSE software (version 7.1; http://drive5.com/uparse/ (accessed on 6 July 2024)), and UCHIME was used to identify and remove the chimeric sequences. The phylogenetic affiliation of each 16S rRNA gene sequence was analyzed using RDP Classifier (http://rdp.cme.msu.edu/ (accessed on 6 July 2024)) against the 16S rRNA database of silva (SSU132) with a 70% confidence threshold [33]. The analysis of microbiota composition was performed in twelve duplicates.

2.7. Statistical Analysis

The data were analyzed using a linear mixed model by SPSS 26.0 software (SPSS Inc., Chicago, IL, USA), with the groups (CON, XP and KLB) as a fixed effect, and the animal as a random effect. When the sampling time is included, the groups and interaction between groups and sampling times were used as the fixed effects, the animal as a random effect, and the sampling time as a repeated measurement. The data of bacterial compositions were analyzed by nonparametric test models, and the differences were judged by the Kruskal–Wallis test. The statistical significance was set at $p < 0.05$, and $p < 0.10$ is considered to have a difference trend.

3. Results

3.1. Nutrient Components and Metabolites Components of YC

The nutrient components of XP and KLB are shown in Table 2. The content of mannan was 90.0 and 140 g/kg of DM, respectively. The metabolites components (the top 10) of XP and KLB are presented in Table 3. Many substances with antioxidant activity and immunity enhancement were found to be abundant in YC, such as acetophenone (12.7%), ascorbic acid (10.3%), citric acid (7.25%), D-(+)-proline (6.42%), succinic acid (5.70%), betaine (5.65%) and DL-malic acid (2.62%) in XP; and ascorbic acid (14.0%), oleamide (9.23%), citric acid

(6.03%), betaine (5.88%), succinic acid (4.42%), indole-3-acrylic acid (2.85%) and DL-malic acid (1.73%) in KLB.

Table 2. Nutrient components of yeast cultures, as dry matter basis.

Items	XP	KLB
OM, g/kg	902	895
CP, g/kg	171	167
Mannan, g/kg	90.0	140
Free amino acid, g/kg	7.40	23.7
Total organic acid, g/kg	88.6	90.9
Acid-soluble protein, g/kg	56.0	55.4
GE, MJ/kg	18.4	17.5

OM: organic matter; CP: crude protein; GE: gross energy.

Table 3. The metabolites components (the top 10) of yeast cultures detected by positive and negative ions mode.

Metabolites of XP	Content, %	Metabolites of KLB	Content, %
	Positive ion		
Acetophenone	12.7	Oleamide	9.23
D-(+)-Proline	6.42	Betaine	5.88
DL-3-Hydroxynorvaline	6.38	L-Pyroglutamic acid	5.43
Saxitoxin	6.05	9-Oxo-10(E),12(E)-octadecadienoic acid	5.21
Betaine	5.65	Indole-3-acrylic acid	2.85
LPC 16:0	3.46	Valine	2.26
N6-Me-dA	3.14	DL-3-Hydroxynorvaline	2.15
Valine	3.12	D-(+)-Proline	2.12
L-Pyroglutamic acid	2.50	L-Tyrosine	1.98
Tyramine	2.41	Deoxycorticosterone 21-glucoside	1.97
	Negative ion		
Ascorbic acid	10.3	Ascorbic acid	14.0
(+/−)12(13)-DiHOME	10.1	(+/−)12(13)-DiHOME	13.8
Citric acid	7.25	Citric acid	6.03
Succinic acid	5.70	2-Isopropylmalic acid	4.44
5-Methoxypsoralen	5.65	Succinic acid	4.42
(+/−)9-HpODE	4.58	4-Oxoproline	3.83
4-Oxoproline	3.69	(+/−)9-HpODE	3.68
DL-Malic acid	2.62	2-Hydroxyvaleric acid	1.76
2-Isopropylmalic acid	2.62	DL-Malic acid	1.73
cis-Aconitic acid	2.43	2-Hydroxycaproic acid	1.56

3.2. *Growth Performance and Apparent Total-Tract Digestibility*

The supplementation of XP and KLB increased ($p < 0.01$) the apparent total-tract digestibility of DM, OM, NDF, ADF and GE of beef cattle, and tended to increase ($p = 0.06$) starch digestibility of beef cattle. Moreover, the supplementation of XP increased ($p < 0.01$) CP digestibility. Consequently, XP and KLB supplementation showed a trend of increased ($p = 0.06$) ADG (Table 4).

Table 4. Effects of yeast cultures supplementation on performance and apparent total-tract digestibility of beef cattle.

Items	Groups			SEM	p-Value
	CON	XP	KLB		
	Performance				
IBW, kg	465	454	460	17.4	0.81
FBW, kg	488	480	493	16.2	0.72
ADG, kg/d	0.63	0.72	0.92	0.11	0.06
DMI, kg/d	10.8	10.7	10.8	0.08	0.30
	Apparent total-tract digestibility, g/kg				
DM	651 ^{b}	710 ^{a}	686 ^{a}	6.00	<0.01
OM	685 ^{b}	743 ^{a}	723 ^{a}	6.19	<0.01
CP	593 ^{b}	669 ^{a}	620 ^{b}	8.81	<0.01
NDF	615 ^{b}	689 ^{a}	691 ^{a}	9.52	<0.01
ADF	397 ^{b}	503 ^{a}	535 ^{a}	17.2	<0.01
Starch	960	968	970	1.85	0.06
GE, KJ/MJ	657 ^{b}	718 ^{a}	694 ^{a}	6.52	<0.01

IBW: initial body weight; FBW: final body weight; ADG: average daily gain; DMI: dry matter intake; DM: dry matter; OM: organic matter; CP: crude protein; NDF: neutral detergent fiber; ADF: acid detergent fiber; GE: gross energy. Different letters indicate significant differences between the groups ($p < 0.05$).

3.3. Dissolved Gases and Ruminal Fermentation Characteristics

Supplementing YC had no effect ($p > 0.10$; Table 5) on rumen dH_2 concentration, whereas XP decreased ($p = 0.04$) the dCH_4 concentration in the rumen. The cattle fed with XP showed a trend ($p = 0.08$) for increasing MCP concentration in the rumen. Both XP and KLB supplementation had no effects ($p > 0.10$) on ruminal pH, acetate to propionate ratio and the molar percentages of acetate, propionate, butyrate and valerate. Interaction ($p = 0.02$) between groups and sampling times was observed for total VFA concentration. The cattle fed with KLB tended to decrease ($p = 0.06$) the total VFA concentration compared with the CON group before the morning feeding, while XP supplementation decreased ($p < 0.01$) the total VFA concentration compared with the CON and KLB groups at 2.5 h after the morning feeding. Interaction ($p < 0.01$) between groups and sampling times was observed for isobutyrate concentration. The supplementation of XP and KLB increased ($p < 0.01$) isobutyrate molar percentages before the morning feeding, but had no effect ($p > 0.10$) at 2.5 h after morning feeding between the groups. Interaction ($p < 0.01$) between groups and sampling times was observed for isobutyrate concentration. The supplementation of XP and KLB increased ($p < 0.01$) isovalerate molar percentages before the morning feeding, but had no effect ($p > 0.10$) at 2.5 h after morning feeding between the groups.

Table 5. Effects of yeast cultures supplementation on dissolved gases and fermentation end-products in the rumen of beef cattle.

Items	Group			Time, h		SEM	p-Value		
	CON	XP	KLB	0	2.5		Group	Time	Group × Time
dH_2, μM	0.14	0.18	0.16	0.14	0.18	0.01	0.31	0.03	0.63
dCH_4, mM	0.26 ^{a}	0.21 ^{b}	0.24 ^{ab}	0.22	0.25	0.01	0.04	0.07	0.11
MCP, g/L	1.49	2.04	1.67	1.94	1.53	0.11	0.08	0.06	0.53
pH	7.16	7.16	7.25	7.51	6.86	0.05	0.23	<0.01	0.54
Total VFA, mM	80.4 ^{a}	69.8 ^{b}	73.3 ^{ab}	57.3	91.7	2.57	0.03	<0.01	0.02
	Molar percentage of individual VFA, mol/100 mol								
Acetate	65.7	66.1	65.8	68.2	63.6	0.58	0.36	<0.01	0.49
Propionate	18.9	19.0	18.7	17.4	20.4	0.29	0.95	<0.01	0.27

Table 5. *Cont.*

Items	Group			Time, h		SEM	*p*-Value		
	CON	XP	KLB	0	2.5		Group	Time	Group × Time
Butyrate	12.0	11.1	11.6	0.34	10.2	0.22	0.10	<0.01	0.26
Isobutyrate	0.94 [b]	1.15 [a]	1.11 [a]	1.41	0.73	0.05	0.01	<0.01	<0.01
Valerate	1.15	1.08	1.24	0.94	1.38	0.04	0.21	<0.01	0.34
Isovalerate	1.31	1.57	1.52	1.89	1.05	0.06	0.05	<0.01	<0.01
Acetate to propionate ratio	3.53	3.55	3.61	3.97	3.15	0.12	0.86	<0.01	0.23

dH_2: dissolved hydrogen; dCH_4: dissolved methane; MCP: microbial protein; VFA: volatile fatty acids. Different letters indicate significant differences between the groups ($p < 0.05$).

3.4. Bacterial Composition in the Rumen

The composition of bacteria in the rumen at phylum level (>1.00%) is shown in Figure 1. No differences between the groups were observed in the relative abundances (CON vs. XP vs. KLB) of Bacteroidetes (56.6% vs. 61.9% vs. and 59.2%), Firmicutes (24.2% vs. 21.7% vs. 23.3%), Proteobacteria (11.3% vs. 10.6% vs. 10.7%), Actinobacteria (1.49% vs. 0.95% vs. 1.30%) and Fibrobacteres (1.05% vs. 0.62% vs. 1.06%) in the rumen ($p > 0.10$).

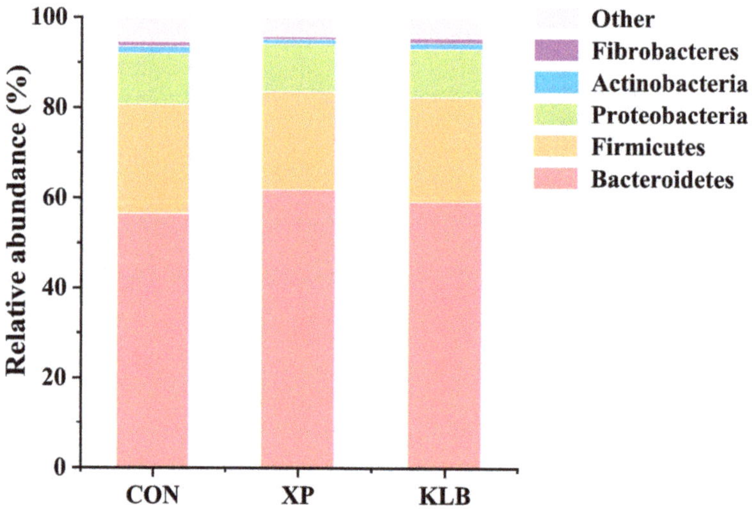

Figure 1. Effect of yeast cultures supplementation on the relative abundance of ruminal bacteria at the phyla level of finishing beef cattle.

The composition of bacteria in the rumen at the genus level (>1.00%) is presented in Figure 2. The supplementation of KLB increased ($p = 0.04$) the relative abundance of *Paraprevotella* (1.87% vs. 2.21% vs. 2.96%; CON vs. XP vs. KLB). However, no differences ($p > 0.10$) between the groups were observed in *Prevotella* (42.4% vs. 48.0% vs. 43.7%), *Stenotrophomonas* (5.08% vs. 4.55% vs. 4.40%), *Sodaliphilus* (3.81% vs. 3.57% vs. 3.61%) and *Succiniclasticum* (3.14% vs. 3.15% vs. 3.02%).

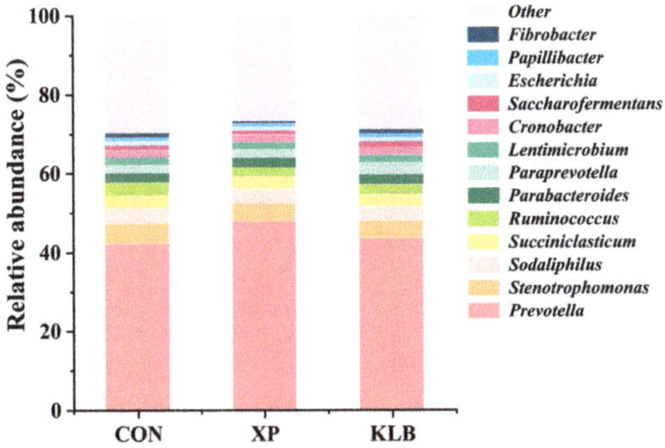

Figure 2. Effect of yeast cultures supplementation on the relative abundance of ruminal bacteria at the genera level of finishing beef cattle.

The relative abundance of Euryarchaeota in the rumen is shown in Figure 3. The cattle fed with XP had a lower relative abundance of ($p = 0.03$) Euryarchaeota compared with the cattle in the CON group.

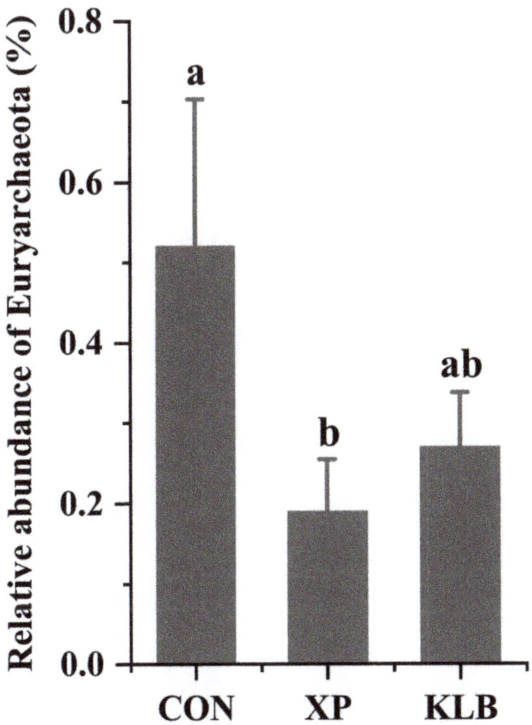

Figure 3. Effect of yeast culture supplementation on the relative abundance of ruminal Euryarchaeota in finishing beef cattle. Different letters indicate significant differences ($p < 0.05$).

3.5. Serum Biochemical Indices

In terms of serum carbohydrate metabolism capacity, supplementing XP and KLB increased ($p \leq 0.01$; Table 6) the concentrations of glucose, α-amylase and pancreatic amylase, but decreased ($p < 0.01$) lactic acid concentration. For serum nitrogen metabolism capacity, the supplementation of XP and KLB decreased ($p = 0.02$) uric acid concentration, whereas it increased ($p < 0.01$) creatinine concentration. Furthermore, the KLB group had higher ($p < 0.01$) serum concentrations of BUN and alanine aminotransferase (ALT) than the CON and XP groups, and had a higher ($p < 0.01$) albumin concentration in serum than the CON group. In terms of serum lipid metabolism capacity, XP supplementation increased ($p = 0.02$) lipase concentration, whereas the supplementation of XP and KLB had no effects ($p \geq 0.10$) on total cholesterol, triglyceride, low-density lipoprotein cholesterol (LDL-C) and high-density lipoprotein cholesterol (HDL-C) concentrations.

Table 6. Effects of yeast cultures supplementation on serum metabolites of finishing beef cattle.

Items	Groups			SEM	p-Value
	CON	XP	KLB		
	Carbohydrate metabolism				
Glucose, mM	3.26 ^{b}	4.57 ^{a}	4.81 ^{a}	0.17	<0.01
Lactic acid, mM	4.80 ^{a}	2.34 ^{b}	2.49 ^{b}	0.30	<0.01
Lactate dehydrogenase, U/L	1105	1180	1181	33.4	0.57
α-amylase, U/L	27.7 ^{b}	40.9 ^{a}	45.6 ^{a}	2.61	<0.01
Pancreatic amylase, U/L	27.3 ^{b}	39.2 ^{a}	43.6 ^{a}	2.47	0.01
	Nitrogen metabolism				
Total protein, g/L	74.3	75.7	79.2	1.20	0.22
Albumin, g/L	34.7 ^{b}	38.2 ^{ab}	39.8 ^{a}	0.69	<0.01
blood ammonia, μM	248	195	226	12.9	0.24
Uric acid, mg/dL	0.60 ^{a}	0.44 ^{b}	0.43 ^{b}	0.03	0.02
BUN, mM	4.33 ^{b}	4.94 ^{b}	5.92 ^{a}	0.18	<0.01
Creatinine, μM	101 ^{b}	140 ^{a}	130 ^{a}	3.85	<0.01
AST, U/L	82.3	70.0	79.7	3.37	0.21
ALT, U/L	24.5 ^{b}	25.2 ^{b}	30.3 ^{a}	0.84	<0.01
	Lipid metabolism				
Total cholesterol, mM	2.60	2.58	2.72	0.16	0.76
Triglyceride, mM	0.28	0.24	0.27	0.01	0.10
LDL-C, mM	0.88	0.80	0.92	0.05	0.59
HDL-C, mM	2.08	2.11	2.23	0.08	0.65
Lipase, U/L	16.9 ^{b}	21.1 ^{ab}	22.7 ^{a}	0.90	0.02

BUN: blood urea nitrogen; AST: aspartate aminotransferase; ALT: alanine aminotransferase; LDL-C: low-density lipoprotein cholesterol; HDL-C: high-density lipoprotein cholesterol. Different letters indicate significant differences between the groups ($p < 0.05$).

3.6. Serum Antioxidant and Immune Indexes

The effects of XP and KLB supplementation on the serum antioxidant and immune capacities of beef cattle are presented in Table 7. With regard to serum antioxidant indexes, the XP group had higher ($p < 0.05$) GSH-Px, catalase and T-AOC concentrations compared with the CON and KLB groups. The cattle fed with XP had a lower ($p = 0.04$) malondialdehyde (MAD) concentration compared with the CON group. However, there was no difference ($p > 0.10$) in superoxide dismutase (SOD) between the CON, XP and KLB groups. In terms of immunoglobulins, the XP group had higher ($p < 0.01$) IgA and IgG concentrations than the CON group, and the cattle fed with KLB had higher ($p < 0.01$) IgA, IgG and IgM concentrations than the CON and XP groups. For immune factors, supplementation with XP and KLB increased ($p < 0.01$) IL-2 concentration, and the cattle fed with KLB had higher ($p < 0.01$) interleukin-10 (IL-10), IFN-γ, complement C3 (C3) and complement C4 (C4) concentrations in serum compared with the CON and XP groups. Furthermore, the cattle fed with XP had higher ($p < 0.01$) IL-10, IFN-γ and C4 concentrations in serum than the CON group.

Table 7. Effects of yeast cultures supplementation on serum antioxidant and immune indexes of finishing beef cattle.

Items	Groups			SEM	p-Value
	CON	XP	KLB		
	Antioxidants				
SOD, U/mL	315	314	307	5.96	0.85
GSH-Px, U/mL	30.0 ^{b}	38.0 ^{a}	28.9 ^{b}	2.69	<0.01
Catalase, U/mL	0.45 ^{b}	0.68 ^{a}	0.47 ^{b}	0.04	<0.01
T-AOC, U/mL	1.18 ^{b}	1.51 ^{a}	1.21 ^{b}	0.07	0.04
MAD, nmol/mL	5.12 ^{a}	3.27 ^{b}	3.88 ^{ab}	0.32	0.04
	Immunoglobulins				
IgA, mg/mL	1.62 ^{c}	1.91 ^{b}	2.52 ^{a}	0.08	<0.01
IgG, mg/mL	3.86 ^{c}	4.72 ^{b}	5.69 ^{a}	0.19	<0.01
IgM, mg/mL	0.98 ^{b}	1.11 ^{b}	1.45 ^{a}	0.05	<0.01
Lysozyme, μg/mL	12.0	11.7	11.7	0.48	0.94
	immune factors				
IL-2, pg/mL	563 ^{b}	726 ^{a}	829 ^{a}	26.9	<0.01
IL-6, pg/mL	130	128	132	4.17	0.91
IL-10, pg/mL	19.8 ^{c}	32.6 ^{b}	41.6 ^{a}	2.07	<0.01
IFN-γ, μg/mL	0.79 ^{c}	1.02 ^{b}	1.29 ^{a}	0.04	<0.01
C3, μg/mL	32.0 ^{b}	38.8 ^{b}	52.9 ^{a}	1.96	<0.01
C4, μg/mL	101 ^{c}	136 ^{b}	174 ^{a}	7.21	<0.01

SOD: superoxide dismutase; GSH-Px: glutathione peroxidase; T-AOC: total antioxidant capacity; MDA: malondialdehyde; IgA: immunoglobulin A; IgG: immunoglobulin G; IgM: immunoglobulin M; IL-2: interleukin-2; IL-6: interleukin-6; IL-10: interleukin-10; IFN-γ: interferon-γ; C3: complement C3; C4: complement C4. Different letters indicate significant differences between the groups ($p < 0.05$).

4. Discussion

Yeast culture has been reported to improve the growth performance, nutrient digestibility, antioxidant capacity and immune function of beef cattle [4,6]. These functions of YC may be related to its rich active substances, such as organic acids, mannan, β-glucan, active peptides, nucleotides and some undiscovered growth factors [1]. In the present study, various metabolites, such as betaine, ascorbic acid, citric acid, succinic acid, malic acid, acetophenone, oleamide and indole-3-acrylic acid were found to be enriched in YC, which may also improve the health and then promote the growth performance of livestock.

Yeast culture can improve the nutrients digestibility, which is related to the improvement of the activities of beneficial microorganisms and digestive enzymes [34,35]. The supplementation of XP and KLB into the diet of beef cattle resulted in a significant increase in the digestibility of NDF and ADF. Consistent with the present study, Wang et al. [2] supplemented YC (10–40 g/d/head) in the diet of fattening sheep and improved the digestibility of NDF and ADF. In addition, supplementation of XP increased the CP digestibility of beef cattle, while KLB supplementation did not, which may be related to the change in rumen microbial community. *Prevotella* is usually the dominant bacteria in the rumen and plays an important role in the degradation of protein [36]. In the present study, the supplementation of XP increased the relative abundance (48.0%) of *Prevotella* in the rumen compared with the CON (42.4%) and KLB (43.7%) groups, which helps uphold the improved CP digestibility. Thus, supplementing YC can promote the digestion of dietary nutrients, which may further improve the growth performance of beef cattle, as demonstrated by the increased ADG.

The total rumen VFA concentration is closely related to the digestion of dietary nutrients. However, the supplementation of XP and KLB did not increase the total VFA concentration as expected, XP even reduced the VFA concentration. One possible reason is that YC promoted the absorption of VFA by rumen epithelium [2,37]. We found that the supplementation of XP and KLB unaltered the molar percentages of propionate, acetate and butyrate and the acetate to propionate ratio. Our results are consistent with Wang et al. [2], who reported that the supplementation of YC (10–40 g/d/head) to sheep did not change the rumen fermentation pathway. However, supplementing XP and KLB to beef cattle increased isobutyrate molar percentages before the morning feeding. Ruminal isobutyrate is mainly derived from the degradation of dietary amino acids, i.e., valine, isoleucine, leucine and proline [38], and the higher contents of valine and proline in XP and KLB could contribute to the increased production of isobutyrate. Furthermore, the cattle fed with XP tended to have a higher MCP concentration compared with the CON group, which is consistent with the fact that XP supplementation enhanced amino acids availability for microbes, and the increased CP digestibility due to XP supplementation can also contribute to the enhanced MCP synthesis.

Ruminal H_2 is produced during the carbohydrates fermentation process and exists as a gaseous or dissolved form, and only the dH_2 can be utilized by the microbes. The dH_2 concentration is determined by the balance between H_2 production pathways (e.g., acetate production from carbohydrate fermentation) and consumption pathways (e.g., CH_4, propionate and MCP formation) in the rumen, and methanogenesis makes the greatest contribution to the H_2 consumption [9–11]. Interestingly, we found that the concentration of dCH_4 in the rumen decreased after XP supplementation, which seems to indicate that methanogenesis was inhibited. Moreover, Euryarchaeota, as the main methanogenic archaea in the rumen [39], was found to be significantly decreased after XP supplementation, which is consistent with the decreased dCH_4 concentration. However, the dH_2 concentration was not increased after methanogenesis inhibition. These results may indicate that the supplementation of XP inhibited methanogenesis, which may lead to more H_2 being redirected to other reduction products, such as the MCP synthesis, which was likely demonstrated by the increased MCP concentration. Therefore, we speculated that XP supplementation inhibited the activity of methanogens, which may lead to more H_2 being utilized to form other reduction products (e.g., MCP), rather than CH_4.

Serum metabolites are an important indicator of nutrient digestion and body metabolism [40]. In the present study, the supplementation of XP and KLB increased serum glucose concentration, which was consistent with the higher digestibility of fiber and starch. In addition, XP and KLB supplementation had no effects on total protein, total cholesterol, blood ammonia, total cholesterol, triglyceride, LDL-C and HDL-C concentrations in serum compared with the CON group, which means that YC had no negative impacts on body metabolism. Although KLB supplementation increased the serum contents of albumin, creatinine and ALT, it was still within the normal physiological range [41].

The animal organism continues to carry out oxidation reactions to meet the needs of normal metabolism of the substances of life. However, when the level of reactive oxygen free radicals exceeds the upper limit of degradation ability (often caused by feed, environment and social factors), it can lead to oxidative stress, endangering animal health and having a negative impact on production efficiency [42]. Serum SOD, GSH-Px, catalase and T-AOC are important indexes to measure the antioxidant performance of animals, while serum MAD concentration can reflect the degree of oxidative damage to the animals [42]. In the present study, the supplementation of XP increased the concentration of GSH-Px, catalase and T-AOC, and decreased the concentration of MAD, indicating that XP supplementation can enhance the antioxidant capacity of beef cattle. Acetophenone, which can improve antioxidant activity, was found to be abundant in XP rather than KLB [43], which may contribute to the improvement of antioxidant capacity in beef cattle.

Serum IgA, IgG and IgM play a pivotal role in both the acquired response and the innate immune systems [44], and IL-2, IL-10, IFN-γ, C3 and C4 play an important role in maintaining physiological balance and innate immunity, regulating inflammation, preventing immune hyperresponsiveness, promoting T cell proliferation and differentiation, and resisting virus invasion [45–48]. In the present study, we found that supplementation with XP increased the concentrations of IgA, IgG, IL-2, IL-10, IFN-γ and C4 in the serum compared with the CON group, while KLB supplementation increased the concentrations of IgA, IgG, IgM, IL-10, IFN-γ, C3 and C4 compared with the CON and XP groups, and increased IL-2 concentration compared with the CON group. These results indicated that YC enhances the immune response of cattle, and KLB is more effective than XP. The metabolites, such as betaine, ascorbic acid, citric acid, succinic acid and malic acid, were abundant in XP and KLB, and have the functions of improving immunity [49–53], which may be the contributing factor to the enhanced immunity observed in cattle. Mannan, oleamide and indole-3-acrylic acid, abundant in KLB rather than XP, have been found to enhance the immunity in animals [54–56], which may promote KLB to be more effective than XP in improving immunity. In addition, *Paraprevotella* in the rumen has been found to be related to immune function [57,58], thus the greater relative abundance of ruminal *Paraprevotella* by KLB supplementation can help explain the improved immunity of beef cattle.

5. Conclusions

Supplementation with YC enhances the nutrient digestion (e.g., fiber), antioxidant capacity and immunity of beef cattle. Compared with KLB, XP improves the CP digestibility and antioxidant capacity and demonstrates the potential to reduce CH_4 production. However, KLB exhibits stronger immune-boosting effects than XP. In summary, YC is an effective additive for improving the production performance and health of beef cattle, with the potential for rumen methanogenesis suppression, which could be considered in the diets of beef cattle to enhance productivity and reduce environmental impact.

Author Contributions: Conceptualization, X.T. and X.Z.; methodology, X.T., D.W., R.W. and M.W.; data curation and formal analysis, S.Y. and X.T.; investigation, X.T., S.G. and R.W.; visualization, S.Y.; validation, X.Q., Y.Z., Z.C. and D.C.; writing—original draft preparation, S.Y.; writing—review and editing, Z.M. and X.Z.; project administration and funding acquisition, X.Q., D.W., X.Z., M.W. and Z.M. All authors have read and agreed to the published version of the manuscript.

Funding: This work was supported by the Science and Technology Innovation Program of Hunan Province (Grant No. 2023RC3206), the Inner Mongolia Major Technology Project (2022JBGS0023), Natural Science Foundation of Hunan Province (2025JJ70628) and National Key R&D Program of China (Grant No. 2023YFD1300900, 2023YFD1300903, 2021YFD1301003).

Institutional Review Board Statement: The animal experiment was conducted in accordance with the guidelines of the Animal Care and Use Committee of the Institute of Subtropical Agriculture, Chinese Academy of Sciences, Changsha, China (Approval NO. ISA 2024-0019).

Informed Consent Statement: Not applicable.

Data Availability Statement: The data presented in this study are available on request from the corresponding authors.

Conflicts of Interest: The authors declare no conflicts of interest.

References

1. Shurson, G.C. Yeast and yeast derivatives in feed additives and ingredients: Sources, characteristics, animal responses, and quantification methods. *Anim. Feed Sci. Technol.* **2018**, *235*, 60–76. [CrossRef]
2. Wang, J.; Zhao, G.; Zhuang, Y.; Chai, J.; Zhang, N. Yeast (*Saccharomyces cerevisiae*) culture promotes the performance of fattening sheep by enhancing nutrients digestibility and rumen development. *Fermentation* **2022**, *8*, 719. [CrossRef]

3. Wang, X.; Li, F.; Zhang, N.; Ungerfeld, E.; Guo, L.; Zhang, X.; Wang, M.; Ma, Z. Effects of supplementing a yeast culture in a pelleted total mixed ration on fiber degradation, fermentation parameters, and the bacterial community in the rumen of sheep. *Anim. Feed Sci. Technol.* **2023**, *296*, 115565. [CrossRef]
4. Batista, L.H.C.; Cidrini, I.A.; Prados, L.F.; Cruz, A.A.C.; Torrecilhas, J.A.; Siqueira, G.R.; Resende, F.D. A meta-analysis of yeast products for beef cattle under stress conditions: Performance, health and physiological parameters. *Anim. Feed Sci. Technol.* **2022**, *283*, 115182. [CrossRef]
5. Chen, H.; Liu, S.; Li, S.; Li, D.; Li, X.; Xu, Z.; Liu, D. Effects of yeast culture on growth performance, immune function, antioxidant capacity and hormonal profile in Mongolian ram lambs. *Front. Vet. Sci.* **2024**, *11*, 1424073. [CrossRef]
6. Li, X.; An, N.; Chen, H.; Liu, D. Effects of yeast culture on growth performance, antioxidant capacity, immune function, and intestinal microbiota structure in Simmental beef cattle. *Front. Vet. Sci.* **2025**, *11*, 1533081. [CrossRef]
7. Zhang, X.; Liang, H.; Xu, L.; Zou, B.; Zhang, T.; Xue, F.; Qu, M. Rumen fermentative metabolomic and blood insights into the effect of yeast culture supplement on growing bulls under heat stress conditions. *Front. Microbiol.* **2022**, *13*, 947822. [CrossRef]
8. Yi, S.; Zhang, X.; Zhang, J.; Ma, Z.; Wang, R.; Wu, D.; Wei, Z.; Tan, Z.; Zhang, B.; Wang, M. *Brittle Culm 15* mutation alters carbohydrate composition, degradation and methanogenesis of rice straw during in vitro ruminal fermentation. *Front. Plant Sci.* **2022**, *13*, 975456. [CrossRef]
9. Yi, S.; Zhang, X.; Chen, X.; Zhou, J.; Gao, C.; Ma, Z.; Wang, R.; Tan, Z.; Wang, M. Fermentation of increasing ratios of grain starch and straw fiber: Effects on hydrogen allocation and methanogenesis through in vitro ruminal batch culture. *PeerJ* **2023**, *11*, e15050. [CrossRef]
10. Li, Q.; Ma, Z.; Huo, J.; Zhang, X.; Wang, R.; Zhang, S.; Jiao, J.; Dong, X.; Janssen, P.H.; Ungerfeld, E.M.; et al. Distinct microbial hydrogen and reductant disposal pathways explain interbreed variations in ruminant methane yield. *ISME J.* **2024**, *18*, wrad016. [CrossRef]
11. Ungerfeld, E.M. Metabolic hydrogen flows in rumen fermentation: Principles and possibilities of interventions. *Front. Microbiol.* **2020**, *11*, 589. [CrossRef]
12. Darabighane, B.; Salem, A.Z.M.; Aghjehgheshlagh, F.M.; Mahdavi, A.; Zarei, A.; Elghandour, M.; Lopez, S. Environmental efficiency of *Saccharomyces cerevisiae* on methane production in dairy and beef cattle via a meta-analysis. *Environ. Sci. Pollut. Res. Int.* **2019**, *26*, 3651–3658. [CrossRef]
13. Garnsworthy, P.C.; Saunders, N.; Goodman, J.R.; Algherair, I.H.; Ambrose, J.D. Effects of live yeast on milk yield, feed efficiency, methane emissions and fertility of high-yielding dairy cows. *Animal* **2025**, *19*, 101379. [CrossRef] [PubMed]
14. Carpinelli, N.A.; Halfen, J.; Trevisi, E.; Chapman, J.D.; Sharman, E.D.; Anderson, J.L.; Osorio, J.S. Effects of peripartal yeast culture supplementation on lactation performance, blood biomarkers, rumen fermentation, and rumen bacteria species in dairy cows. *J. Dairy Sci.* **2021**, *104*, 10727–10743. [CrossRef]
15. Wang, W.; Wei, Z.; Li, Z.; Ren, J.; Song, Y.; Xu, J.; Liu, A.; Li, X.; Li, M.; Fan, H.; et al. Integrating genome- and transcriptome-wide association studies to uncover the host-microbiome interactions in bovine rumen methanogenesis. *iMeta* **2024**, *3*, e234. [CrossRef]
16. *NY/T 815*; Feeding Standard of Beef Cattle. Ministry of Agriculture and Rural Affairs, PRC: Beijing, China, 2004.
17. AOAC International. *Official Methods of Analysis*, 18th ed.; AOAC International: Gaithersburg, MD, USA, 2005.
18. Van Soest, P.; Robertson, J.; Lewis, B. Methods for dietary fiber, neutral detergent fiber, and nonstarch polysaccharides in relation to animal nutrition. *J. Dairy Sci.* **1991**, *74*, 3583–3597. [CrossRef] [PubMed]
19. Kartchner, R.J.; Theurer, B. Comparison of hydrolysis methods used in feed, digesta, and fecal starch analysis. *J. Agric. Food Chem.* **1981**, *29*, 8–11. [CrossRef]
20. *GB/T 23742*; Animal Feeding Stuffs-Determination of Ash Insoluble in Hydrochloric Acid. Standards Press of China: Beijing, China, 2009.
21. Van Keulen, J.; Young, B.A. Evaluation of acid-insoluble ash as a natural marker in ruminant digestibility studies. *J. Anim. Sci.* **1977**, *44*, 282–287. [CrossRef]
22. *T/QBAA 001*; Determination of Mannan Content in Yeast Culture (Saccharomyces cerevisiae)—High Performance Liquid Chromatography. Standards Press of China: Beijing, China, 2023.
23. *GB/T 18246;* Determination of Amino Acids in Feeds. Standards Press of China: Beijing, China, 2019.
24. *NY/T 3801*; Determination of Acid-Soluble Protein in Feed Materials. Ministry of Agriculture and Rural Affairs, PRC: Beijing, China, 2020.
25. *GB/T 12456;* Determination of Total Acid in Foods. Standards Press of China: Beijing, China, 2021.
26. Want, E.J.; Wilson, I.D.; Gika, H.; Theodoridis, G.; Plumb, R.S.; Shockcor, J.; Holmes, E.; Nicholson, J.K. Global metabolic profiling procedures for urine using UPLC-MS. *Nat. Protoc.* **2010**, *5*, 1005–1018. [CrossRef]
27. Dunn, W.B.; Broadhurst, D.; Begley, P.; Zelena, E.; Francis-McIntyre, S.; Anderson, N.; Brown, M.; Knowles, J.D.; Halsall, A.; Haselden, J.N.; et al. Procedures for large-scale metabolic profiling of serum and plasma using gas chromatography and liquid chromatography coupled to mass spectrometry. *Nat. Protoc.* **2011**, *6*, 1060–1083. [CrossRef]

28. Dai, W.D.; Xie, D.C.; Lu, M.L.; Li, P.L.; Lv, H.P.; Yang, C.; Peng, Q.H.; Zhu, Y.; Guo, L.; Zhang, Y.; et al. Characterization of white tea metabolome: Comparison against green and black tea by a nontargeted metabolomics approach. *Food Res. Int.* **2017**, *96*, 40–45. [CrossRef] [PubMed]
29. Wang, M.; Wang, R.; Janssen, P.H.; Zhang, X.M.; Sun, X.Z.; Pacheco, D.; Tan, Z.L. Sampling procedure for the measurement of dissolved hydrogen and volatile fatty acids in the rumen of dairy cows. *J. Anim. Sci.* **2016**, *94*, 1159–1169. [CrossRef] [PubMed]
30. Zinn, R.A.; Owens, F.N. A rapid procedure for purine measurement and its use for estimating net ruminal protein synthesis. *Can. J. Anim. Sci.* **1986**, *66*, 157–166. [CrossRef]
31. Ma, Z.Y.; Zhang, X.M.; Wang, R.; Wang, M.; Liu, T.; Tan, Z.L. Effects of chemical and mechanical lysis on microbial DNA yield, integrity, and downstream amplicon sequencing of rumen bacteria and protozoa. *Front. Microbiol.* **2020**, *11*, 581227. [CrossRef]
32. Zakrzewski, M.; Goesmann, A.; Jaenicke, S.; Jünemann, S.; Eikmeyer, F.; Szczepanowski, R.; Al-Soud, W.A.; Sørensen, S.; Pühler, A.; Schlüter, A. Profiling of the metabolically active community from a production-scale biogas plant by means of high-throughput metatranscriptome sequencing. *J. Biotechnol.* **2012**, *158*, 248–258. [CrossRef]
33. Amato, K.R.; Yeoman, C.J.; Kent, A.; Righini, N.; Carbonero, F.; Estrada, A.; Gaskins, H.R.; Stumpf, R.M.; Yildirim, S.; Torralba, M.; et al. Habitat degradation impacts black howler monkey (*Alouatta pigra*) gastrointestinal microbiomes. *ISME J.* **2013**, *7*, 1344–1353. [CrossRef]
34. Wagner, J.J.; Engle, T.E.; Belknap, C.R.; Dorton, K.L. Meta-analysis examining the effects of *Saccharomyces cerevisiae* fermentation products on feedlot performance and carcass traits. *Pro. Anim. Sci.* **2016**, *32*, 172–182. [CrossRef]
35. Zhang, J.C.; Chen, P.; Zhang, C.; Khalil, M.M.; Zhang, N.Y.; Qi, D.S.; Wang, Y.W.; Sun, L.H. Yeast culture promotes the production of aged laying hens by improving intestinal digestive enzyme activities and the intestinal health status. *Poult. Sci.* **2020**, *99*, 2026–2032. [CrossRef] [PubMed]
36. Patra, A.K.; Yu, Z. Genomic Insights into the Distribution of Peptidases and Proteolytic Capacity among *Prevotella* and *Paraprevotella* Species. *Microbiol. Spectr.* **2022**, *10*, e0218521. [CrossRef]
37. Baaske, L.; Gabel, G.; Dengler, F. Ruminal epithelium: A checkpoint for cattle health. *J. Dairy Res.* **2020**, *87*, 322–329. [CrossRef]
38. Liu, Q.; Wang, C.; Yang, W.Z.; Zhang, B.; Yang, X.M.; He, D.C.; Zhang, P.; Dong, K.H.; Huang, Y.X. Effects of isobutyrate on rumen fermentation, lactation performance and plasma characteristics in dairy cows. *Anim. Feed Sci. Technol.* **2009**, *154*, 58–67. [CrossRef]
39. Malik, P.K.; Trivedi, S.; Kolte, A.P.; Mohapatra, A.; Biswas, S.; Bhattar, A.V.K.; Bhatta, R.; Rahman, H. Comparative analysis of rumen metagenome, metatranscriptome, fermentation and methane yield in cattle and buffaloes fed on the same diet. *Front. Microbiol.* **2023**, *14*, 1266025. [CrossRef]
40. Puppel, K.; Kuczyńska, B. Metabolic profiles of cow's blood; a review. *J. Sci. Food Agric.* **2016**, *96*, 4321–4328. [CrossRef] [PubMed]
41. Motta, G.A.; Neto, P.S.M.; Nociti, R.P.; Santana, A.E. Hematological normality, serum biochemistry, and acute phase proteins in healthy beef calves in the brazilian savannah. *Animals* **2023**, *13*, 2398. [CrossRef]
42. Hao, Y.; Xing, M.; Gu, X. Research progress on oxidative stress and its nutritional regulation strategies in pigs. *Animals* **2021**, *11*, 1384. [CrossRef]
43. Liu, T.; Zheng, W.; Wang, L.L.; Wang, L.; Zhang, Y.K. TLR4/NF-κB signaling pathway participates in the protective effects of apocynin on gestational diabetes mellitus induced placental oxidative stress and inflammation. *Reprod. Sci.* **2020**, *27*, 722–730. [CrossRef] [PubMed]
44. Vattepu, R.; Sneed, S.L.; Anthony, R.M. Sialylation as an important regulator of antibody function. *Front. Immunol.* **2022**, *13*, 818736. [CrossRef]
45. Copenhaver, M.; Yu, C.Y.; Hoffman, R.P. Complement components, C3 and C4, and the metabolic syndrome. *Curr. Diabetes Rev.* **2019**, *15*, 44–48. [CrossRef]
46. Bendickova, K.; Fric, J. Roles of IL-2 in bridging adaptive and innate immunity, and as a tool for cellular immunotherapy. *J. Leukoc. Biol.* **2020**, *108*, 427–437. [CrossRef]
47. Fard, N.E.; Xiao, M.; Sehmi, R. Regulatory ILC2—Role of IL-10 producing ILC2 in Asthma. *Cells* **2023**, *12*, 2556. [CrossRef]
48. Casanova, J.L.; MacMicking, J.D.; Nathan, C.F. Interferon-γ and infectious diseases: Lessons and prospects. *Science* **2024**, *384*, eadl2016. [CrossRef] [PubMed]
49. Yang, Z.; Shao, Y.; Yang, J.; Xing, X.; Yang, H.; Wang, Z. Betaine enhances hepatic antioxidant activity and thymus-associated immunity in lipopolysaccharide-challenged goslings. *BMC Vet. Res.* **2025**, *21*, 77. [CrossRef]
50. Hieu, T.V.; Guntoro, B.; Qui, N.H.; Quyen, N.T.K.; Hafiz, F.A.A. The application of ascorbic acid as a therapeutic feed additive to boost immunity and antioxidant activity of poultry in heat stress environment. *Vet. World* **2022**, *15*, 685–693. [CrossRef] [PubMed]
51. Fikry, A.M.; Attia, A.I.; Ismail, I.E.; Alagawany, M.; Reda, F.M. Dietary citric acid enhances growth performance, nutrient digestibility, intestinal microbiota, antioxidant status, and immunity of Japanese quails. *Poult. Sci.* **2021**, *100*, 101326. [CrossRef]
52. Farsani, M.N.; Ahani, S.; Ahani, S.; Ghafarifarsani, H. Interactive effects of dietary probiotic and succinic acid on the growth performance, digestive enzyme activities, immunomodulation, antioxidative capacity, and disease resistance in rainbow trout (*Oncorhynchus mykiss*) juveniles. *Ann. Anim. Sci.* **2024**, *24*, 1237–1249. [CrossRef]

53. Qiu, K.; He, W.; Zhang, H.; Wang, J.; Qi, G.; Guo, N.; Zhang, X.; Wu, S. Bio-fermented malic acid facilitates the production of high-quality chicken via enhancing muscle antioxidant capacity of broilers. *Antioxidants* **2022**, *11*, 2309. [CrossRef]
54. Capó, X.; Martorell, M.; Tur, J.A.; Sureda, A.; Pons, A. 5-Dodecanolide, a compound isolated from pig lard, presents powerful anti-inflammatory properties. *Molecules* **2021**, *26*, 7363. [CrossRef]
55. Angthong, P.; Uengwetwanit, T.; Uawisetwathana, U.; Koehorst, J.J.; Arayamethakorn, S.; Schaap, P.J.; Santos, V.M.D.; Phromson, M.; Karoonuthaisiri, N.; Chaiyapechara, S.; et al. Investigating host-gut microbial relationship in *Penaeus monodon* upon exposure to *Vibrio harveyi*. *Aquaculture* **2023**, *567*, 739252. [CrossRef]
56. Wang, A.; Archile, A.; Patterson, R. Feeding a novel mannan-rich yeast carbohydrate product improves production performance and humoral immunity of broiler chickens. *Animals* **2024**, *14*, 1667. [CrossRef]
57. Pan, M.; Kong, F.; Xing, L.; Yao, L.; Li, Y.; Liu, Y.; Li, C.; Li, L. The structural characterization and immunomodulatory activity of polysaccharides from *Pleurotus abieticola* fruiting bodies. *Nutrients* **2022**, *14*, 4410. [CrossRef]
58. Gu, F.; Zhu, S.; Tang, Y.; Liu, X.; Jia, M.; Malmuthuge, N.; Valencak, T.G.; McFadden, J.W.; Liu, J.X.; Sun, H.Z. Gut microbiome is linked to functions of peripheral immune cells in transition cows during excessive lipolysis. *Microbiome* **2023**, *11*, 40. [CrossRef] [PubMed]

animals

MDPI

Article

Effects of Yeast Culture on Lamb Growth Performance, Rumen Microbiota, and Metabolites

Jinlong Xu [1,2,3], Xiongxiong Li [4], Qingshan Fan [1,2], Shengguo Zhao [3,4] and Ting Jiao [1,2,3,*]

1 College of Prataculturai Science, Gansu Agricultural University, Lanzhou 730070, China; xjl_gsau@163.com (J.X.); fanqsh18@lzu.edu.cn (Q.F.)
2 Key Laboratory for Grassland Ecosystem of Ministry of Education, Gansu Agricultural University, Lanzhou 730070, China
3 Provincial R&D Institute of Ruminants in Gansu, Lanzhou 730070, China; zhaosg@gsau.edu.cn
4 College of Animal Science and Technology, Gansu Agricultural University, Lanzhou 730070, China; lxx_gsau@163.com
* Correspondence: jiaoting207@126.com; Tel.: +86-136-0930-4144; Fax: +86-0931-642-468

Simple Summary: Modern feeding methods and high-grain diets have negative effects on the rumen health of ruminants. Yeast culture (YC) is a kind of green additive probiotic rich in yeast polysaccharides, mannan oligosaccharides, vitamins, peptides, and organic acids. These components can be used as fermentation substrates and growth factors to promote the growth of beneficial bacteria in the gastrointestinal tract of ruminants to effectively improve their gastrointestinal health. Therefore, this study aimed to scientifically verify the theoretical basis for the effect of YC in maintaining the rumen of lambs and promote partial replacement of antibiotics for lamb fattening and health maintenance in practice. The lambs were divided into five groups, fed 0%, 3%, 6%, 9%, and 12% YC. The basic diet was the same, and the test period was 70 days. By measuring the production performance, health indicators, and rumen environment of lambs, it was found that the 9% YC changed the rumen fermentation environment of lambs, promoted the development of rumen tissue, improved feed utilization, and ultimately improved the fattening effect. The results suggest that adding 9% YC to pelleted diets at high concentrations helps promote the development of rumen epithelium and regulates the fermentation pattern of the rumen, thereby maintaining rumen health in sheep.

Abstract: The effects of incorporating yeast culture (YC) into pelleted feeds on sheep production and the potential impact on rumen microbial populations, microbial metabolism, and fermentation have not been extensively studied. This study aimed to evaluate the effect of YC on growth performance, rumen tissue development, rumen fermentation, and rumen microflora in sheep and to explore the potential microbial mechanisms involved. Fifty healthy 3-month-old male lambs of small-tailed Han sheep, with an average weight of 28.44 ± 0.63 kg, were randomly divided into five groups: control (0% YC), 3% YC, 6% YC, 9% YC, and 12% YC. The pre-feeding period lasted for 15 days, followed by an official feeding period of 60 days. On the last day of the formal feeding period, six lambs that exhibited the best growth performance were randomly selected from the control group and the 9% YC group. These sheep were slaughtered, then the rumen epithelial tissue and rumen contents were collected for the measurement of rumen fermentation, microbial populations, and metabolites. Compared to the control group, the YC-treated groups showed higher daily and final body weight gains, as well as increased levels of propionic acid, butyric acid, and total volatile fatty acids ($p < 0.05$). YC supplementation also enhanced rumen papilla length and width ($p < 0.05$). Additionally, YC increased the relative abundance of certain microbial species ($p < 0.05$). These results suggest that

 https://doi.org/10.3390/ani15050738

supplementing 9% YC in pelleted diets for small-tailed Han sheep may enhance growth performance and improve the rumen environment.

Keywords: lamb; yeast culture; growth performance; rumen tissue; rumen microorganisms; microbial metabolism

1. Introduction

The rumen is a unique digestive organ in ruminants, playing a critical role in the digestion of fiber and the absorption of nutrients by the host. This organ houses a complex and diverse microbiota that interacts synergistically to form a highly efficient "natural fermenter", capable of breaking down cellulose in food [1]. The health and function of the rumen not only directly impact the physiological status of ruminants but are also closely linked to their production performance [2]. Therefore, it is essential to maintain the rumen's internal environment. However, modern intensive feeding practices and the widespread use of high-grain diets have negatively affected sheep rumen health [3]. The use of antibiotics in animal husbandry to deal with the production problems of sheep promotes the development of antibiotic-resistant strains, and imbalance in the rumen microbiome may also pose a risk to human health [4]. In recent years, the application of plant-derived natural substances and probiotics have gained recognition as effective strategies to replace antibiotics in maintaining ruminant health [5,6]. Among these, yeast culture (YC), a green additive probiotic, is rich in yeast polysaccharides, mannan oligosaccharides, vitamins, peptides, and organic acids. These components can serve as fermentation substrates and growth factors, promoting the growth of beneficial bacteria in the gastrointestinal tract of ruminants, thereby effectively improving their gastrointestinal health [7,8].

Years of research have shown that YC has a positive impact on the health and growth of ruminants [9,10]. It can promote the development of the rumen epithelium of male fattening sheep, improve nitrogen utilization, and improve the rumen microbial environment by increasing the relative abundance of *Succiniclasticum* and *Fibrobacter* [11]. Furthermore, Pi et al. [12] found that adding YC can stabilize the pH value of the rumen by reducing lactic acid-producing bacteria and increasing lactic acid-utilizing bacteria or rumen protozoa, thereby determining the rumen environment. Further results from Li et al. [13] indicated that the addition of YC improved fermentation parameters by modulating the Anaplasma phage, *Prevotella_7*, and *Lachnospiraceae*_NK3A20 groups in lamb rumen, thereby enhancing the flavor of Aohu sheep. Supplementation with yeast culture (YC) also enhanced the abundance of fibrinolytic microbiota and enriched metabolites in rumen associated with amino acid and lipid metabolism pathways of male Hu lambs, thereby maintaining rumen homeostasis under high-concentrate dietary conditions [14]. Liu et al. [15] found that the concentrations of acetic acid and total volatile fatty acids (TVFAs) in the rumen of small-tailed Han sheep increased, the concentration of pH decreased, and the relative abundance of *Lachnospiraceae*_NK3A20 and *Olsenella* increased when using a high-concentration diet. The polysaccharide components in yeast cultures modulate the structure and metabolism of the rumen microbiota, thereby increasing nutrient utilization in ruminants and improving the quality of livestock-derived products [16]. Although existing studies have demonstrated the benefits of yeast culture (YC) in regulating rumen microbial communities, nitrogen utilization efficiency, and volatile fatty acid profiles across various ruminant species, the feasibility of incorporating YC into pelleted diets and its mechanistic impacts on energy metabolism pathways through functional gene–metabolite interactions remain less explored. This limits the widespread adoption of YC as a green natural additive with properties of health maintenance and regulation of the rumen environment, integrated into pellet feed, in intensive lamb production systems.

This study hypothesized that supplementing pelleted feed with brewer's yeast culture (YC) might improve growth performance and maintain rumen health by altering rumen bacteria and metabolites, which could positively affect rumen tissue morphology in mature sheep. Therefore, this study aimed to evaluate the impact of YC supplementation in pelleted feed on growth performance, rumen tissue development, rumen fermentation indices, rumen bacterial composition, and metabolomics in small-tailed Han sheep.

2. Materials and Methods

2.1. Yeast Culture

Yeast culture (YC) is produced using distiller grains as the fermentation substrate, which is inoculated with high-activity probiotics such as *Bacillus subtilis*, *Saccharomyces cerevisiae*, and *Aspergillus cerevisiae*. The production process involves using composite strains inoculated with liquid culture, followed by secondary solid fermentation for 48 to 72 h. The YC was provided by Sichuan Runge Biotechnology Co., Ltd., Mianzhu City, China. The number of viable bacteria reached 1×10^8 CFU/g, and the inoculation rate of composite strains was 10%. The nutritional active ingredients of YC include crude protein (dry matter basis) $\geq$ 17%, crude fiber $\leq$ 20%, crude ash $\leq$ 12%, crude fat $\geq$ 2.6%, and total organic acid $\geq$ 5%.

The study was conducted at Yingqian Agricultural and Animal Husbandry Development Co., Ltd. in Linxia City, China. This experiment has been approved by the Animal Ethics Committee of Gansu Agricultural University (GSAU-2nd-AST-2023-053). Fifty healthy 3-month-old male small-tailed Han lambs (average weight: 28.44 $\pm$ 0.63 kg) were randomly divided into five groups: control (CON) and 3%, 6%, 9%, and 12% YC. Each treatment group (YC) consisted of 10 sheep, which were housed in separate single pens with one replicate per sheep.

The trial utilized a nutritionally balanced total mixed ration (TMR) formulated in compliance (PRC NY/T 816-2004), with detailed nutritional composition presented in Table 1. Animals received TMR rations twice daily (08:00 and 18:00) ad libitum, with graded supplementation of *Saccharomyces cerevisiae* culture (YC; viable count $\geq 1 \times 10^8$ CFU/g) at 3%, 6%, 9%, and 12% (dry matter basis). The 15-day adaptation phase was followed by a 60-day test period. At the end of the test, the 9% YC group exhibited the greatest daily weight gain, the fastest growth rate, and the greatest final weight. Consequently, six lambs were randomly selected from both the 9% YC experimental and control groups for slaughter and further analysis.

Table 1. Basic diet formulation (dry matter basis).

Items	Diets (%)	Nutrient Levels	
Corn	32.0	Digestive energy DE/(MJ/kg)	11.02
Wheat bran	6.0	Moisture content/%	11.76
Grass meal	10.0	CP/%	16.43
Wheatgrass	10.0	Ash/%	8.16
Soybean meal (43%)	5.0	Salinity/%	0.96
Cottonseed meal (46%)	5.0	Ca/%	1.23
Bran	8.0	P/%	0.49
Corn germ meal	15.0	NDF/%	37.84
DDGS (distiller dried grains with solubles)	5.0	ADF%	26.28
Premix [a]	4.0	NFE (g/kg) [b]	617.92

[a] Composition (per kg of dry matter): 100,000–500,000 IU of vitamin A, 50,000–200,000 IU of vitamin D3, ≥500 IU of vitamin E, Fe 1500–7000 mg, Cu 300–750 mg, Mn 1000–5000 mg, Zn 1500–4000 mg, I 20–30 mg, Se 5–20 mg, Co 8–35 mg. [b] NFE (g/kg DM) = 1000 − (EE + CP + CF + Ash) [17,18]. CP: crude protein; NDF: neutral detergent fiber; ADF: acid detergent fiber.

2.2. Sample Collection and Processing

Rumen fluid and epithelial tissues were collected from the rumen after lambs were slaughtered. A total of 70 mL of rumen content was obtained from each lamb and filtered through four layers of gauze. The collected rumen fluid was then divided into three portions: one portion was transferred to a 10 mL cryotube for DNA extraction, another portion was allocated to a separate 10 mL cryotube for metabolite identification, and the final portion was placed into a 50 mL centrifuge tube for the assessment of rumen fermentation parameters. Once dispensed, the samples were transported to the laboratory in liquid nitrogen tanks and stored in an ultralow-temperature freezer at −80 °C for future analysis.

2.3. Product Performance

Throughout the treatment phase, both the initial individual body weight and the final body weight were recorded. A daily log was maintained to track the amount of feed consumed, from which the average daily feed intake (ADFI), average daily gain (ADG), and feed-to-gain ratio (F/G) were calculated.

2.4. Determination of Rumen Histomorphometry

Five tissue samples, each measuring 2.5 × 2.5 cm, were excised from the abdominal capsule region of each sheep. The rumen and abdominal tissues were immersed in 4% paraformaldehyde for 24 h, after which paraffin sections were prepared by orienting the specimens in a water-based medium within a ventilated environment. The process for sectioning comprised sampling, dehydration, waxing, embedding, and sectioning, all conducted in a ventilated setting. After hematoxylin–eosin (HE) staining, the sections were dehydrated and sealed with a mixture of anhydrous ethanol and xylene. The rumen papilla's length and width and the muscle layer's thickness were then measured using Case Viewer software (Case Viewer 2.4.0), followed by image acquisition using a Panoramic 250 digital slide scanner (3DHISTECH, Budapest, Hungary).

2.5. Rumen Fermentation Characteristics

Rumen pH was determined using a calibrated digital pH meter (P611, Shanghai Leici Instrument Co., Ltd., Shanghai, China) equipped with a temperature-compensated electrode, with triplicate measurements recorded within 2 min post-sampling to minimize atmospheric exposure. The concentration of ammonia nitrogen (NH_3-N) in the rumen was quantified using established colorimetric methods, performed with the assistance of a 721-type spectrophotometer (TU-1901). Additionally, volatile fatty acids (VFAs) were analyzed using gas chromatography with a GC-2010 Plus from Shimadzu (Kyoto, Japan). These methodologies ensure accurate and reliable assessments of the chemical composition of the rumen fluid [19].

2.6. High-Throughput Sequencing and Analysis

Microbial DNA was extracted from rumen content samples using an MN NucleoSpin 96 Soil kit (Macherey-Nagel, Düren, Germany), and the concentration and purity were assessed using a NanoPhotometer (N60, Düren, Germany). The community structure of the rumen microbiota was determined by PCR amplification of the V3–V4 region of the 16S rRNA gene. The primers used were 338F (5′-ACTCCTACGGGAGGCA GCAG-3′) and 806R (5′-GGACTACHVGGGTWTCTAAT-3′). The library obtained from PCR amplification was sequenced on an Illumina MiSeq platform (Illumina, San Diego, CA, USA), and bioinformatic analysis was performed using BMKCloud.

The raw data produced by the Illumina MiSeq platform underwent a series of processes, including the merging of paired-end reads (using FLASH v1.2.7), filtering (via Trimmomatic v0.33), and removal of chimeric sequences (with UCHIME v4.2) to derive optimized sequences, known as tags. Subsequently, Usearch software (v11 64bit) [20] was employed to cluster these tags at a similarity threshold of 97%, generate operational taxonomic units (OTUs), and taxonomic annotation of the OTUs was performed using the Silva (bacteria) taxonomic database. The OTU assessment results were analyzed at various taxonomic levels, resulting in community structure maps for each sample by phylum, class, order, family, genus, and species. Alpha diversity metrics were applied to evaluate species diversity within individual samples, yielding indices such as Ace, Chao1, Shannon, and Simpson. Sample dilution curves [21] and rank abundance curves were also constructed. Beta diversity analyses were carried out to investigate variations in species diversity, particularly in the composition and structure of the microbiota. A PCA map was generated based on the distance matrix [22], and biomarkers with statistically significant differences between groups were identified through intergroup significance analysis (LEfSe analysis) [23]. Metastats software (QIIME2 2023.2) was used for t-tests on species abundance data across groups to derive p-values [24], which were subsequently adjusted to yield q-values. Ultimately, species responsible for compositional differences between the two sample groups were identified based on the q-values. Functional analysis of the 16S gene was conducted using the Kyoto Encyclopedia of Genes and Genomes (KEGG) framework and the Clusters of Orthologous Groups (COGs).

2.7. Metabolomic Analysis of Rumen Fluid

Rumen fluid specimens were passively thawed at 4 °C following extraction from a −80 °C cryostorage system. Aliquots (500 μL) were homogenized with an equivalent volume of acetonitrile–methanol (1:1, v/v) solvent mixture and vortex-mixed for 30 sec, followed by 10 min sonication (PS-60AL apparatus) and 60 min incubation at −20 °C. Particulate removal was achieved through dual-phase centrifugation (13,000 rpm, 4 °C, 15 min). Chromatographic separation was performed on a Shimadzu LC-30 platform equipped with an Acquity UPLC HSS T3 column (2.1 × 100 mm, 1.8 μm), employing the aqueous mobile phase at a 0.3 mL/min flow rate with column thermostating at 50 °C, utilizing 2 μL injection volumes. High-resolution mass spectrometric detection was implemented using a Triple TOF 5600 system operated in dual ionization modes. Raw spectral data processing through MassLynx V4.2 incorporated peak annotation and spectral alignment, with metabolite identification via Progenesis QI cross-referenced against the METLIN and Biomark proprietary databases, maintaining stringent mass accuracy thresholds (<100 ppm). Pathway mapping and compound classification were executed through KEGG, HMDB, and Lipid MAPS resources, with differential analysis performed using two-tailed T-tests to determine statistical significance (p-value computation for all analytes).

2.8. Statistical Analysis

The statistical analyses were conducted utilizing SPSS v.27.0 (IBM, Armonk, NY, USA) and R v.4.0.2 (R Foundation, Vienna, Austria). This investigation systematically assessed multiple variables encompassing forage nutritional composition, rumen fermentation characteristics, microbial community structure (analyzed at phylum and genus taxonomic levels), fibrinolytic enzyme kinetics, CAZyme gene expression profiles, and rumen histomorphometric parameters. Comparative analyses between experimental groups were performed using Student's t-test, with statistical significance defined at $\alpha = 0.05$ (two-tailed). Nonparametric Wilcoxon rank-sum tests were implemented for pathway enrichment analysis of differentially expressed genes in metabolic pathways. Bivariate associations between microbial taxa (top 20 operational taxonomic units) and differentially abundant metabolites were quantified using Spearman's rank correlation analysis (ρ coefficients ranging −1 to +1). Data visualization was executed using ggplot2 and related packages within the R programming environment. This multiparametric analytical framework facilitated a comprehensive evaluation of interrelationships among nutritional, microbial, enzymatic, and histological parameters in the rumen ecosystem.

3. Results

3.1. Growth Performance Analysis

As shown in Table 2, the final body weight and average daily gain in the 9% YC group were significantly higher than those of the control group ($p < 0.05$). At the same time, the average daily weight gain was the highest in the 9% YC group, with the lowest feed-to-gain ratio.

Table 2. Effect of adding yeast culture on the growth performance of lambs.

Items	CON	3% YC	6% YC	9% YC	12% YC	p Value
Initial weight (kg)	27.33 ± 0.58	27.77 ± 0.81	28.23 ± 0.82	28.47 ± 0.72	27.5 ± 0.57	0.764
Final weight (kg)	42.9 ± 0.25 ^{b}	44.87 ± 1.11	45 ± 0.89	46.57 ± 0.80 ^{a}	45.02 ± 0.39	0.040
ADG (g/d)	277.98 ± 24.13	305.36 ± 27.50	299.40 ± 25.59	323.21 ± 28.19	312.80 ± 29.18	0.232
ADFI (kg/d)	1.61 ± 0.09	1.62 ± 0.07	1.59 ± 0.1	1.67 ± 0.13	1.66 ± 0.09	0.977
F/G	5.79 ± 0.31	5.31 ± 0.24	5.32 ± 0.35	5.17 ± 0.41	5.30 ± 0.28	0.706

Note: CON = control group; YC = yeast culture group. ^{a,b} Values within a row with different letters are significantly different ($p < 0.05$).

3.2. Differences in VFAs in Sheep Rumen

The concentrations of TVFA, propionic acid, and butyric acid were significantly higher in the 9% YC group than those of the control group, while the ethyl–propyl ratio was considerably lower in the 9% YC group compared to the control group (Table 3; $p < 0.05$).

Table 3. Effect of the addition of yeast culture on rumen fermentation.

Items	CON	9% YC	p Value
pH	6.65 ± 0.16	6.44 ± 0.15	0.346
NH_3-N (mg/dL)	8.70 ± 2.50	5.93 ± 0.55	0.051
TVFA (mmol/L)	67.41 ± 2.30 ^{b}	81.91 ± 3.47 ^{a}	0.006
Acetate (mmol/L)	41.86 ± 0.73	47.79 ± 3.13	0.119
Propionate (mmol/L)	14.30 ± 1.59 ^{b}	19.11 ± 0.99 ^{a}	0.002
Butyrate (mmol/L)	8.69 ± 1.13 ^{b}	11.91 ± 0.76 ^{a}	0.039
Isobutyric acid (mmol/L)	0.73 ± 0.08	0.78 ± 0.05	0.644
Valerate (mmol/L)	0.711 ± 0.02	0.78 ± 0.04	0.151
Isovalerate (mmol/L)	1.12 ± 0.20	1.54 ± 0.14	0.124
A/P	2.96 ± 0.13	2.51 ± 0.12	0.028

Note: CON = control group; YC = yeast culture group. TVFA: total volatile fatty acid. A/P: acetic acid-to-propionic acid ratio. ^{a,b} Values within a row with different letters are significantly different ($p < 0.05$).

3.3. Differences in Rumen Organization

The papilla height and width in the rumen of sheep in the 9% YC group were significantly higher than those in the control group (Figure 1 and Table S1).

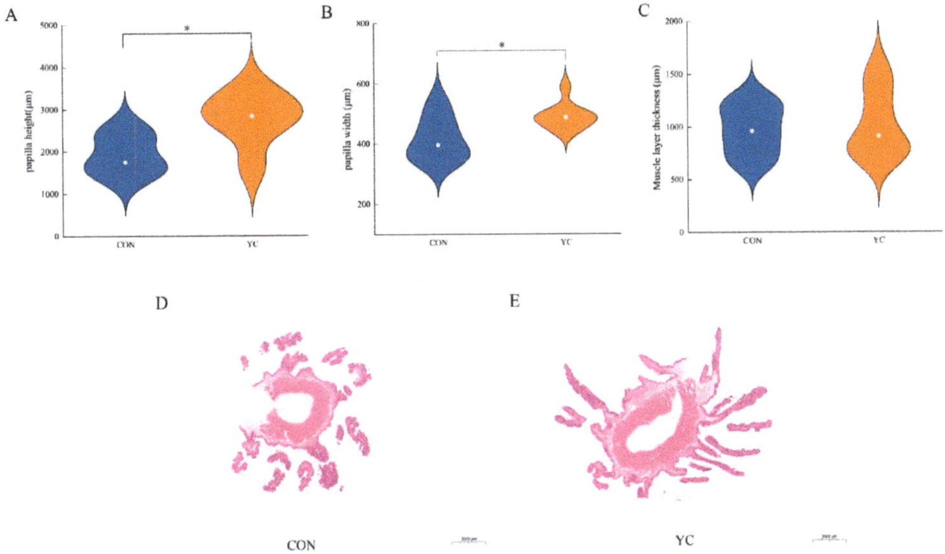

Figure 1. (**A–C**) Measurement of indices related to rumen morphology. (**D**,**E**) H&E staining of the ruminal wall. CON = control group; YC = yeast culture group. * Significant difference between groups ($p < 0.05$).

3.4. *Analysis of the Alpha Diversity of Rumen Microorganisms*

There was no significant difference between the two groups in rumen microbiota alpha diversity analysis, including the Shannon, Simpson, ACE, and Chao1 indices ($p > 0.05$) (Figure 2).

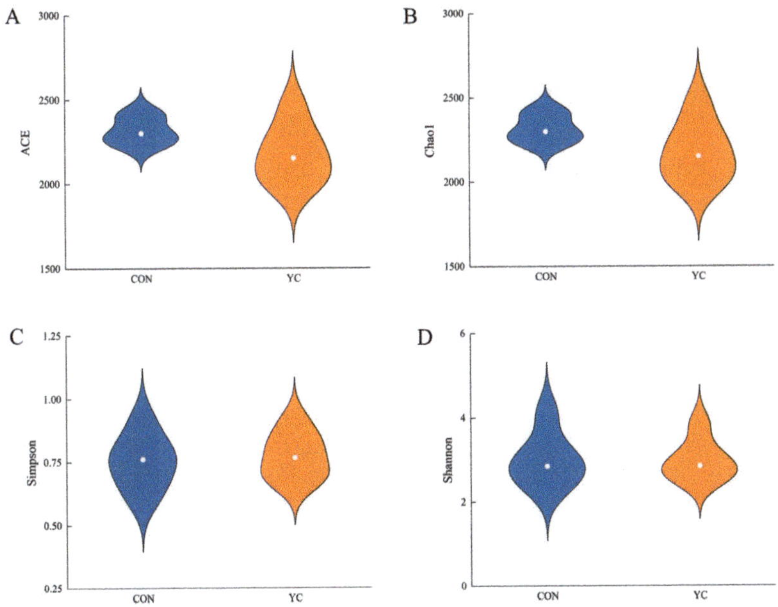

Figure 2. Alpha diversity of rumen microorganisms in both groups. (**A**), ACE index. (**B**), Chao 1 index. (**C**), Simpson index. (**D**), Shannon index. CON = control group; YC = yeast culture group.

3.5. Differences in the Composition of Rumen Microbial Communities in Sheep

At the phylum level, rumen bacteria mainly included Bacteroidetes, Firmicutes, Proteobacteria, and Fibrobacteres (Figure 3A). The relative abundance of Fibrobacteres in the 9% YC group was significantly higher than that in the control group (Table S2, $p < 0.05$).

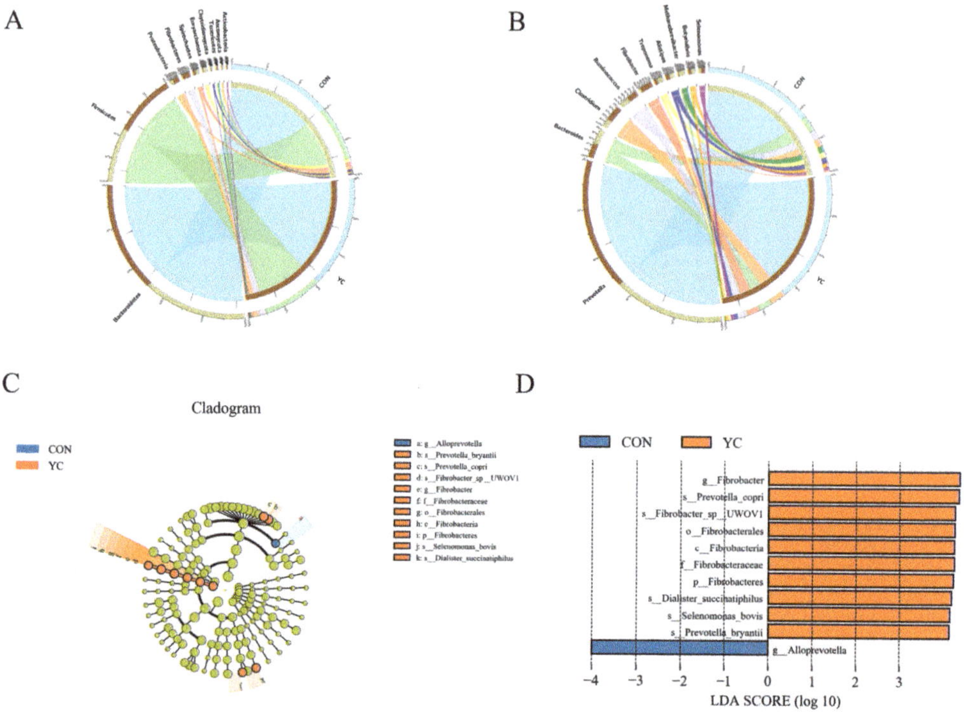

Figure 3. Relative abundance of phylum (**A**) and genus (**B**) level species. (**C**) LEfSe analysis of branching diagrams. (**D**) Significantly different bacterial taxa were identified by linear discriminant analysis effect size (LEfSe) (threshold value set to 3.0). CON = control group; YC = yeast culture group.

At the genus level, rumen bacteria primarily consist of *Prevotella*, *Bacteroides*, *Clostridium*, *Ruminococcus*, and *Fibrobacter* (Figure 3B). The relative abundance of *Clostridium*, *Chlamydia*, and *Fibrobacter* was significantly higher in the 9% YC group than that in the control group (Supplementary Table S3, $p < 0.05$). Conversely, *Alloprevotella*, *Candidatus_Nanosyncoccus*, and *Tannerella* were significantly higher in the control group than those in the YC group (Table S3, $p < 0.05$).

LEfSe difference analysis was performed to examine the rumen microorganisms identified in the two groups (Figure 3C). The results of the LEfSe analysis (Figure 3D) revealed that the genera with higher abundance in the control group included Alloprevotella. In contrast, the genera with higher abundance in the YC group were Fibrobacteres, Dialister_succinatiphilus, Selenomonas_bovis, Prevotella_bryantii, and Prevotella_copri.

3.6. Differences in CAZy and KEGG Pathways of Sheep Rumen

The concentrations of glycoside hydrolases (GHs), glycosyltransferases (GTs), polysaccharide lyases (PLs), and carbohydrate esterases (CEs) were higher in the rumen of sheep in the YC group than in the control group ($p > 0.05$) (Supplementary Table S3). The GH, GT, PL, and CBM families associated with polysaccharide degradation were compared. The relative abundance of CBM50, GH23, GH73, GH24, GT9, GT83, GT32, GH4, GH84,

CBM41, CBM88, CBM26, GT49, GH108, GH100, and GH108 was significantly higher in the rumen of sheep in the YC group compared to the control group. Conversely, the relative abundance of GH171, CBM12, CBM66, CBM77, and PL4 in the rumen of control sheep was considerably higher than that in the YC group (Figure S1).

At level 2 (Figure 4), the global and overview maps showed pathways including carbohydrate metabolism, amino acid metabolism, translation, nucleotide metabolism, metabolism of cofactors and vitamins, replication and repair, energy metabolism, membrane transport, folding, sorting and degradation, glycan biosynthesis and metabolism, lipid metabolism, metabolism of other amino acids, cellular community—prokaryotes, signal transduction, biosynthesis of other secondary metabolites, metabolism of terpenoids and polyketides, transport and catabolism, transcription, and cell motility ($p > 0.05$).

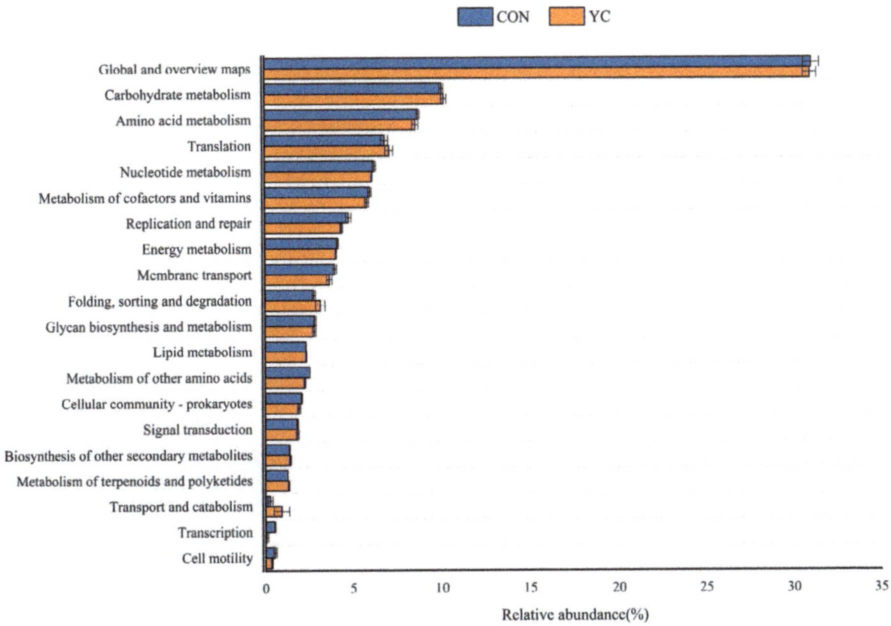

Figure 4. Functional genes related to the KEGG secondary metabolic pathway in both groups. CON = control group; YC = yeast culture group.

3.7. Differences in Rumen Metabolites

In the OPLS-DA model, the R^2Y value for comparison approached 1 between the two groups, while the slope of the fitted regression line for Q^2Y was positive. This indicated that the established model was both stable and reliable, making it suitable for comparing the differences between the two groups (Figure 5A). Using a significance threshold of $p < 0.05$, we identified 1232 differential metabolites between the two groups, of which 771 were significantly upregulated and 461 were significantly downregulated (Figure 5B).

KEGG enrichment analysis was conducted to elucidate the metabolic pathways associated with the significantly different metabolites. The key differential metabolites predominantly pertained to amino acid metabolism, digestion, lipid metabolism, and energy metabolism pathways (Figure 5D). Notably, the primary differential metabolic pathways identified between the two groups included arachidonic acid metabolism, arginine and proline metabolism, neurodegenerative pathways related to multiple diseases, serotonergic synapses, and oxidative phosphorylation (Figure 5C).

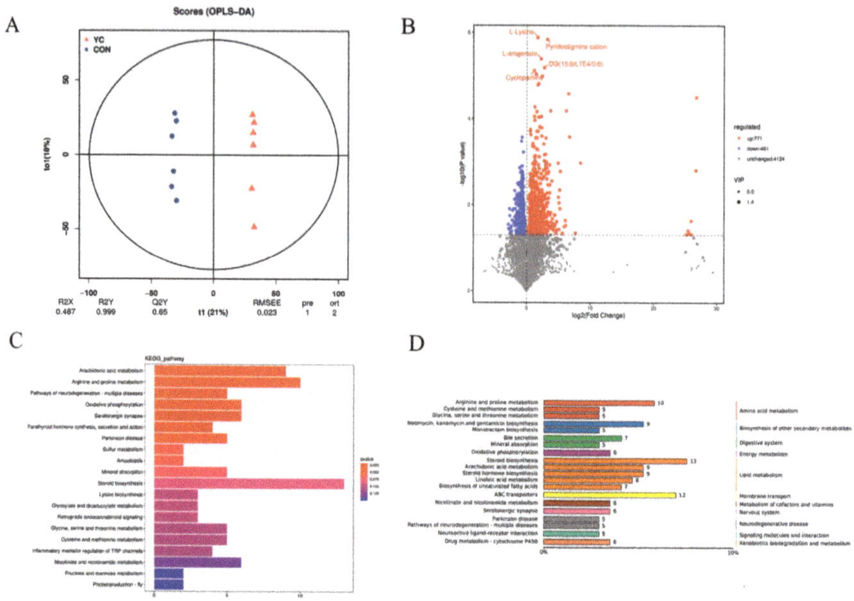

Figure 5. (**A**) OPLS-DA model score plots in the two groups. (**B**) Volcano maps of differential metabolites in the two groups. (**C**) KEGG map of enriched pathways. (**D**) Classification diagram of differential metabolites. CON = control group; YC = yeast culture group.

3.8. Correlation Analysis

Fibrobacter was significantly and positively correlated with thromboxane B2 ($p < 0.05$). Parabacteroides were significantly and positively correlated with 15-OxoETE and 2,3-dinor-8-iso prostaglandin F1 alpha ($p < 0.05$). Additionally, 15-keto-prostaglandin E2 was positively correlated with Bacteroides, Phocaeicola, and Parabacteroides, but negatively correlated with Eubacterium ($p < 0.05$) (Figure 6).

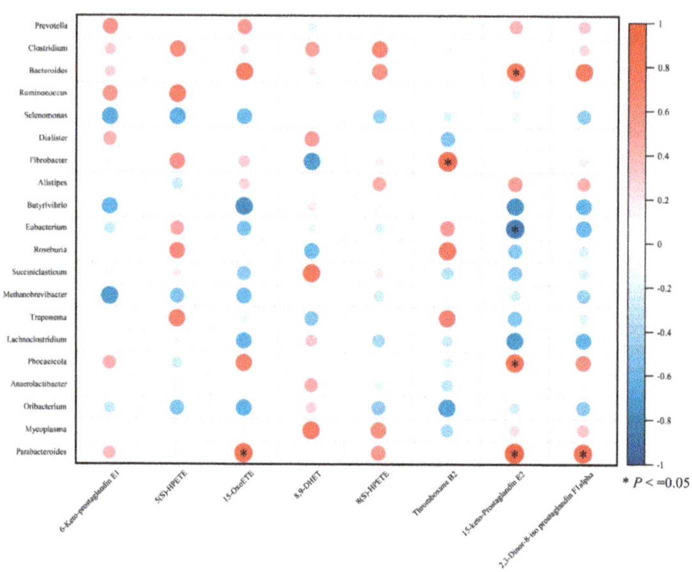

Figure 6. Microbiome–metabolome correlation heatmap.

4. Discussion

The YC used in this study reached $\geq 1 \times 10^8$ CFU/g and produced abundant cellular metabolites, including organic acids, oligosaccharides, and vitamins. During the experiment, the addition of 9% YC to the pelleted feed significantly enhanced the sheep's daily gain and final body weight. However, it is noteworthy that feed intake did not result in significant differences between the two experimental groups, suggesting that the improvement in growth performance in ruminants by adding YC to pelleted feeds is not dependent on changes in feed intake. This finding is consistent with the results reported by Qi and Chen et al. [12,25]. We hypothesize that this effect may be closely related to the role of YC in regulating rumen fermentation, rumen microbial communities, and their metabolites.

Ruminants degrade cellulose, hemicellulose, and carbohydrates in their feed through rumen microbes to produce volatile fatty acids (VFAs), which are essential for energy and growth [26]. Previous studies have shown that adding yeast cultures to lamb rations improves rumen fermentation performance and VFA absorption, while also enhancing the rumen fermentation environment and promoting rumen development [14,27]. In this study, we found that supplementing yeast cultures in pelleted feeds significantly increased the concentrations of propionic acid, butyric acid, and total short-chain fatty acids (SCFAs), while also decreasing the acetic acid-to-propionic acid ratio. The entry of yeast fermentation metabolites into the sheep rumen significantly increased the abundance of fibrinolytic bacteria and improved the breakdown and utilization efficiency of neutral and acidic detergent fibers. Yeast cultivation promotes the growth of lactic acid-utilizing bacteria by altering the structure of the rumen microbiota, thereby utilizing the lactic acid produced in the rumen to increase rumen pH and promote rumen fermentation while maintaining a stable rumen environment [15]. Most of the VFAs produced by fermentation are absorbed through the rumen epithelium and play a key role in regulating the energy supply for the host's life activities [28]. The addition of 9% yeast culture to pelleted feed significantly increased butyric and propionic acid content, which in turn stimulated rumen tissue development, significantly increasing the height and width of rumen papillae [29,30]. This suggests that yeast culture supplementation may enhance the efficient absorption, utilization, and transport of VFAs in sheep.

Macro-genomic sequencing was employed to gain deeper insights into the impact of adding yeast cultures to pelleted feed on the profile and regulation of rumen bacteria in sheep. The findings revealed that the predominant bacterial populations at the phylum level in the rumen of both experimental groups were primarily from Bacteroidetes and Firmicutes. The yeast culture contains a cellulase system, hemicellulase (xylanase), and neutral protease. Hemicellulase and cellulase have higher activity, so they can help improve the digestion of the cellulose components in the diet [14]. In this study, the relative abundance of Fibrobacter in the 9% YC group significantly increased, producing various cellulases and hemicellulases that helped to break down cellulose and hemicellulose in plant cell walls. These can break down plant cell walls into fermentable sugars (such as glucose), thereby increasing the content of acetic acid in the rumen fermentation parameter of the 9% YC group [31–33]. *Clostridium* generates substantial short-chain fatty acids (SCFAs), particularly butyric acid, through carbohydrate fermentation [34]. This observation explains the marked increase in butyric acid levels following yeast culture supplementation, which subsequently promotes mitotic activity in rumen wall tissues [35]. The enhanced relative abundance of *Prevotella* in the 9% YC group led to increased propionate production. The liver uses propionic acid as a raw material for gluconeogenesis, providing more energy to the body, resulting in higher daily weight gain in the 9% YC group compared to other groups [36,37]. Meanwhile, *Prevotella* utilizes nitrogen sources such as ammonia nitrogen and amino acids to convert non-protein nitrogen in the rumen into high-quality microbial protein that can be utilized by animals, thereby improving the utilization rate of feed nitrogen [1]. This may indicate the effective utilization of nitrogen in microbial protein synthesis and reduction of nitrogen loss and also explains the level of histidine nitrogen in the 9% YC group being lower than that in the control group.

The influence of yeast cultures on the abundance of rumen bacteria may be attributed to the presence of mannan oligosaccharides and glucans in the yeast cell wall. These components are readily degraded in the rumen and serve as sources of fermentable metabolites, minerals, and enzymes, significantly altering the characteristics and physiological functions of rumen bacteria. This alteration can lead to increased proliferation of bacteria specializing in the hydrolysis of cellulose, starch, and protein, thereby enhancing the overall efficiency of nutrient utilization during digestion [10,38]. Yeast cultures play a vital role in fostering the growth of lactic acid-utilizing bacteria, essential for maintaining a stable rumen pH. This stabilization creates an optimal environment for the development of bacteria involved in cellulose breakdown, as well as those that produce acetic and propionic acids. Previous research has shown that YC significantly increases the population of cellulolytic bacteria, lactate-utilizing bacteria, and enzymes involved in carbohydrate metabolism in the rumen. Additionally, another study indicated that the metabolites present in yeast cultures can promote the growth of cellulolytic bacteria [39,40]. The findings from these studies align with our research, reinforcing the beneficial effects of yeast cultures on rumen microbiota. In an intriguing development, our study demonstrated that the incorporation of yeast cultures led to a decrease in the relative abundance of Pseudomonas *Methanobrevibacter*, which consequently resulted in lower methane production in the rumen of sheep. This reduction not only mitigated energy loss for the host but also contributed to a decrease in greenhouse gas emissions. Fibrobacter, comprising core fiber-degrading bacteria in the rumen, efficiently breaks down difficult-to-digest fibers, thereby reducing fiber retention time and subsequently decreasing total methane production [41]. Additionally, Clostridium species utilize the butyrate kinase pathway, which consumes H_2 during butyrate formation, potentially reducing the availability of methane precursors and consequently decreasing methane emissions [42,43]. Previous studies have revealed a strong negative correlation between the number of *Prevotella* species and methane emissions. It has been suggested that members of the *Prevotella* genus can utilize hydrogen to produce propionic acid, thereby reducing methane output [44,45]. Our findings are consistent with these studies, further emphasizing the potential of yeast cultures to enhance rumen efficiency while reducing environmental impacts.

A metabolomic approach based on untargeted LC-MS was used to analyze changes in rumen metabolites. Our study showed that the addition of YC affected rumen metabolites, which mainly consisted of amino acids and their metabolites, as well as lipids and their metabolites. YC modulates rumen metabolite profiles primarily by altering rumen microorganisms, ultimately impacting the host animal [14]. We further analyzed the functional enrichment of differential metabolites between the two groups of sheep. The YC group showed higher enrichment scores in the arachidonic acid and arginine and proline metabolic pathways. Further analysis of the key metabolites in these pathways revealed that the arachidonic acid metabolic pathway in the YC group exhibited higher enrichment scores for metabolites such as 6-keto-prostaglandin E1, 15-OxoETE, 5(S)-HPETE, 8,9-DHET, 8(S)-HPETE, thromboxane B2, 15-keto-prostaglandin E2, and 2,3-dinor-8-iso prostaglandin F1 alpha, which were significantly upregulated. Previous studies have shown that arachidonic acid enhances animal immunity and energy metabolism [33,46,47]. Previous studies have demonstrated that YC supplementation significantly impacts rumen metabolism. In dairy goats, YC primarily influenced pathways related to lipid metabolism, glycan biosynthesis, amino acid metabolism, and vitamin-cofactor metabolism [39]. Similarly, in sheep fed high-concentrate diets, YC supplementation altered the profile of key metabolites, including amino acids, organic acids, carbohydrates, and glycerophosphates [14]. These metabolic changes reflect the fundamental relationship between feed composition and rumen microbial community structure, where microbial cross-feeding interactions determine the overall metabolic capacity of the rumen ecosystem [29]. Our findings align with these previous observations, particularly in the enhanced amino acid and lipid metabolism pathways observed in the 9% YC group. In this experiment, some metabolites belonging to the arachidonic acid signaling pathway were significantly elevated in sheep fed YC-supplemented pelleted diets, suggesting that YC can promote fatty acid and amino acid metabolism through rumen microorganisms and metabolites. This in turn enhances the ability of sheep to utilize arachidonic acid, an essential fatty acid, thereby improving energy utilization and production performance.

5. Conclusions

In summary, YC can significantly promote the development of rumen tissues in sheep, improve rumen fermentation and microbial flora structure, decrease the abundance of flora associated with starch degradation, and increase the abundance of flora associated with fiber catabolism. These changes in the microbiota further affected rumen metabolism, leading to differences in metabolites related to amino acid and lipid metabolism. The results suggest that adding 9% YC to pelleted diets at high concentrations helps promote the development of rumen epithelia and regulates the fermentation pattern of the rumen, thereby maintaining rumen health in sheep.

Supplementary Materials: The following supporting information can be downloaded at: https://www.mdpi.com/article/10.3390/ani15050738/s1, Figure S1: Top 10 carbohydrate differences; Table S1: Differences in rumen tissue structure; Table S2: Differences in rumen bacterial phyla (Abundance >1%) and genera (top 20); Table S3: Differences in carbohydrate-active enzymes.

Author Contributions: Conceptualization, S.Z. and T.J.; methodology, T.J.; software, J.X. and Q.F.; validation, X.L. and Q.F.; formal analysis, X.L.; investigation, J.X.; resources, T.J.; data curation, T.J.; writing—original draft preparation, J.X.; writing—review and editing, T.J.; visualization, T.J.; supervision, T.J.; project administration, T.J.; funding acquisition, T.J. and S.Z. All authors have read and agreed to the published version of the manuscript.

Funding: This research was funded by the Major Cultivation Project of Higher Education Research and Innovation Platform of Gansu Provincial Department of Education (2024CXPT-07); Southern Xinjiang Key Industry Innovation and Development Support Plan Project (2022DB017); Agricultural Science and Technology Project of Gansu Province (GNKJ-2021-41); Technology Innovation Guidance Plan of Gansu Province Science and Technology Plan Project —Lugan Science and Technology Cooperation Topic (22CX2NA005); "Flying Wings" College Student Science and Technology Volunteer Service Project (20230608ZZ10070098) and Guanghe County Cattle and Sheep Industry Quality and Efficiency Technology Demonstration Project (LXZC21520230202).

Institutional Review Board Statement: All animal experiments in this study were performed according to animal welfare requirements and were approved by the Animal Committee of Gansu Agricultural University (Approval No. GSAU-Eth-AST-2023-035) on 10 January 2024.

Informed Consent Statement: Not applicable.

Data Availability Statement: The data presented in this study are available on request from the corresponding author.

Acknowledgments: The authors would like to thank Management personnel of the breeding farm and laboratory management personnel of the College of Grassland Science, Gansu Agricultural University for their help in the successful completion of this study.

Conflicts of Interest: The authors declare no conflict of interest.

References

1. Zhang, R.; Zhang, L.; An, X.; Li, J.; Niu, C.; Zhang, J.; Geng, Z.; Xu, T.; Yang, B.; Xu, Z.; et al. Hybridization promotes growt performance by altering rumen microbiota and metabolites in sheep. *Front. Vet. Sci.* **2024**, *11*, 1455029. [CrossRef] [PubMed]
2. Duan, P.P.; Rehemujiang, H.; Zhang, L.D.; Lu, M.L.; Li, C.C.; Hu, L.H.; Wang, Y.L.; Diao, Q.Y.; Xu, G.S. *Lycium barbarum* (Wolfberry Branches and Leaves Enhance the Growth Performance and Improve the Rumen Microbiota in Hu Sheep. *Animals* **2024**, *14*, 161 [CrossRef] [PubMed]
3. Chen, K.; Shui, Y.; Deng, M.; Guo, Y.; Sun, B.; Liu, G.; Liu, D.; Li, Y. Effects of different dietary energy levels on growt performance, meat quality and nutritional composition, rumen fermentation parameters, and rumen microbiota of fattenin Angus steers. *Front. Microbiol.* **2024**, *15*, 1378073. [CrossRef] [PubMed]
4. Du, J.; Wang, Y.; Su, S.; Wang, W.; Guo, T.; Hu, Y.; Yin, N.; An, X.; Qi, J.; Xu, X. Supplementation with Complex Phytonutrient Enhances Rumen Barrier Function and Growth Performance of Lambs by Regulating Rumen Microbiome and Metabolom *Animals* **2025**, *15*, 228. [CrossRef]
5. Liu, T.; Pang, R.; Huang, L.; Mao, T.; Yu, J.; Hua, J.; Zhong, Y.; Ren, C.; Zhang, Z.; Zhu, W. Effects of allicin addition o growth performance, rumen microbiome, and ruminal epithelial proteome of high-grain-fed goats. *Anim. Feed Sci. Techno.* **2024**, *310*, 115944. [CrossRef]
6. Gao, K.; Geng, C. Alterations in the rumen bacterial communities and metabolites of finishing bulls fed high-concentrate diet supplemented with active dry yeast and yeast culture. *Front. Microbiol.* **2022**, *13*, 908244. [CrossRef]
7. Gao, K.; Geng, C. Comparison of rectum fecal bacterial community of finishing bulls fed high-concentrate diets with active dr yeast and yeast culture supplementation. *Anim. Biosci.* **2023**, *36*, 63–74. [CrossRef]
8. Halfen, J.; Carpinelli, N.; Del Pino, F.; Chapman, J.; Sharman, E.; Anderson, J.; Osorio, J. Effects of yeast culture supplementatio on lactation performance and rumen fermentation profile and microbial abundance in mid-lactation Holstein dairy cows. *J. Dair Sci.* **2021**, *104*, 11580–11592. [CrossRef]
9. Suntara, C.; Cherdthong, A.; Wanapat, M.; Uriyapongson, S.; Leelavatcharamas, V.; Sawaengkaew, J.; Chanjula, P.; Foiklang S. Isolation and Characterization of Yeasts from Rumen Fluids for Potential Use as Additives in Ruminant Feeding. *Vet. Sc* **2021**, *8*, 52. [CrossRef]
10. Li, Z.; Hu, Y.; Li, H.; Lin, Y.; Cheng, M.; Zhu, F.; Guo, Y. Effects of yeast culture supplementation on milk yield, rumen fermentatio metabolism, and bacterial composition in dairy goats. *Front. Vet. Sci.* **2024**, *11*, 1447238. [CrossRef]
11. Wang, J.; Zhao, G.; Zhuang, Y.; Chai, J.; Zhang, N. Yeast (*Saccharomyces cerevisiae*) Culture Promotes the Performance of Fattenin Sheep by Enhancing Nutrients Digestibility and Rumen Development. *Fermentation* **2022**, *8*, 719. [CrossRef]
12. Qi, P.; Wang, L. Effect of Adding Yeast Cultures to High-Grain Conditions on Production Performance, Rumen Fermentatio Profile, Microbial Abundance, and Immunity in Goats. *Animals* **2024**, *14*, 1799. [CrossRef] [PubMed]
13. Li, X.; Wang, Y.; Xu, J.; Yang, Q.; Sha, Y.; Jiao, T.; Zhao, S. Effects of yeast cultures on meat quality, flavor composition and rume microbiota in lambs. *Curr. Res. Food Sci.* **2024**, *9*, 100845. [CrossRef] [PubMed]
14. Wang, H.; Su, M.; Wang, C.; Li, D.; Li, Q.; Liu, Z.; Qi, X.; Wu, Y.; Zhao, Y.; Li, T.; et al. Yeast culture repairs rumen epithelial injur by regulating microbial communities and metabolites in sheep. *Front. Microbiol.* **2023**, *14*, 1305772. [CrossRef]
15. Liu, Y.-Z.; Chen, X.; Zhao, W.; Lang, M.; Zhang, X.-F.; Wang, T.; Farouk, M.H.; Zhen, Y.-G.; Qin, G.-X. Effects of yeast cultur supplementation and the ratio of non-structural carbohydrate to fat on rumen fermentation parameters and bacterial-communit composition in sheep. *Anim. Feed Sci. Technol.* **2019**, *249*, 62–75. [CrossRef]
16. Majewska, M.P.; Miltko, R.; Bełżecki, G.; Kowalik, B. Population of protozoa and carbohydrate-digesting enzymes in the rume of sheep fed a diet supplemented with yeast Saccharomyces cerevisiae. *Small Rumin. Res.* **2021**, *205*, 106544. [CrossRef]
17. Boukrouh, S.; Noutfia, A.; Moula, N.; Avril, C.; Hornick, J.-L.; Chentouf, M.; Cabaraux, J.-F. Effects of Sulla Flexuosa Hay a Alternative Feed Resource on Goat's Milk Production and Quality. *Animals* **2023**, *13*, 709. [CrossRef]
18. Boukrouh, S.; Noutfia, A.; Moula, N.; Avril, C.; Louvieaux, J.; Hornick, J.-L.; Cabaraux, J.-F.; Chentouf, M. Growth performance carcass characteristics, fatty acid profile, and meat quality of male goat kids supplemented by alternative feed resources: Bitte vetch and sorghum grains. *Arch. Anim. Breed.* **2024**, *67*, 481–492. [CrossRef]
19. Li, X.; Wang, Z.; Qin, W.; Gao, X.; Wu, J.; Zhao, S.; Jiao, T. Effects of oregano essential oil, cobalt and synergistic of both of them o rumen degradation rate and fermentation characteristics for corn silage. *Ital. J. Anim. Sci.* **2022**, *21*, 1476–1488. [CrossRef]
20. Edgar, R.C. UPARSE: Highly accurate OTU sequences from microbial amplicon reads. *Nat. Methods* **2013**, *10*, 996–998. [CrossRef
21. Wang, Y.; Sheng, H.-F.; He, Y.; Wu, J.-Y.; Jiang, Y.-X.; Tam, N.F.-Y.; Zhou, H.-W. Comparison of the Levels of Bacterial Dive sity in Freshwater, Intertidal Wetland, and Marine Sediments by Using Millions of Illumina Tags. *Appl. Environ. Microbio* **2012**, *78*, 8264–8271. [CrossRef] [PubMed]

22. Dubois, P.C.; Trynka, G.; Franke, L.; Hunt, K.A.; Romanos, J.; Curtotti, A.; Zhernakova, A.; Heap, G.A.; Adány, R.; Aromaa, A.; et al. Multiple common variants for celiac disease influencing immune gene expression. *Nat. Genet.* **2010**, *42*, 295–302. [CrossRef] [PubMed]
23. Segata, N.; Izard, J.; Waldron, L.; Gevers, D.; Miropolsky, L.; Garrett, W.S.; Huttenhower, C. Metagenomic biomarker discovery and explanation. *Genome Biol.* **2011**, *12*, R60. [CrossRef] [PubMed]
24. White, J.R.; Nagarajan, N.; Pop, M. Statistical methods for detecting differentially abundant features in clinical metagenomic samples. *PLoS Comput. Biol.* **2009**, *5*, e1000352. [CrossRef]
25. Chen, H.; Liu, S.; Li, S.; Li, D.; Li, X.; Xu, Z.; Liu, D. Effects of yeast culture on growth performance, immune function, antioxidant capacity and hormonal profile in Mongolian ram lambs. *Front. Vet. Sci.* **2024**, *11*, 1424073. [CrossRef]
26. Aschenbach, J.R.; Penner, G.B.; Stumpff, F.; Gäbel, G. Ruminant Nutrition Symposium: Role of fermentation acid absorption in the regulation of ruminal pH. *J. Anim. Sci.* **2011**, *89*, 1092–1107. [CrossRef]
27. Xu, Z.; Yang, L.; Chen, H.; Liu, S.; Li, X.; Li, S.; Ying, C.; Li, X.; Du, R.; Liu, D. *Saccharomyces cerevisiae* and *Kluyveromyces marxianus* yeast co-cultures modulate the ruminal microbiome and metabolite availability to enhance rumen barrier function and growth performance in weaned lambs. *Anim. Nutr.* **2024**, *19*, 139–152. [CrossRef]
28. Tun, H.M.; Li, S.; Yoon, I.; Meale, S.J.; Azevedo, P.A.; Khafipour, E.; Plaizier, J.C. *Saccharomyces cerevisiae* fermentation products (SCFP) stabilize the ruminal microbiota of lactating dairy cows during periods of a depressed rumen pH. *BMC Vet. Res.* **2020**, *16*, 237. [CrossRef]
29. Xue, M.-Y.; Sun, H.-Z.; Wu, X.-H.; Liu, J.-X.; Guan, L.L. Multi-omics reveals that the rumen microbiome and its metabolome together with the host metabolome contribute to individualized dairy cow performance. *Microbiome* **2020**, *8*, 64. [CrossRef]
30. Xue, M.-Y.; Xie, Y.-Y.; Zhong, Y.; Ma, X.-J.; Sun, H.-Z.; Liu, J.-X. Integrated meta-omics reveals new ruminal microbial features associated with feed efficiency in dairy cattle. *Microbiome* **2022**, *10*, 32. [CrossRef]
31. Shao, P.; Sha, Y.; Liu, X.; He, Y.; Wang, F.; Hu, J.; Wang, J.; Li, S.; Chen, X.; Yang, W.; et al. Supplementation with *Astragalus* Root Powder Promotes Rumen Microbiota Density and Metabolome Interactions in Lambs. *Animals* **2024**, *14*, 788. [CrossRef] [PubMed]
32. Lu, Z.; Kong, L.; Ren, S.; Aschenbach, J.R.; Shen, H. Acid tolerance of lactate-utilizing bacteria of the order Bacteroidales contributes to prevention of ruminal acidosis in goats adapted to a high-concentrate diet. *Anim. Nutr.* **2023**, *14*, 130–140. [CrossRef] [PubMed]
33. Ren, Y.; Zhaxi, Y.; Ciwang, R.; Wang, Z.; Liu, M. Responses of rumen microorganisms and metabolites to different roughage of domesticated Tibetan sheep. *Front. Microbiol.* **2023**, *14*, 1247609. [CrossRef] [PubMed]
34. Li, Q.; Wu, Y.; Qi, X.; Liu, Z.; Wang, C.; Ma, X.; Ma, Y. Prickly Ash Seeds Improve the Ruminal Epithelial Development and Growth Performance of Hu Sheep by Modulating the Rumen Microbiota and Metabolome. *Microorganisms* **2024**, *12*, 2242. [CrossRef]
35. Wang, K.; Jiang, M.; Chen, Y.; Huang, Y.; Cheng, Z.; Datsomor, O.; Jama, S.M.; Zhu, L.; Li, Y.; Zhao, G.; et al. Changes in the rumen development, rumen fermentation, and rumen microbiota community in weaned calves during steviol glycosides treatment. *Front. Microbiol.* **2024**, *15*, 1395665. [CrossRef]
36. De Vadder, F.; Kovatcheva-Datchary, P.; Zitoun, C.; Duchampt, A.; Bäckhed, F.; Mithieux, G. Microbiota-Produced Succinate Improves Glucose Homeostasis via Intestinal Gluconeogenesis. *Cell Metab.* **2016**, *24*, 151–157. [CrossRef]
37. Betancur-Murillo, C.L.; Aguilar-Marín, S.B.; Jovel, J. Prevotella: A Key Player in Ruminal Metabolism. *Microorganisms* **2022**, *11*, 1. [CrossRef]
38. Carpinelli, N.; Halfen, J.; Trevisi, E.; Chapman, J.; Sharman, E.; Anderson, J.; Osorio, J. Effects of peripartal yeast culture supplementation on lactation performance, blood biomarkers, rumen fermentation, and rumen bacteria species in dairy cows. *J. Dairy Sci.* **2021**, *104*, 10727–10743. [CrossRef]
39. Sirisan, V.; Pattarajinda, V.; Vichitphan, K.; Leesing, R. Isolation, identification and growth determination of lactic acid-utilizing yeasts from the ruminal fluid of dairy cattle. *Lett. Appl. Microbiol.* **2013**, *57*, 102–107. [CrossRef]
40. Ogunade, I.; Schweickart, H.; McCoun, M.; Cannon, K.; McManus, C. Integrating 16S rRNA Sequencing and LC–MS-Based Metabolomics to Evaluate the Effects of Live Yeast on Rumen Function in Beef Cattle. *Animals* **2019**, *9*, 28. [CrossRef]
41. Ma, H.; Dong, A.; Xu, Y.; Wu, Q.; Lambo, M.T.; Zhang, Y.; Dou, X.; Li, Y. Regulatory effects of high concentrate diet synergistically fermented with cellulase and lactic acid bacteria: Ruminal fermentation, methane production, and rumen microbiome. *Anim. Feed Sci. Technol.* **2025**, *319*, 116194. [CrossRef]
42. Bature, I.; Xiaohu, W.; Ding, X. The roles of phytogenic feed additives, trees, shrubs, and forages on mitigating ruminant methane emission. *Front. Vet. Sci.* **2024**, *11*, 1475322. [CrossRef] [PubMed]
43. Ungerfeld, E.M. Metabolic Hydrogen Flows in Rumen Fermentation: Principles and Possibilities of Interventions. *Front. Microbiol.* **2020**, *11*, 589. [CrossRef] [PubMed]
44. Aguilar-Marin, S.B.; Betancur-Murillo, C.L.; Isaza, G.A.; Mesa, H.; Jovel, J. Lower methane emissions were associated with higher abundance of ruminal Prevotella in a cohort of Colombian buffalos. *BMC Microbiol.* **2020**, *20*, 364. [CrossRef] [PubMed]

45. Wallace, R.J.; Sasson, G.; Garnsworthy, P.C.; Tapio, I.; Gregson, E.; Bani, P.; Huhtanen, P.; Bayat, A.R.; Strozzi, F.; Biscarini, F.; et al. A heritable subset of the core rumen microbiome dictates dairy cow productivity and emissions. *Sci. Adv.* **2019**, *5*, eaav8391. [CrossRef]
46. Bai, J.; Tang, L.; Bi, Y.; Li, M. Multi-omics insights into the energy compensation of rumen microbiota of grazing yaks in cold season. *Front. Microbiol.* **2024**, *15*, 1467841. [CrossRef]
47. Liu, X.; Sha, Y.; Lv, W.; Cao, G.; Guo, X.; Pu, X.; Wang, J.; Li, S.; Hu, J.; Luo, Y. Multi-Omics Reveals That the Rumen Transcriptome, Microbiome, and Its Metabolome Co-regulate Cold Season Adaptability of Tibetan Sheep. *Front. Microbiol.* **2022**, *13*, 859601. [CrossRef]

animals

MDPI

Article

Hotspot Analysis of Rumen Microbiota and Methane Mitigation in Ruminants: A Bibliometric Analysis from 1998 to 2023

Xueyue Zheng [1], Lijie Tang [1], Rong Wang [2,*], Xiumin Zhang [2], Min Wang [2] and Duanqin Wu [3,*]

1 College of Animal Science and Technology, Hunan Agricultural University, Changsha 410128, China; zhengxueyue1111@163.com (X.Z.); 19191018407@163.com (L.T.)
2 CAS Key Laboratory for Agro-Ecological Processes in Subtropical Region, National Engineering Laboratory for Pollution Control and Waste Utilization in Livestock and Poultry Production, Institute of Subtropical Agriculture, the Chinese Academy of Sciences, Changsha 410125, China; xmzhang@isa.ac.cn (X.Z.); mwang@isa.ac.cn (M.W.)
3 Institute of Bast Fiber Crops, Chinese Academy of Agricultural Sciences, Changsha 410205, China
* Correspondence: rongwang@isa.ac.cn (R.W.); wuduanqin@caas.cn (D.W.)

Simple Summary: Enteric methane emissions from livestock are a significant source of anthropogenic greenhouse gases and a contributor to climate change. A significant portion of this methane is derived from the digestive activities of ruminants. The present study determined research trends on rumen microbiota and methane mitigation from livestock over the last two decades. This study evaluated 189 articles published between 1998 and 2023, identifying hotspots and research gaps in the field of enteric methane mitigation. The number of studies increased exponentially from 2008, with most studies being conducted in China and North America, where livestock farming is widespread. The data revealed that although some progress has been made, further research is still needed, particularly on sustainable and effective CH_4 mitigation strategies. This information can be used to reduce the environmental impacts of livestock production, contributing to sustainable food production systems and climate change mitigation.

Abstract: Enteric methane (CH_4) is mainly produced by methanogens during the anerobic fermentation of ingested carbohydrates (fiber and starch) in the rumen. The aim of this study was to investigate the research and dynamic trends regarding rumen microbiota and CH_4 emissions. Using the Web of Science (WoS) database, we conducted a comprehensive review of 189 relevant papers published from 1998 to 2023 and analyzed them using CiteSpace and VOSviewer. The results showed that most of the papers were written by biologists, and dairy products and animal science accounted for a relatively small number of publications. Chinese scholars contributed the most in this field (22.2%, 42 papers), ranking first in the world. The analysis of research interest found that the use of exogenous additives to regulate CH_4 emissions from ruminant livestock has received increasing attention in recent years. In terms of research trends, traditional animal nutrition studies accounted for most of the research before 2015, but with the development of microbiome technology, the research focus gradually shifted to understanding the CH_4 generation pathway mechanism. From the perspective of keywords, the focus of enteric CH_4 research has gradually shifted from phenotype indicators such as growth performance and digestibility to the level of microbial mechanisms such as protozoa and methanogens. Despite the advancements that have been made, there remains a pressing need for further investigation into sustainable and effective CH_4 mitigation strategies.

Keywords: greenhouse gas; methane emission; rumen microbiota; bibliometrics

1. Introduction

Methane (CH_4) is the second-most abundant greenhouse gas, following carbon dioxide (CO_2), and has a warming potential 28 times greater than CO_2. It contributes up to 25% to global warming [1,2]. The Intergovernmental Panel on Climate Change (IPCC) has consistently emphasized the significance of CH_4 mitigation in decelerating global warming, and issued the Kyoto Protocol in 1997, which formally took effect from February 2005. This marked a historic milestone, as it was the first instance where emissions were regulated to limit their impact on climate change [3]. Consequently, in November 2015, the IPCC introduced the Paris Agreement, which stands as a significant milestone in international legal frameworks after the Kyoto Protocol. The Paris Agreement set out detailed global actions to combat climate change and stipulated that the world should achieve net-zero greenhouse gas emissions in the second half of this century. The goal of the Paris Agreement is to reduce the ecological risks of climate change and prevent climate change from posing a potentially existential threat to humankind [4,5].

Atmospheric CH_4 emissions come from natural ecosystems and human activities [6,7], of which livestock emissions contribute about 33% to the total emissions and are the largest single agricultural source [8]. Cattle and sheep are the main contributors (80%) to all livestock CH_4 emissions [9]. About 90% of CH_4 is produced in the rumen, and thus the reduction of ruminal CH_4 production is a key approach to CH_4 mitigation [10]. CH_4 emissions are also considered a major pathway of feed energy loss (about 2% to 12% of the total digestible energy) [8,11]. Therefore, studies on the reduction of ruminal CH_4 emissions can potentially bring benefits to both alleviating greenhouse effect and improving feed energy utilization efficiency [12].

Nutritional strategies for reducing CH_4 emissions in ruminants include dietary composition, feed quality improvement, rumen passage rate increase, oil, and CH_4 inhibitors [11,13,14]. It has been shown that a reduction in methanogen copy number and the suppression of ruminal hydrogen in CH_4 synthesis can effectively reduce CH_4 emissions [15] and that changes in rumen microbial community structure and fermentation patterns can reduce hydrogen production, thus reducing CH_4 emissions [16]. These studies show that changes in rumen microbiota are associated with a reduction in CH_4 emissions. Although many authors have reported on the methods, advantages, and disadvantages of CH_4 mitigation in ruminants [17–19], there is little literature on the progress and future research directions of CH_4 mitigation related to rumen microbiota. Therefore, this paper analyzes the current research status on rumen microbiota in relation to CH_4 emissions by comparing journals and evaluates the international cooperation and research focus in this field. The purpose of this study was to understand the current research progress and the role of rumen microbiota in CH_4 mitigation and to provide scientific guidance for future research in this field.

2. Materials and Methods

2.1. Data Sources and Research Strategies

Web of Science (WoS) is a multidisciplinary literature database developed by Thomson Reuters Corporation within the open Internet environment. It contains the most significant scientific search engines and pertinent bibliometric data retrieval databases. It is widely acknowledged as the foremost quality-oriented database globally, offering a standardized record for retrieving global scientific literature across diverse research fields [16]. WoS provides bibliometric and network analysis, with important metadata such as abstract, type of document, timeline, citation, list of scholars, country information, university affiliation, and journal impact factor. WoS is an open access database that does not involve human subjects and does not require institutional review board approval [20,21]. This study utilizes the WoS database as its primary source of scientific data. The databases employed encompass various indices, such as the Science Citation Index Expanded (Expanded), Social Sciences Citation Index (SSCI), Emerging Sources Citation Index (ESCI), Conference Proceedings Citation Index-Science (CPCI-S), Book Citation Index—Social Sciences and Humanities (BKCI-SSH), and Arts and Humanities Citation Index.

Advanced search was used to find publications related to rumen microbiota and CH_4 emissions in ruminants. Selection of keywords is very important in analysis, as the choice of keywords directly affects the results of a study. Therefore, we used different combinations of keywords in several cycles [22]. Finally, we used the following retrieval strategy to find appropriate search terms using Boolean logic (i.e., OR and AND): TS = ("microorganism" OR "germ" OR "microbe" OR "Microbiology" OR "Microbial") AND ("cow" OR "Cattle" OR "Bovine" OR "steer" OR "sheep" OR "goats" OR "cud chewer" OR "ruminant" or "ruminates" or "OX" or "bull" or "heifer" or "buffalo") AND ("methane" or "CH_4"). The retrieval period was from 1 January 1998 to 16 December 2023. The literature search resulted in a total of 1638 references after screening for articles written in English within the specified time frame. Among these, we found 191 studies related to rumen microbes and CH_4 emissions based on a review of the title and abstract. Then, according to the data source of the article, the bibliometric scoring method was used to conduct quantitative and qualitative analysis of the development trends and patterns of research in this field (Figure 1).

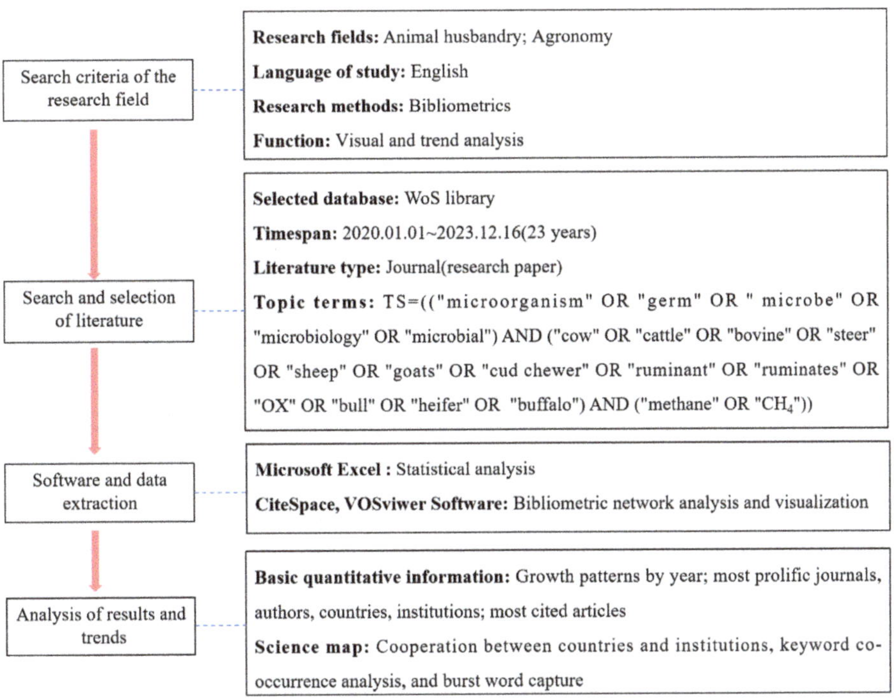

Figure 1. Flowchart of research procedure.

2.2. *Statistical Analysis*

This study categorized the relevant research into three distinct phases, each corresponding to significant international policies concerning CH_4 emissions from ruminants: January 1998–January 2005 (issuance of the Kyoto Protocol), February–October 2015 [23], and November 2015–December 2023 (proposed Paris Agreement). These phases represent the primary clusters within the trend of research on rumen microbiota and CH_4 mitigation. An analysis of shifts in research trends within this field has been conducted, encompassing a comprehensive examination of impact factors and citation rates associated with relevant journals. The assessment of research in this study included: (1) publication output and subject category according to the WoS database (from 1998 to 2023); (2) the most influential among the top ten journals; (3) the most cited articles; (4) the research productivity of the top ten countries; and (5) screening of field keywords. Additionally, impact factors, derived from an average number of citations from Journal Citation Reports (JCR) published in 2022, were utilized to evaluate journal performance via SCIMago Journal Rank (SJR).

Publication trends from 1998 to 2023 were assessed using Excel 2019, with a focus on publication volumes, and bibliometrics was subsequently employed to examine the existing number of publications, citation frequency, and author distribution associated with the rumen microbiota and methane mitigation, in order to discern current research trends. Furthermore, the assessment of citation frequency and impact factor contributed to an objective foundation for academic evaluation. Author credibility was determined through the evaluation of total citations (TCs) and average citations per paper (TCs/P), while TCs were also used as a criterion for gauging the reputation of the source institution. Social network analysis was subsequently employed to discern dynamic patterns in research focus and collaborations among leading authors, countries, and institutions. Both VOSviewer 1.6.20 and CiteSpace 6.2.R6 software enabled quantitative analysis and visualization of literature units and keywords [24,25], with "Country" depicting institutional cooperation networks nationally and "Keyword" representing high-frequency keyword co-occurrence map analysis [26]. Additionally, co-citation analysis was leveraged to explore the interconnections between cited documents and the evolution of study content and dynamics [27].

3. Results and Discussion

3.1. *Number of Published Articles*

3.1.1. Publication Output

The present study analyzed a total of 191 papers, which were contributed by 889 authors affiliated with 313 organizations from 46 countries. These publications appeared in 63 journals and cited a total of 6012 papers from 1767 distinct journals. Figure 2 depicts the publication trends concerning rumen microbiota and CH_4 mitigation. During the initial phase (1998–2005), the annual number of publications on these selected topics did not exceed two, resulting in a cumulative total of four publications. This suggests that research in this domain was still at an early stage, possibly due to the recent inception of the Kyoto Protocol. Although nations began addressing concerns related to CH_4 emissions during this period, no relevant policies were introduced. In the subsequent phase (2006–2015), there was a significant surge in publications focusing on these subjects, reaching a count of 134 papers, which constituted approximately 70.9% of the overall output. The citation counts also experienced substantial growth, amounting to approximately 5500 citations. This increase can be attributed to heightened scholarly interest following the implementation of the Kyoto Protocol. In the final stage (2016–2023), although there was no further increase in publication numbers indicating sustained focus on research topics, the momentum seemed to have stabilized compared to previous years. This stabilization could be attributed to an evolution over a decade within dedicated teams researching rumen microbiota and CH_4 mitigation. However, the annual citation counts continued to decline, possibly because more articles were available for citation and researchers had more options when choosing which articles to cite.

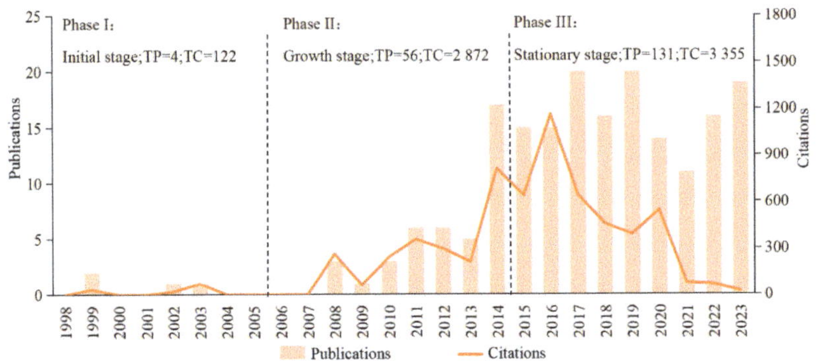

Figure 2. Number of publications and citations per year from 1998 to 2023 on Web of Science.

3.1.2. Number of Articles Published by Journals

The analysis of the literature sources revealed that apart from a few journals in veterinary and food science technology, many publications over the past 26 years are primarily associated with animal science and microbiology. This concentration is evident in terms of publication volume. Specifically, four journals have published more than 10 papers each—*Journal of Dairy Science* (impact factor: 3.5), *Frontiers in Microbiology* (impact factor: 5.2), *Journal of Animal Science* (impact factor: 3.3), and *Animal Feed Science and Technology* (impact factor: 3.2). These journals focus on key areas such as nutrition, physiology, feed science, microbiology and animal production, predominantly targeting beef cattle, dairy cows, sheep, and other livestock and poultry species (Table 1). According to Bradford's law in bibliometrics [28,29], a certain field's journal can be divided into three parts, where one-third of articles concentrate on a few specific journals—consistent with this study's findings. Through analyzing the citation patterns among these journals, it was found that the *Journal of Dairy Science* received the highest number of citations per article—an average of 53.42. This prominence is likely due to its reputation as a leading journal in the field of animal science, attracting considerable industry attention. Furthermore, it may be attributed to the fact that the dairy industry bears the brunt of CH_4 emissions originating from livestock. The second most frequently cited journal was *Frontiers in Microbiology*, which had an average citation rate of 32.26 per article. This may be attributed to the journal's high impact factor and its open access policy.

Table 1. The 10 most frequently published journals in the field of rumen microbiota and methane mitigation in ruminants from 1998 to 2023 with the relative number of articles.

No.	Journal	IF	Publications	Citations	Average Citations/Publication	JCR Categories
1	*Journal of Dairy Science*	3.5	19	1015	53.42	Dairy & Animal Science (Q1/9/62) Food Science & Technology (Q2/57/142)
2	*Frontiers in Microbiology*	5.2	19	613	32.26	Microbiology (Q2/38/135)
3	*Journal of Animal Science*	3.3	14	522	37.29	Dairy & Animal Science (Q1/10/62)
4	*Animal Feed Science and Technology*	3.2	13	454	34.92	Dairy & Animal Science (Q1/11/62)
5	*Asian-Australasian Journal of Animal Sciences*	2.7	8	159	19.88	Dairy & Animal Science (Q2/17/62)

Table 1. *Cont.*

No.	Journal	IF	Publications	Citations	Average Citations/Publication	JCR Categories
6	*Animals*	3.0	8	117	14.63	Dairy & Animal Science (Q1/12/62) Veterinary Science (Q1/13/144)
7	*Journal of Animal Physiology and Animal Nutrition*	2.7	7	139	19.86	Dairy & Animal Science (Q1/13/62) Veterinary Science (Q1/19/144)
8	*Animal*	3.6	7	95	13.57	Dairy & Animal Science (Q1/7/62) Veterinary Science (Q1/9/144)
9	*Livestock Science*	1.8	6	253	42.17	Dairy & Animal Science (Q2/29/62)
10	*Animal Production Science*	2.2	6	76	12.67	Dairy & Animal Science (Q3/37/62)

3.1.3. Most Cited Research Article

The SCI mainly uses citation count as a key indicator to measure the quality and influence of scientific papers [30,31]. Here, the 191 papers had a total of 6349 citations, with an average citation frequency of 33.24 times per paper. The 10 most cited papers on enteric CH_4 emission regulation by rumen microbiota during the last 26 years are summarized in Table 2. The paper "Specific microbiome-dependent mechanisms underlie the energy harvest efficiency of ruminants" by Ben Shabat SK et al. (2016) had the highest citation frequency (409 times) and the highest annual citation frequency (51.13 times/year) [32]. In this study, the taxonomic composition, gene content, microbial activity, and metabolite composition of rumen microbiota were characterized through diversity + metagenomics + metabolomics technology in 146 dairy cows with different feed conversion rates. The potential mechanism of different feed conversion rates from various angles was explored, and the results showed that the complex relationship between microbial genes, low species richness, and high feed conversion rate phenotype can accurately predict animal feed conversion rates. This study indicated that the enrichment of specific microorganisms and their metabolic pathways could enhance the utilization of energy and carbon sources in animals and reduce CH_4 emissions. This is of great significance for the rational use of feed resources by rumen microbiota to improve the ecological environment of livestock farming.

Table 2. Detailed information on the top 10 research publications about rumen microbiota and methane mitigation based on the total cited from 1998 to 2023.

No.	Title	Journal and Year	Average Citations per Year	Impact Factor and Category Quartile	Country	Total Cited	Ref.
1	Specific microbiome-dependent mechanisms underlie the energy harvest efficiency of ruminants	*ISME Journal*, 2016	51.13	11.0 (Q1)	Israel	409	[32]
2	The rumen microbial metagenome associated with high methane production in cattle	*BMC Genomics*, 2015	23.67	4.70 (Q1)	UK	213	[33]

Table 2. *Cont.*

No.	Title	Journal and Year	Average Citations per Year	Impact Factor and Category Quartile	Country	Total Cited	Ref.
3	Nitrate and sulfate: Effective alternative hydrogen sinks for mitigation of ruminal methane production in sheep	*Journal of Dairy Science*, 2010	14.57	3.50 (Q1)	The Netherlands	204	[34]
4	Bovine host genetic variation influences rumen microbial methane production with best selection criterion for low methane emitting and efficiently feed converting hosts based on metagenomic gene abundance	*PLOS Genetics*, 2016	24.88	4.50 (Q1)	UK	199	[35]
5	Two different bacterial community types are linked with the low-methane emission trait in sheep	*PLoS ONE*, 2014	18	3.70 (Q1)	New Zealand	180	[36]
6	Rumen metagenome and metatranscriptome analyses of low methane yield sheep reveals a Sharpea-enriched microbiome characterised by lactic acid formation and utilisation	*Microbiome*, 2016	21.75	15.5 (Q1)	New Zealand	174	[37]
7	Host genetics and the rumen microbiome jointly associate with methane emissions in dairy cows	*PLOS Genetics*, 2018	25.67	11.0 (Q1)	Denmark	154	[38]
8	Methane production in dairy cows correlates with rumen methanogenic and bacterial community structure	*Frontiers in Microbiology*, 2017	25.5	6.20 (Q2)	Sweden	153	[39]
9	Effects of addition of tea saponins and soybean oil on methane production, fermentation and microbial population in the rumen of growing lambs	*Livestock Science*, 2010	10.14	1.80 (Q2)	China	142	[18]
10	Methane emission by goats consuming diets with different levels of condensed tannins from lespedeza	*Animal Feed Science and Technology*, 2008	8.38	3.60 (Q1)	US.	134	[40]

Five of the ten most-cited articles were published in 2014–2018, suggesting that this period was particularly important for the development of the field. Interestingly, half of the most-cited papers were not published in journals with a high number of publications, and they came from biology journals more frequently than animal science journals. This is likely because impact factors are generally higher in biology journals, and many have interdisciplinary sections. Another possible explanation is that researchers prefer to publish their work in high-impact interdisciplinary journals to accelerate the field's progress.

3.2. Analysis of Authors' Countries

3.2.1. Number of Articles by Country

As shown in Figure 3, the literature involving rumen microbial and CH_4 mitigation originates from 46 countries around the world. The figure shows the number of published papers by country or region. China (42), the United States (23), Thailand (23), Canada (21), France (15), Australia (14), New Zealand (12), Spain (12), the United Kingdom (11), and Switzerland (9) are the 10 countries with the most publications in the field of rumen microbial and CH_4 mitigation over the past 23 years. The number of papers published by scholars in China accounts for 21.99% of all papers on this topic, indicating that China has carried out more research in this field than any other country. This is related to China's long-term development of carbon reduction regulations and policies in animal husbandry. For example, since 2001, China has issued carbon reduction policies for animal husbandry, including The Animal Husbandry Law of the People's Republic of China, The Standard for Pollutant Emissions from Livestock Farming Industry, and so on. In 2007, the State Council issued Opinions on Promoting Sustainable and Healthy Development of Animal Husbandry, which was the first time that the State Council put forward the requirements of carbon neutrality in animal husbandry in the form of normative documents during China's reform and opening up. Since 2015, the Animal Husbandry Law of the People's Republic of China and the Environmental Protection Law of the People's Republic of China have been formally implemented one after another, further forming a complete legal system of animal husbandry and providing legal guarantees for the sustainable development of animal husbandry. The number of policy texts concerning carbon reduction in animal husbandry has increased significantly, including training, subsidies, pilot projects, assessments, and other aspects.

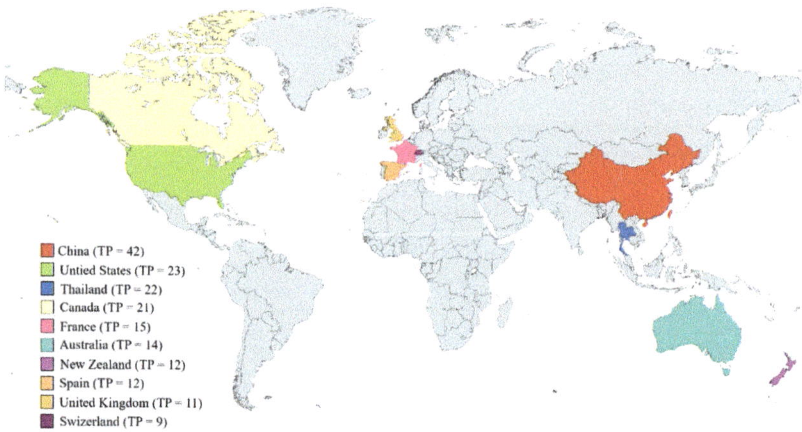

Figure 3. The number of publications for each geographic region in the world (based on the nationality of the first author).

3.2.2. Co-Authorship Among Countries

Figure 4 illustrates the extensive collaboration among national authors in the field of rumen microbial and CH_4 mitigation, with 46 nations contributing to published papers and a total of 148 links established. Notably, China emerges as a prominent player in this research area, exhibiting the highest density of links (with a total link strength of 28). This prominence can be attributed to President Xi's initiatives since the 18th National Congress of the Communist Party of China, aiming to foster international cooperation for building a beautiful Earth and a community with shared future for mankind through One Belt and One Road, as well as accelerating ecological civilization reform domestically to create a beautiful China. Through effective diplomatic efforts, China has actively engaged in global initiatives focused on reducing CH_4 emissions. Numerous domestic universities and research institutions have initiated international cooperation platforms, such as the International Research Center on Dairy Science at the Institute of Animal Sciences within the Chinese Academy of Agricultural Sciences. These endeavors have progressively positioned China as an indispensable partner within the international scientific research cooperation network.

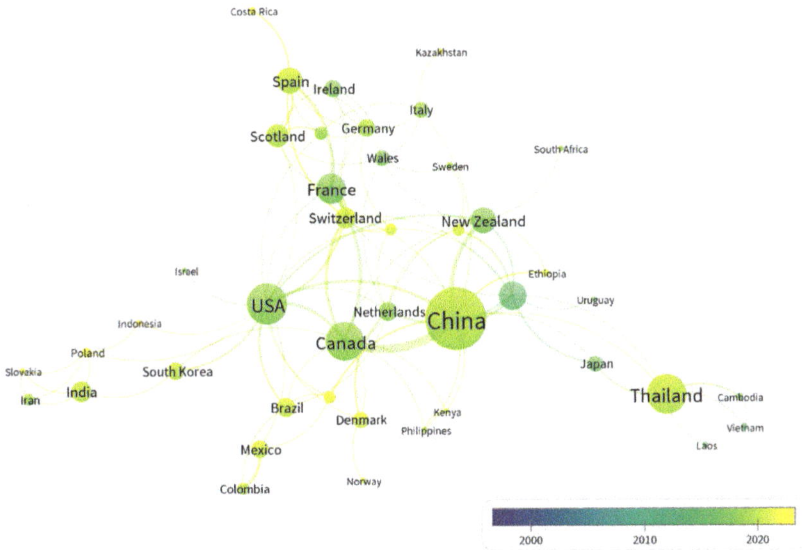

Figure 4. Research cooperation network among authors from different countries. Thicker lines indicate stronger partnerships.

International cooperation, particularly that facilitated by the United States (which has 17 links with a total link strength of 25), is marginally surpassed by China's efforts. This is followed by Canada (with 11 links and a total link strength of 24) and Australia (which also has 11 links, but with a total link strength of 16). In relation to temporal development, countries including Canada, the United States, France, Australia, and the Netherlands commenced research on rumen microbial and CH_4 mitigation comparatively early, around 2015. More recently, India and Costa Rica have entered this field within the last three years (2021–2023). The Global Inventory Report, released by the United Nations, confirms that all countries' gas emissions are significantly deviating from the targets set by the Paris Agreement. Consequently, it is incumbent upon all nations to perpetuate their efforts towards system transformation to integrate carbon reduction into every facet of human life.

3.3. Research Hotspots and Emerging Trends Based on Keyword Co-Occurrence Analysis

3.3.1. Research Hotspot Trends of Change

Keywords can clearly show the research status and development trends of scholarly articles. The Network Visualization view of VOSviewer 1.6.20 was used for keyword co-occurrence analysis, and keywords with co-occurrences greater than five were filtered to construct a network diagram of research hotspot changes (Figure 5). Fewer keywords were found in the early literature, which was basically after 2010, and was consistent with the distribution of publication volumes. Before 2015, the research focused on feed additives such as coconut oil [41,42], condensed tannins [43,44] and monensin [45]. Since 2020, microbiota (microbial synthesis, 16s metabarcoding) have gradually increased, indicating that the research focus has changed from macronutrients to molecular nutrition [46,47]. In addition, an increasing number of studies have begun to explore the regulatory effects of rumen microbiota on CH_4 emissions.

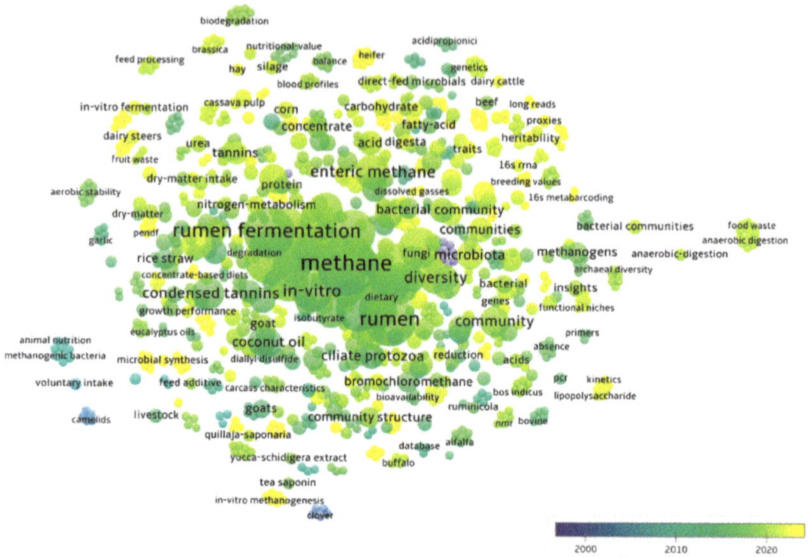

Figure 5. Research hotspots trends of change. Each small circle signifies a keyword. The size of the circle is directly proportional to the frequency of its occurrence. Furthermore, the thickness of the line connecting these circles indicates an increased co-occurrence frequency between keywords, signifying a closer relationship.

The co-occurrence analysis showed that besides the keywords CH_4, methanogenesis, rumen, and ruminant (dairy cows, sheep, and cattle), the frequency of rumen fermentation (or ruminal fermentation) and diversity also increased (Figure 5). This indicates that more and more researchers studying the effect of rumen microbiota on CH_4 emission in ruminants have focused on the indicators related to rumen fermentation. The main reason is that the VFAs produced by rumen fermentation are often accompanied by hydrogen production, which is the main precursor of ruminal methanogenesis and closely related to CH_4 emissions [48,49]. More and more in vivo experiments were conducted to study the effects of changes in rumen microbial diversity on CH_4 emissions.

3.3.2. Keyword Cluster Analysis

Keyword clustering analysis is a process that distills the co-occurrence network relationship into a more manageable number of clusters, relying on cluster statistics. In this study, we employed CiteSpace 6.2.R6 software to systematically conduct keyword clustering, enabling us to discern the relationships and primary research areas across three distinct stages (1998–2005, 2006–2015, and 2016–2023). It is noteworthy that the recurring presence of certain keywords underscores the significance of vocabulary in the current research phase.

In the keyword network map of phase I (1998–2005), two clusters were found (Figure 6A). In this phase, microbial protein synthesis was the focus, as reviewed by [50,51]. The studies used different feed additions (e.g., lotus corniculatus) or feed treatments and studied their effect on CH_4 emissions [52]. Most studies in this first phase were performed with traditional animal nutrition. This was due to a lack of technology at that time and knowledge of background mechanisms, which led to research on a more macro scale.

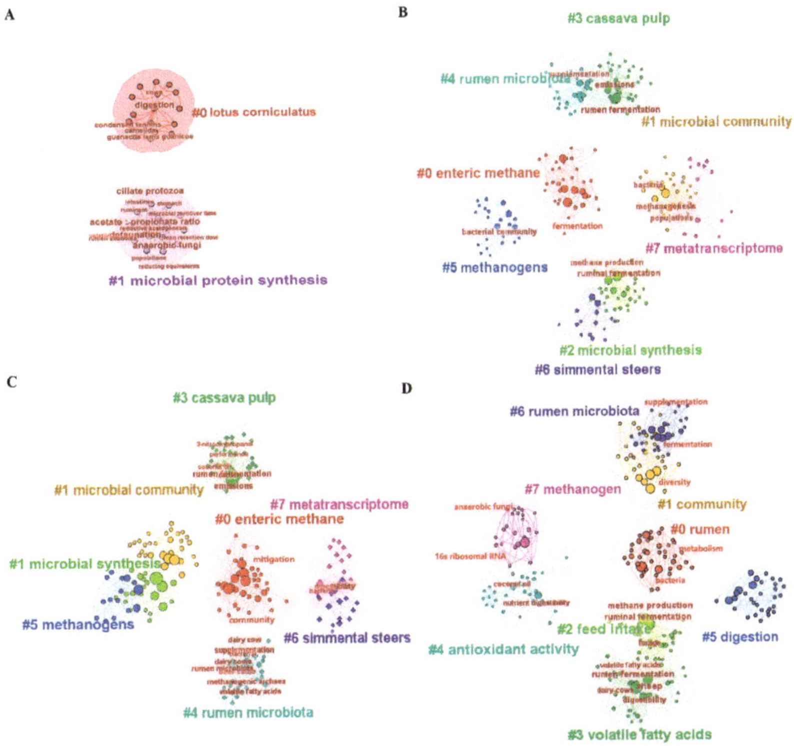

Figure 6. Keyword cluster analysis chart. (**A**) Phase I: 1998–2005; (**B**) phase II: 2006–2020; (**C**) phase III: 2021–2023; (**D**) all phases: 1998–2023.

In phase II (2006–2015), a keyword network map was generated, resulting in the identification of eight distinct clusters (Figure 6B). This contrasts with phase I, where novel feed additives such as condensed tannins and eucalyptus leaf meal were introduced [53,54]. Steers were predominantly used as the experimental animal. Metatranscriptome analysis emerged as a significant technique for measurement purposes. The research focus during this phase was primarily on methanogens, microbial community dynamics, and synthesis processes [55]. These results indicated that the regulation of ruminant feed was still the main research focus for controlling CH_4 emissions in this field.

In phase III (2016–2023), eight clusters were obtained from the keyword network map (Figure 6C). In contrast to the previous two stages, the research in this stage mainly used steers and dairy cows as subjects, and the metatranscriptome and 16S sequencing continued to be the main research methods [56]. At the same time, new research content such as feed intake, heritability, and dissolved hydrogen began to appear [57]. The increase in the number of research topics and the cross-disciplinary characteristics are consistent with the current research trends. It can also be seen that in recent years, research on the mechanisms of CH_4 generation has been gradually strengthened, and the reduction in CH_4 emissions at the source is the focus of current research, which is further developed on the basis of the previous two stages [58]. Based on the three phases, it can be seen that scholars have paid more attention to the regulation of feed additives on rumen microbiota and CH_4 emissions in ruminants over the past 26 years (Figure 6D). In addition to CH_4 emission, feed intake, digestibility, and microflora were also important when choosing research content [59].

3.3.3. Generation and Analysis of Burst Word

Burst words are keywords with a high frequency in a given period. They reflect the latest research hotspots and the development of theories and subjects at the forefront of research. CiteSpace's burst detection function was used to analyze the high-frequency words that have appeared in recent years so as to further understand the research hotspots in this field. Figure 7 show several obvious periods of burst words. The use of microbial protein synthesis increased significantly from 1999 to 2008. After 2008, focus turned to the study of feed additives such as coconut oil, and then gradually turned to fiber, metabolism, enteric CH_4, supplementation, methanogenic archaea, and so on [60,61]. In recent years, the dynamics of CH_4 production, methods of evaluating nutrient digestibility, and ruminal fermentation parameters are also important topics.

Figure 7. Top 25 keywords with the strongest citation bursts. Red lines indicate the years when the keyword was used frequently. Blue lines indicate the years when the keyword was not used frequently between 1998 and 2023.

The initial phase of research in ruminant nutrition primarily emphasized the enhancement of rumen microbial protein production [62]. Given that microbial proteins can supply 70% to 100% of the amino acids necessary for ruminants, the augmentation of microbial protein production could decrease the requirement for additional crude protein supplementation, thereby reducing feeding costs [63]. As a result, coconut oil has become a new field of research interest due to its cost-effectiveness, palatability, and accessibility as a feed resource. Furthermore, coconut oil is abundant in saturated medium- and long-chain fatty acids, which have the potential to diminish the ruminal protozoa population [64]. This reduction could boost rumen microbial protein synthesis, offset potential deficiencies in microbial protein availability, and inhibit the growth of rumen methanogens, consequently decreasing CH_4 generation [65]. The emergence of terms such as "rumen fermentation", "metabolism", and "methanogenic archaea" predominantly occurred between 2015 and 2020, indicating a shift in research focus towards unraveling the pathway of CH_4 generation. Methanogens play a crucial role within this context. This trend also suggests a transition in research content towards the microbial domain.

4. Future Research Directions

The increasing focus on CH_4 emissions from ruminant livestock has led to a surge in research exploring the relationship between ruminal microorganisms in ruminants and CH_4 emissions [66,67]. Most current studies in this field concentrate on short-term animal feeding experiments (lasting 21 to 42 days) employing a variety of direct measurement methods, such as respiration chambers, sulfur hexafluoride tracing technology, and the GreenFeed system to assess the impact on emission reduction. However, there is a relative paucity of research on long-term strategies. Additionally, the use of significantly different measurement methods across various institutions introduces a degree of uncertainty and inconsistency in evaluating CH_4 emission reduction effects.

To address these limitations, future research needs to delve more deeply into the disparities between various measurement techniques to enhance measurement precision and data trustworthiness. Concurrently, it is imperative to amalgamate current research insights to scrutinize the temporal and spatial dynamics of CH_4 emissions. This would bolster the regional CH_4 emission dataset and furnish theoretical and empirical backing for the development of effective emission reduction strategies. As research attention transitions from macro-phenotypes to rumen microorganisms and their CH_4 production pathways, there will be a growing emphasis on studies that employ additives for targeted regulation to curtail CH_4 emissions. This approach is poised to become a dominant research trend, thereby elevating the efficiency of CH_4 emission reduction.

Nonetheless, effectively mitigating CH_4 emissions without compromising animal production performance continues to pose a significant challenge in practical production. Consequently, future endeavors should encompass: firstly, a comprehensive exploration of the specific mechanisms by which feed additives and ingredients impact rumen microorganisms; secondly, ensuring the long-term viability of economic benefits and emission reduction effects in actual production settings; finally, synergistically employing multiple emission reduction strategies—such as selecting optimal feeds and utilizing effective additives—to maximize the emission reduction effect.

5. Conclusions

This study conducted a comprehensive analysis of relevant research papers in the field of rumen microbiota and CH_4 emission spanning from 1998 to 2023. The analysis was performed using CiteSpace and VOSviewer software, providing a systematic review of the developmental trends within this field. The findings reveal an escalating trend in the number of pertinent publications, predominantly centered in Europe, America, and China. An analysis of keywords indicates a discernible shift in research focus from phenotypic indicators, such as growth performance and digestibility, towards mechanistic explorations of microorganisms, including protozoa and methanogens. The reduction in CH_4 emissions from the gastrointestinal tract of ruminant livestock fundamentally hinges on encouraging hydrogen utilization within the rumen and preventing its use by methanogenic bacteria to produce CH_4. Present nutritional regulation strategies encompass diet optimization, feed quality enhancement, increased rumen flow rate, addition of hydrogen pools, and methanogenic bacterial inhibition. However, the enduring effects of certain CH_4 reduction techniques, such as oil and nitrate addition, require further investigation. Furthermore, pasture management and genetic breeding also serve as significant CH_4 emission reduction tools. In a production context, it is crucial to consider both current production efficiency and potential CH_4 reduction capabilities when devising a feasible mitigation plan. Future research should focus on synergizing different nutritional regulation strategies, assessing the sustainability of CH_4 reduction outcomes, developing low-CH_4-emitting livestock breeds, evaluating economic implications of livestock production systems, ensuring food safety, and addressing consumer preferences.

Author Contributions: X.Z. (Xueyue Zheng): conceptualization, methodology, formal analysis, data curation, writing—original draft preparation. L.T., D.W., R.W., M.W. and X.Z. (Xiumin Zhang): writing—review and editing. D.W. and R.W.: supervision. All authors have read and agreed to the published version of the manuscript.

Funding: This research was funded by the National Natural Science Foundation of China (grant 32372927), Industrial Innovation Chain Project of Hunan Province (2021NK1050), Science and Technology Innovation Program of Hunan Province (grant 2024RC3244), Agricultural Science and Technology Innovation Program (ASTIP-IBFC), and the Chinese Academy of Agricultural Sciences through the Natural Science Foundation of Hunan Province (grant 2023JJ50056).

Institutional Review Board Statement: Not applicable.

Informed Consent Statement: Not applicable.

Data Availability Statement: Dataset available on request from the authors.

Conflicts of Interest: The authors declare that they have no known competing financial interests or personal relationships that could have appeared to influence the work reported in this paper.

References

1. Jackson, R.B.; Saunois, M.; Bousquet, P.; Canadell, J.G.; Poulter, B.; Stavert, A.R.; Bergamaschi, P.; Niwa, Y.; Segers, A.; Tsuruta, A. Increasing anthropogenic methane emissions arise equally from agricultural and fossil fuel sources. *Environ. Res. Lett.* **2020**, *15*, 071002. [CrossRef]
2. Shashank, T.; Chhatarpal, S.; Jay, S.S. *Wetlands: A Major Natural Source Responsible for Methane Emission*; Springer: Singapore, 202
3. Kim, Y.; Tanaka, K.; Matsuoka, S. Environmental and economic effectiveness of the Kyoto Protocol. *PLoS ONE* **2020**, *15*, e023629 [CrossRef] [PubMed]
4. Meinshausen, M.; Lewis, J.; Mcglade, C.; Gütschow, J.; Nicholls, Z.; Burdon, R.; Cozzi, L.; Hackmann, B. Realization of Paris Agreement pledges may limit warming just below 2 °C. *Nature* **2022**, *604*, 304–309. [CrossRef] [PubMed]
5. Nisbet, E.G.; Fisher, R.E.; Lowry, D.; France, J.L.; Allen, G.; Bakkaloglu, S.; Broderick, T.J.; Cain, M.; Coleman, M.; Fernandez, J. et al. Methane mitigation: Methods to reduce emissions, on the path to the paris agreement. *Rev. Geophys.* **2020**, *58*, e2019RG00067 [CrossRef]
6. Lei, Z.; Hanqin, T.; Hao, S.; Shufen, P.; Xiaoyu, Q.; Naiqing, P.; Shree, R.S.D. Methane emissions from livestock in East Asia during 1961–2019. *Ecosyst. Health Sust.* **2021**, *7*, 215–226.
7. Basu, S.; Lan, X.; Dlugokencky, E.; Michel, S.; Schwietzke, S.; Miller, J.; Bruhwiler, L.; Oh, Y.; Tans, P.; Apadula, F. Estimating emissions of methane consistent with atmospheric measurements of methane and δ^{13}C of methane. *Atmos. Chem. Phys.* **2022**, *22*, 15351–15377. [CrossRef]
8. Dangal, S.R.S.; Tian, H.; Zhang, B.; Pan, S.; Lu, C.; Yang, J. Methane emission from global livestock sector during 1890–2014: Magnitude, trends and spatiotemporal patterns. *Glob. Change Biol.* **2017**, *23*, 4147–4161. [CrossRef] [PubMed]
9. Mirzaei-Aghsaghali, A.; Maheri-Sis, N. Factors affecting mitigation of methane emission from ruminants: Microbiology and biotechnology strategies. *J. Anim. Behav. Biometeorol.* **2015**, *4*, 22–31. [CrossRef]
10. Opio, C.I.; Gerber, P.J.; Mottet, A.; Falcucci, A.; Tempio, G.; MacLeod, M.; Vellinga, T.V.; Henderson, B.B.; Steinfeld, H. *Greenhouse Gas Emissions from Ruminant Supply Chains—A Global Life Cycle Assessment*; Food and Agriculture Organization: Rome, Italy, 2013
11. Tseten, T.; Sanjorjo, R.A.; Kwon, M.; Kim, S.W. Strategies to mitigate enteric methane emissions from ruminant animals. *J. Microbiol. Biotechnol.* **2022**, *32*, 269–277. [CrossRef] [PubMed]
12. Cottle, D.J.; Nolan, J.V.; Wiedemann, S.G. Ruminant enteric methane mitigation: A review. *Anim. Prod. Sci.* **2011**, *51*, 491–514. [CrossRef]
13. Bačėninaitė, D.; Džermeikaitė, K.; Antanaitis, R. Global warming and dairy cattle: How to control and reduce methane emission *Animals* **2022**, *12*, 2687. [CrossRef]
14. Meale, S.J.; Popova, M.; Saro, C.; Martin, C.; Bernard, A.; Lagree, M.; Yáñez-Ruiz, D.R.; Boudra, H.; Duval, S.; Morgavi, D.P. Early life dietary intervention in dairy calves results in a long-term reduction in methane emissions. *Sci. Rep.* **2021**, *11*, 3003. [CrossRef] [PubMed]
15. Martínez-lvaro, M.; Auffret, M.D.; Stewart, R.D.; Dewhurst, R.J.; Duthie, C.A.; Rooke, J.A.; Wallace, R.J.; Shih, B.; Freeman, T.C. Watson, M. Identification of complex rumen microbiome interaction within diverse functional niches as mechanisms affecting the variation of methane emissions in bovine. *Front. Microbiol.* **2020**, *11*, 659. [CrossRef]
16. Pereira, A.M.; Dapkevicius, M.D.L.N.E.; Borba, A.E.S. Alternative pathways for hydrogen sink originated from the ruminal fermentation of carbohydrates: Which microorganisms are involved in lowering methane emission? *Anim. Microbiome* **2022**, *4*, 5 [CrossRef] [PubMed]
17. Honan, M.; Feng, X.; Tricarico, J.M.; Kebreab, E. Feed additives as a strategic approach to reduce enteric methane production in cattle: Modes of action, effectiveness and safety. *Anim. Prod. Sci.* **2022**, *62*, 1303–1317. [CrossRef]
18. Huiling, M.; Wang, J.; Zhou, Y.Y.; Liu, J. Effects of addition of tea saponins and soybean oil on methane production, fermentation and microbial population in the rumen of growing lambs. *Livest. Sci.* **2010**, *129*, 56–62.
19. Bekele, W.; Guinguina, A.; Zegeye, A.; Simachew, A.; Ramin, M. Contemporary methods of measuring and estimating methane emission from ruminants. *Methane* **2022**, *1*, 82–95. [CrossRef]
20. Butt, N.S.; Malik, A.A.; Shahbaz, M.Q. Bibliometric analysis of statistics journals indexed in web of science under emerging source citation index. *Sage Open* **2021**, *11*, 88870. [CrossRef]

21. Xu, X.; Mishra, G.D.; Jones, M. Mapping the global research landscape and knowledge gaps on multimorbidity: A bibliometric study. *J. Glob. Health* **2017**, *7*, 010414. [CrossRef]
22. Sweileh, W.M. Research trends on human trafficking: A bibliometric analysis using Scopus database. *Glob. Health* **2018**, *14*, 106. [CrossRef]
23. Ait, S. *Report of the Conference of the Parties on Its Third Session, Held at Kyoto from 1 to 11 December 1997 Addendum Part Two: Action Taken By the Conference of the Parties*; UNFCCC: Bonn, Germany, 1998.
24. Chen, C. A glimpse of the first eight months of the COVID-19 literature on microsoft academic graph: Themes, citation contexts, and uncertainties. *Front. Res. Metr. Anal.* **2020**, *5*, 607286. [CrossRef] [PubMed]
25. Feng, X.W.; Hadizadeh, M.; Cheong, J.P.G. Global trends in physical-activity research of autism: Bibliometric analysis based on the web of science database (1980–2021). *Int. J. Environ. Res. Public Health* **2022**, *19*, 7278. [CrossRef]
26. Pauna, V.H.; Buonocore, E.; Renzi, M.; Russo, G.F.; Franzese, P.P. The issue of microplastics in marine ecosystems: A bibliometric network analysis. *Mar. Pollut. Bull.* **2019**, *149*, 110612. [CrossRef]
27. Andrade, D.F.; Romanelli, J.P.; Pereira-Filho, E.R. Past and emerging topics related to electronic waste management: Top countries, trends, and perspectives. *Environ. Sci. Pollut. Res.* **2019**, *26*, 17135–17151. [CrossRef]
28. Brookes, B.C. Bradford's Law and the bibliography of science. *Nature* **1969**, *224*, 953–956. [CrossRef] [PubMed]
29. Desai, N.; Veras, L.; Gosain, A. Using Bradford's law of scattering to identify the core journals of pediatric surgery. *J. Surg. Res.* **2018**, *229*, 90–95. [CrossRef] [PubMed]
30. Qian, J.; Yuan, Z.; Li, J.; Zhu, H. Science citation index (SCI) and scientific evaluation system in China. *Humanit. Soc. Sci. Commun.* **2020**, *7*, 108. [CrossRef]
31. Yi, B.; Ludo, W.; Yong, H. A multi-dimensional framework for characterizing the citation impact of scientific publications. *Quant. Sci. Stud.* **2020**, *2*, 155–183.
32. Shabat, S.K.; Sasson, G.; Doron-Faigenboim, A.; Durman, T.; Yaacoby, S.; Berg Miller, M.E.; White, B.A.; Shterzer, N.; Mizrahi, I. Specific microbiome-dependent mechanisms underlie the energy harvest efficiency of ruminants. *ISME J.* **2016**, *10*, 2958–2972. [CrossRef] [PubMed]
33. Wallace, R.J.; Rooke, J.A.; McKain, N.; Duthie, C.A.; Hyslop, J.J.; Ross, D.W.; Waterhouse, A.; Watson, M.; Roehe, R. The rumen microbial metagenome associated with high methane production in cattle. *BMC Genom.* **2015**, *16*, 839. [CrossRef] [PubMed]
34. van Zijderveld, S.M.; Gerrits, W.J.; Apajalahti, J.A.; Newbold, J.R.; Dijkstra, J.; Leng, R.A.; Perdok, H.B. Nitrate and sulfate: Effective alternative hydrogen sinks for mitigation of ruminal methane production in sheep. *J. Dairy Sci.* **2010**, *93*, 5856–5866. [CrossRef] [PubMed]
35. Roehe, R.; Dewhurst, R.J.; Duthie, C.A.; Rooke, J.A.; McKain, N.; Ross, D.W.; Hyslop, J.J.; Waterhouse, A.; Freeman, T.C.; Watson, M.; et al. Bovine host genetic variation influences rumen microbial methane production with best selection criterion for low methane emitting and efficiently feed converting hosts based on metagenomic gene Abundance. *PLoS Genet.* **2016**, *12*, e1005846. [CrossRef] [PubMed]
36. Kittelmann, S.; Pinares-Patiño, C.S.; Seedorf, H.; Kirk, M.R.; Ganesh, S.; McEwan, J.C.; Janssen, P.H. Two different bacterial community types are linked with the low-methane emission trait in sheep. *PLoS ONE* **2014**, *9*, e103171. [CrossRef]
37. Kamke, J.; Kittelmann, S.; Soni, P.; Li, Y.; Tavendale, M.; Ganesh, S.; Janssen, P.H.; Shi, W.; Froula, J.; Rubin, E.M. Rumen metagenome and metatranscriptome analyses of low methane yield sheep reveals a Sharpea -enriched microbiome characterised by lactic acid formation and utilisation. *Microbiome* **2016**, *4*, 56. [CrossRef] [PubMed]
38. Difford, G.F.; Plichta, D.R.; Løvendahl, P.; Lassen, J.; Noel, S.J.; Højberg, O.; Wright, A.G.; Zhu, Z.; Kristensen, L.; Nielsen, H.B.; et al. Host genetics and the rumen microbiome jointly associate with methane emissions in dairy cows. *PLoS Genet.* **2018**, *14*, e1007580. [CrossRef]
39. Rebecca, D.; Johan, D.; Li, S.; Horacio, G.; Bettina, M.; Anna, S.; Jan, B. Methane production in dairy cows correlates with rumen methanogenic and bacterial community structure. *Front. Microbiol.* **2017**, *8*, 226.
40. Animut, G.; Puchala, R.; Goetsch, A.L.; Patra, A.K.; Sahlu, T.; Varel, V.H.; Wells, J. Methane emission by goats consuming diets with different levels of condensed tannins from lespedeza. *Anim. Feed Sci. Tech.* **2008**, *144*, 212–227. [CrossRef]
41. Hollmann, M.; Powers, W.J.; Fogiel, A.C.; Liesman, J.S.; Beede, D.K. Response profiles of enteric methane emissions and lactational performance during habituation to dietary coconut oil in dairy cows. *J. Dairy Sci.* **2013**, *96*, 1769–1781. [CrossRef] [PubMed]
42. Ding, X.Z.; Long, R.J.; Zhang, Q.; Huang, X.D.; Guo, X.S. Reducing methane emissions and the methanogen population in the rumen of Tibetan sheep by dietary supplementation with coconut oil. *Trop. Anim. Health Prod.* **2012**, *44*, 1541–1545. [CrossRef] [PubMed]
43. Piñeiro-Vázquez, A.; Canul-Solís, J.; Alayón-Gamboa, J.; Chay-Canul, A.; Ayala-Burgos, A.; Aguilar-Pérez, C.; Solorio-Sánchez, F.; Ku-Vera, J. Potential of condensed tannins for the reduction of emissions of enteric methane and their effect on ruminant productivity. *Arch. Med. Vet.* **2015**, *47*, 263–272. [CrossRef]

44. Puchala, R.; Min, B.R.; Goetsch, A.L.; Sahlu, T. The effect of a condensed tannin-containing forage on methane emission by goats. *J. Anim. Sci.* **2005**, *83*, 182–186. [CrossRef]
45. Grainger, C.; Auldist, M.J.; Clarke, T.; Beauchemin, K.A.; McGinn, S.M.; Hannah, M.C.; Eckard, R.J.; Lowe, L.B. Use of monensin controlled-release capsules to reduce methane emissions and improve milk production of dairy cows offered pasture supplemented with grain. *J. Dairy Sci.* **2008**, *91*, 1159–1165. [CrossRef]
46. Fu, Y.; Yao, S.; Wang, T.; Lu, Y.; Han, H.; Liu, X.; Lv, D.; Ma, X.; Guan, S.; Yao, Y.; et al. Effects of melatonin on rumen microorganisms and methane production in dairy cow: Results from in vitro and in vivo studies. *Microbiome* **2023**, *11*, 196.
47. Ramayo-Caldas, Y.; Zingaretti, L.M.; Popova, M.; Estellé, J.; Bernard, A.; Pons, N.; Bellot, P.; Mach, N.; Rau, A.; Roume, H. Identification of rumen microbial biomarkers linked to methane emission in Holstein dairy cows. *J. Anim. Breed. Genet.* **2020**, *137*, 49–59. [CrossRef]
48. Wang, R.; Wang, M.; Lin, B.; Ungerfeld, E.M.; Tan, Z.L. Associations of ruminal hydrogen and pH with fiber digestibility and microbiota composition induced by increasing starch intake in beef cattle. *Anim. Feed Sci. Technol.* **2021**, *278*, 114980. [CrossRef]
49. Tian, X.; Gao, C.; Hou, Z.P.; Wang, R.; Zhang, X.M.; Li, Q.S.; Wei, Z.S.; Wu, D.Q.; Wang, M. Comparisons of ramie and corn stover silages: Effects on chewing activity, rumen fermentation, microbiota and methane emissions in goats. *Fermentation* **2022**, *8*, 432. [CrossRef]
50. Karsli, M.A.; Russell, J.R. Effects of some dietary factors on ruminal microbial protein synthesis. *Turk. J. Vet. Anim. Sci.* **2001**, *28*, 681–686.
51. McSweeney, C.S.; Palmer, B.; Bunch, R.; Krause, D.O. Effect of the tropical forage calliandra on microbial protein synthesis and ecology in the rumen. *J. Appl. Microbiol.* **2001**, *90*, 78–88. [CrossRef] [PubMed]
52. Woodward, S.L.; Waghorn, G.C.; Laboyrie, P.G. Condensed tannins in birdsfoot trefoil (*Lotus corniculatus*) reduce methane emissions from dairy cows. *Proc. N. Z. Soc. Anim. Prod.* **2004**, *64*, 160–164.
53. Kamra, D.N.; Pawar, M.; Singh, B. *Effect of Plant Secondary Metabolites on Rumen Methanogens and Methane Emissions by Ruminants*; Springer: Dordrecht, The Netherlands, 2012; pp. 351–370.
54. Thao, N.T.; Wanapat, M.; Cherdthong, S.K. Effects of supplementation of eucalyptus (*E. Camaldulensis*) leaf meal on feed intake and rumen fermentation efficiency in swamp buffaloes. *Anim. Biosci.* **2015**, *28*, 951–957. [CrossRef] [PubMed]
55. Zhou, M.; Chung, Y.H.; Beauchemin, K.A.; Holtshausen, L.; Oba, M.; McAllister, T.A.; Guan, L.L. Relationship between rumen methanogens and methane production in dairy cows fed diets supplemented with a feed enzyme additive. *J. Appl. Microbiol.* **2011**, *111*, 1148–1158. [CrossRef]
56. Tapio, I.; Snelling, T.J.; Strozzi, F.; Wallace, R.J. The ruminal microbiome associated with methane emissions from ruminant livestock. *J. Anim. Sci. Biotechnol.* **2017**, *8*, 7. [CrossRef] [PubMed]
57. Li, Q.S.; Wang, R.; Ma, Z.Y.; Zhang, X.M.; Jiao, J.Z.; Zhang, Z.G.; Ungerfeld, E.M.; Yi, K.L.; Zhang, B.Z.; Long, L.; et al. Dietary selection of metabolically distinct microorganisms drives hydrogen metabolism in ruminants. *ISME J.* **2022**, *16*, 2535–2546. [CrossRef] [PubMed]
58. Arndt, C.; Hristov, A.; Price, W.; Mcclelland, S.; Pelaez, A.M.; Cueva, S.F.; Oh, J.; Dijkstra, J.; Bannink, A.; Bayat, A. Full adoption of the most effective strategies to mitigate methane emissions by ruminants can help meet the 1.5 °C target by 2030 but not 2050. *Proc. Natl. Acad. Sci. USA* **2022**, *119*, e2111294119. [CrossRef]
59. Khurana, R.; Brand, T.; Tapio, I.; Bayat, A.R. Effect of a garlic and citrus extract supplement on performance, rumen fermentation, methane production, and rumen microbiome of dairy cows. *J. Dairy Sci.* **2023**, *106*, 4608–4621. [CrossRef] [PubMed]
60. Crompton, L.A.; Mills, J.A.N.; Reynolds, C.K. Effect of feeding frequency and replacing calcium salts of palm oil with crushed rapeseed or coconut oil on methane emissions in lactating dairy cows. *Proc. Nutr. Soc.* **2010**, *69*, E329. [CrossRef]
61. Liu, H.; Vaddella, V.; Zhou, D. Effects of chestnut tannins and coconut oil on growth performance, methane emission, ruminal fermentation, and microbial populations in sheep. *J. Dairy Sci.* **2011**, *94*, 6069–6077. [CrossRef]
62. Chumpawadee, S.; Sommart, K.; Vongpralub, T.; Pattarajinda, V. Effects of Synchronizing the Rate of Dietary Energy and Nitrogen Release on Ruminal Fermentation, Microbial Protein Synthesis, Blood Urea Nitrogen and Nutrient Digestibility in Beef Cattle. *Anim. Biosci.* **2005**, *19*, 181–188. [CrossRef]
63. Harmon, D.L.; Swanson, K.C. Review: Nutritional regulation of intestinal starch and protein assimilation in ruminants. *Animal* **2020**, *14* (Suppl. S1), s17–s28. [CrossRef]
64. Shi, L.; Zhang, Y.; Wu, L.; Xun, W.; Zhou, H. Moderate coconut oil supplement ameliorates growth performance and ruminal fermentation in Hainan black goat kids. *Front. Vet. Sci.* **2020**, *7*, 622259. [CrossRef]
65. Lee, C.; Hristov, A.N.; Heyler, K.S.; Cassidy, T.W.; Long, M.; Corl, B.A.; Karnati, S.K. Effects of dietary protein concentration and coconut oil supplementation on nitrogen utilization and production in dairy cows. *J. Dairy Sci.* **2011**, *94*, 5544–5557. [CrossRef] [PubMed]

6. Zhang, L.; Tian, H.; Shi, H.; Pan, S.; Chang, J.; Dangal, S.R.S.; Qin, X.; Wang, S.; Tubiello, F.N.; Canadell, J.G.; et al. A 130-year global inventory of methane emissions from livestock: Trends, patterns, and drivers. *Glob. Change Biol.* **2022**, *28*, 5142–5158. [CrossRef] [PubMed]
7. Wang, L.; Li, Y.; Zhang, Y.; Wang, L. The effects of different concentrate-to-forage ratio diets on rumen bacterial microbiota and the structures of Holstein cows during the feeding cycle. *Animals* **2020**, *10*, 957. [CrossRef] [PubMed]

animals

MDPI

Article

Effect of Hemp Seed Oil on Milk Performance, Blood Parameters, Milk Fatty Acid Profile, and Rumen Microbial Population in Milk-Producing Buffalo: Preliminary Study

Qichao Gu [1,2], Bo Lin [1,2], Dan Wan [1,2], Zhiwei Kong [1,2], Qinfeng Tang [1,2], Qi Yan [1,2], Xinghua Cai [1,2], Hao Ding [1,2], Guangsheng Qin [3,*] and Caixia Zou [1,2,*]

1 College of Animal Science and Technology, Guangxi University, Nanning 530004, China; guqchao@gmail.com (Q.G.); linbo@gxu.edu.cn (B.L.); wan89dan@163.com (D.W.); kongzhiwei2013@126.com (Z.K.); 899qinghui@163.com (Q.T.); yanqi7798@163.com (Q.Y.); 13367811557@163.com (X.C.); haodingxj@163.com (H.D.)

2 Guangxi Key Laboratory of Animal Breeding, Disease Control and Prevention, Nanning 530004, China

3 Buffalo Research Institute Chinese Academy of Agricultural Sciences and Guangxi Zhuang Nationality Autonomous Region, Nanning 530001, China

* Correspondence: qinguangsheng2004@126.com (G.Q.); zou.cx@gxu.edu.cn (C.Z.)

Simple Summary: Vegetable oils containing unsaturated fatty acids play an important role in improving milk quality and enhancing buffaloes' overall health. Hemp seed oil (HSO), a native vegetable oil from Bama in Guangxi, China, which is known as "the hometown of longevity", is particularly rich in unsaturated fatty acid (e.g., linoleic acid). In this study, our results show improvements in the serum antioxidant capacity and lipid metabolism regulation following dietary supplementation with HSO. Specifically, we found that HSO supplementation increased the proportion of polyunsaturated fatty acids (e.g., omega-6 and omega-3), as well as milk C18 saturated (C18:0) and unsaturated fatty acids (e.g., C18:2n6c, C18:3n3, and C18:2n9c) in milk samples. This study provides a basis for using HSO as a natural feed additive to improve the milk quality and health of milk-producing buffalo.

Abstract: Vegetable oils rich in unsaturated fatty acids have been shown to improve animal health and enrich milk with functional fatty acids in various studies. This study investigates the effects of dietary supplementation with hemp seed oil (HSO), a native vegetable oil from the "longevity village" of Bama (Guangxi, China), on the milk performance, milk fatty acid composition, blood indicators, and rumen bacterial community of milk-producing buffalo. Seventeen healthy, four-year-old, crossbred, milk-producing buffaloes with the same parity (three), as well as similar body weights (BW = 580 ± 25 kg), number of days producing milk (DIM, 153 ± 10 d), and milk yields (8.56 ± 0.89 kg/d) were divided into three groups (n = 6, 5, and 6) and assigned to the following diets: (1) no HSO supplement (H0, n = 6), (2) a supplement of 100 g/d of HSO (H1, n = 5), and (3) a supplement of 200 g/d of HSO (H2, n = 6). The total experimental period was 42 days (including a 14-day adaptation period and a 28-day treatment period). The data were statistically analyzed by repeated measures analysis of variance. The results showed that compared to that of no HSO supplement group, the dry matter intake (DMI) showed a decreasing tendency (p = 0.06), while feed efficiency and rumen fermentation remained similar across all the groups (p > 0.05) with dietary HSO supplementation. Moreover, with dietary HSO supplementation, the total antioxidant capacity (T-AOC) (p = 0.05) and catalase (CAT) (p < 0.01) and glutathione peroxidase (GSH-Px) (p = 0.02) contents in the serum were greatly increased, with the highest levels observed in the H2 group (increased by 1.16 U/mL, 1.15 U/mL, and 134.51 U/mL, respectively). In contrast, the malondialdehyde (MDA) content was significantly decreased with dietary HSO supplementation (p = 0.02) and was

 https://doi.org/10.3390/ani15040514

the lowest in the H1 group (decreased by 0.72 nmol/mL). The high-density lipoprotein cholesterol (HDL-C) content in the blood showed an increasing tendency with dietary HSO supplementation ($p = 0.09$). Moreover, with dietary HSO supplementation, the proportions of C18:0 ($p = 0.02$), C18:1n9t ($p = 0.02$), C18:2n6c ($p = 0.02$), C18:3n3 ($p < 0.01$), C18:2n9c ($p = 0.04$), omega-3 ($p = 0.02$), and omega-6 ($p = 0.02$) were significantly increased, with the highest levels observed in the H2 group (increased by 5.29 g/100 g FA, 1.81 g/100 g FA, 0.55 g/100 g FA, 0.14 g/100 g FA, 0.75 g/100 g FA, 0.17 g/100 g FA, and 0.56 g/100 g FA, respectively). Additionally, rumen *Acetobacter* abundance was significantly affected by HSO addition ($p = 0.03$), with rumen *Acetobacter* abundance decreasing in the H1 group (by 0.55%) and increasing in the H2 group (by 0.73%). These results suggest that adding HSO to milk-producing buffalo diets does not affect feed efficiency or rumen fermentation, although it decreases the DMI. Meanwhile, it can improve the nutritional quality of milk, enhance the antioxidant status, and regulate blood lipid metabolism in milk-producing buffaloes.

Keywords: milk-producing buffalo; oil supplementation; natural additive; antioxidants; milk fatty acid profiles; rumen microbiota

1. Introduction

In modern society, people are becoming increasingly conscious of eating a healthy diet. Hemp seed oil is a lipid vegetable oil extracted from the dried and mature seeds of hemp (*Cannabis sativa* L.). Hemp is a specialty of Bama (Guangxi, China), which is also known as the "hometown of longevity", with a cultivation area of more than 2300 hectares, and a hemp seed yield of about 1000–2000 kg/hectare [1,2]. Notably, this oil is characterized by being the only natural vegetable oil that can be dissolved in water and has an extremely high unsaturated fatty acid (UFA) content, among which more than 80% of the total fatty acid composition is polyunsaturated fatty acids (PUFAs); additionally, over 50% of the total fatty acid composition is 18:2n-6 (linoleic acid, LA), and ~20% is 18:3n-3 (linolenic acid, ALA) [3]. It can be an important source of omega-6 and omega-3 PUFAs, which are present in an optimal nutrient ratio of 3:1 (LA to ALA) [3]. Other lipid vegetable oils (e.g., palm oil, adlay seed oil, soybean oil, coconut oil, and blended seed oil) are also rich in UFAs. When added to animal diets, these oils can improve the antioxidant capacity of animal serum by increasing antioxidant enzyme activities (the total antioxidant capacity increased by 0.57–6.56 U/mL, while the malondialdehyde and catalase contents decreased by 1.92–23.46 nmol/mL and 1.50–4.97 U/mL, respectively) [4,5]. Additionally, these oils can regulate lipid metabolism by increasing the high-density lipoprotein cholesterol content (increased by ~22.5 mg/dL) and reducing the total cholesterol and triglyceride levels in the blood (decreased by ~5 mg/dL) [6]. Hemp seed oil not only has higher levels of antioxidant properties than those of many other seed oils (e.g., linseed oil, olive oil, and niger seed oil) [7], but it has also demonstrated the ability to lower the total cholesterol (decreased by 1.28 mmol/mL) and triglyceride (decreased by 2.65 mmol/mL) levels in rats [8].

The previous studies have shown that adding sunflower seed oil (130 g/d), linseed oil (130 g/d), and babassu oil (90 g/d) to diets can improve feed intake, with the dry matter intake decreasing by 0.15 kg/d, 0.18 kg/d, and 0.46 kg/d respectively [9,10]. Moreover, linseed oil supplementation increases the yields of milk fat and milk lactose, with increases of 15 g/d and 13 g/d, respectively [9]. In addition, sunflower seed oil or linseed oil (130 g/d) supplementation can effectively regulate the milk fatty acid (FA) composition of ruminants, including a decrease in saturated fatty acids in milk, with those in milk 10:0, milk 11:0, milk 12:0, milk 13:0, milk 14:0, milk 15:0, and milk 16:0 decreasing by 3.51 g/100 g FA or 2.60 g/100 g FA, 0.08 g/100 g FA or 0.06 g/100 g FA, 2.50 g/100 g FA or 2.12 g/100 g FA, 0.15 g/100 g FA or 0.13 g/100 g FA, 4.14 g/100 g FA or 3.92 g/100 g FA, 0.50 g/100 g FA or 0.44 g/100 g FA, and 10.38 g/100 g FA or 10.72 g/100 g FA, respectively [9]. Meanwhile, regular sunflower oil (48 g/d) [11], linseed oil (48 g/d or 66 g/d) [11,12], and echium oil (~40 g/d) [13] supplementation increases the amount of functional unsaturated fatty acids in milk, such as conjugated linoleic acid (CLA), with increases of 1.22 g/100 g FA, 1.04 g/100 g FA, 1.64 g/100 g FA, and 15.51 g/100 g FA, respectively. Cozma et al., who fed dairy goats diets supplemented with 93 g/d hemp seed oil, also reported an increase in the CLA content in goat milk, with an increase of 0.18 g/100 g FA [14]. Moreover, Delavaud et al. found that both cows and goats fed diets containing hydrogenated palm oil (3.0% of total DMI; 639.3 g/d for cows and 79.5 g/d for goats) showed an increase in hydrogenated palm oil content in milk C18:0, with increases of 2.70 g/100 g FA and 3.10 g/100 g FA in the cows and the goats, respectively [15]. Rumen microbiota hydrogenate unsaturated fatty acids into saturated fatty acids through biohydrogenation, which is an important process that affects the composition of milk fatty acids [16,17]. This is an important source of saturated fatty acids in milk fat, and some biohydrogenation intermediates produced during the hydrogenation process are the source of unsaturated fatty acids in milk fat [18–20]. Several recent studies indicated that the addition of vegetable oil (such as olive oil, sunflower oil, coconut oil, and palm oil) affected the composition of ruminal bacteria [21–23]. It is unclear whether hemp seed oil, another lipid vegetable oil, can also induce changes in the bacterial composition of the dairy buffalo rumens, thereby affecting bacterial milk fatty acid metabolism.

Therefore, the aim of this study was to preliminarily evaluate the effects of dietary HSO from Bama, known as the world's "hometown of longevity", on milk-producing buffaloes, hypothesizing that dietary HSO could improve the milk performance, alter the blood antioxidant properties and lipid levels, modify the rumen bacterial structure, and increase the proportion of functional milk fatty acids (such as milk C18 saturated and unsaturated fatty acids) in the milk of milk-producing buffaloes.

2. Materials and Methods

2.1. Experimental Material

HSO was self-squeezed from Bama hemp seeds purchased in Bama, Guangxi, China. The hemp seeds (150 kg) were first cleaned and dried, after which they were crushed and shelled to obtain hemp kernels. These kernels (about 38 kg) were then pressed at room temperature using a press (Weiruima 120, Wuhan Weiruima Machinery Manufacturing Co., Ltd., Wuhan, China) without steaming or frying to obtain HSO. The fatty acid composition of HSO was determined according to the method described by Montserrat-de et al., with some modifications [3]. Briefly, 0.2 g of HSO was placed into a 15 mL stoppered test tube, followed by the addition of 0.5 mL of 2 mg/mL C19:0 as an internal standard and 4 mL of n-hexane. The mixture was gently shaken to dissolve the oil. Then, 5 mL of 0.5 mol/L of sodium methoxide solution was added, and the mixture was gently shaken and allowed to stand at room temperature for 10 min. A small amount of distilled water was added to the mixture, which was then gently shaken until an emulsion formed. This mixture was allowed to stand and clarify, allowing all the n-hexane solution to rise to the top. The supernatant was transferred to a dry test tube, and a small amount of anhydrous sodium sulfate was added to dry the tube. The supernatant was subsequently taken and analyzed in terms of fatty acid composition using an Agilent-7890A gas chromatograph equipped with a flame ionization detector (FID) and an HP-INNOWAX 19091N-133 capillary column (30.0 m × 0.25 mm × 0.25 μm). The fatty acid composition of HSO is shown in Table 1.

Table 1. Composition of fatty acids in HSO (g/100 g of fatty acid).

Fatty Acid	Content
C14:0	0.012
C16:0	6.301
C16:1n7	0.094
C18:0	2.918
C18:1n9t	0.001
C18:1n11t	0.021
C18:1n9c	10.238
C18:2n6c	58.321
C18:3n6	0.476
C20:0	0.769
C18:3n3	20.243
C18:2n9c	0.365
C22:0	0.226
C20:4n6	0.015
SFA [1]	10.226
USFA [2]	86.774

[1] SFA, saturated fatty acid, [2] USFA, unsaturated fatty acid.

2.2. Experimental Design

A randomized block design was used in this research. The total experimental period lasted 42 days, including a 14-day adaptation period, followed by a 28-day formal experiment period. Seventeen healthy, four-year-old crossbred milk-producing buffalo, all of which exhibited the same parity (3) and had similar weights (BW = 580 ± 25 kg), number of days producing milk (DIM, 153 ± 10 d), and milk yields (8.56 ± 0.89 kg/d), were used in this study. The milk-producing buffalo were housed in separate fenced pens (9 m length × 4 m width) at the Buffalo Research Institute of the Chinese Academy of Agricultural Sciences in Nanning, located within China's Guangxi Province. The milk-producing buffaloes were then divided into three dietary treatment groups (n = 6, 5, and 6, respectively): (1) with the addition of 0 g/d HSO (H0, n = 6), (2) with the addition of 100 g/d HSO (H1, n = 5), and (3) with the addition of 200 g/d HSO (H2, n = 6). HSO was added to concentrate, mixed thoroughly, and then fed individually to each milk-producing buffalo at 6:00 a.m. in the HSO treatment groups. The diets all had the same protein and energy contents. The diets, which were formulated according to the recommendations of Borghese et al. [24] (suitable for a milk yield of 8 kg/d), were provided to the milk-producing buffalo in equal amounts twice daily at 6:00 a.m. and 2:30 p.m. To facilitate the feeding of the buffaloes, balance nutrition, and improve feed utilization, the forage (including elephant grass, distilled brewer's grain, and cassava pulp) and concentrates of the diets (forage/concentrate = 50:50) were provided separately. Each milk-producing buffalo was fed individually and had free access to water at all times. The ingredients and chemical composition of the basic diet are shown in Table 2. The samples of feed from the buffalo's basic diet were analyzed for dry matter (method 935.29), crude protein (method 990.03), ash (method 942.05), and ether extract (method 920.39) contents [25]. The crude fiber, neutral detergent fiber, and acid detergent fiber contents were determined using an ANKOM A2000i fiber analyzer (ANKOM Technology, NY, USA). The maximum ingestion capacity of each buffalo was estimated by ensuring 10% feed refusal of a standard milk-production diet. During the last seven days of the experiment, the remaining feed in the trough was removed 30 min before daily feeding. The leftover feed was weighed, and this value was subtracted from the total weight of each feed ingredient allocated the previous day to determine the daily feed intake (subsequently converted to DMI) for each milk-producing buffalo. The DMI data are shown in Table S1.

Table 2. Composition (%, as fed) and nutrients (%, DM) in basal diets.

Item	Basal Diet
Ingredients	
Maize grain	13.31
Wheat bran	3.33
Soy bean meal	4.99
Cottonseed meal	2.77
Rapeseed meal	1.66
Stone powder	0.28
$CaHPO_4 \cdot 2H_2O$	0.28
$NaHCO_3$	0.42
NaCl	0.42
Premix [1]	0.28
Distilled brewer's grain	23.90
Elephant grass	11.31
Cassava pulp	37.05
Nutritional level of diet	
Dry matter	46.60
Crude protein	17.65
Crude fiber	45.55
Acid detergent fiber	18.20
Neutral detergent fiber	37.14
Ether extract	3.34
Ash	8.01
NE_L, MJ/kg [2]	4.57

NE_L = net energy for lactation. [1] The composition of the premix is as follows: VA 500 KIU/kg; VD3150 KIU/kg; VE 3000 IU/kg; Fe 4.0 g/kg; Cu 1.3 g/kg; Mn 3.0 g/kg; Zn 6.0 g/kg; I 80 mg/kg; Se 50 mg/kg; and Co 80 mg/kg. [2] NEL was calculated according to the Feed Database in China (2022). The formula for calculating NEL is as follows: NEL (MJ/kg) = 0.1025 × TDN% − 0.502, TDN% = 1.15 × CP% + 1.75 × EE% + 0.45 × CF% + 0.0085 × NDF%2 + 0.25 × NFE% − 3.4. The other nutritional contents of the diet are measured values.

2.3. Milk Sampling and Analysis

Milk samples were collected throughout the 28–42 days of the total experiment; the milk yield is shown in Supplementary Table S1. Samples (100 mL) were collected from each experimental milk-producing buffalo at 8:30 a.m. and 3:30 p.m. The two samples from each animal were mixed thoroughly to form a mixed milk sample of 200 mL. The mixed milk sample was divided into two parts. One part (100 mL) was immediately tested for milk composition (e.g., milk fat, milk protein, total solid, solid non-fat, and lactose contents). For the different types of milk, energy-corrected milk (ECM) = milk (kg/d) + [(fat (kg/d)−40) + (protein (kg/d)−31)] × 0.01155 [26,27] and 4% fat-corrected milk (4% FCM) = 0.4 × milk (kg/d) + 15 × fat (kg/d) [26]. The milk composition and feed efficiency (milk yield/DMI; ECM/DMI) are shown in Table S1; the other part (100 mL) was stored in a refrigerator at −80 °C to determine the fatty acid composition of milk.

Fatty acids in the milk samples were extracted for methyl esterification and gas phase detection. Based on the fatty acid determination method established by Marín et al. [13], the column selection and operation steps were followed with some modifications. An Agilent-7890A gas chromatograph was used to determine fatty acids, with a special capillary column for fatty acid methyl ester determination (Agilent HP88 (100 m × 0.25 mm × 0.20 μm)), nitrogen as the carrier gas, and a flow rate of 1.1 mL/min. The temperature of both the injection port and the detector was 250 °C. The temperature of the storage box was programmed as follows: the initial temperature was held at 150 °C for 5 min, and then increased to 175 °C at a rate of 2 °C/min and held for 15 min, increased to 200 °C at a rate of 7 °C/min and held for 20 min, and finally increased to 220 °C at a rate of 5 °C/min and held for 25 min. The sample injection volume was 1 μL, including the internal standard, and the split ratio was 20:1. The detector parameters were a hydrogen flow of 30 mL/min and an air flow 400 of mL/min.

2.4. Blood Sampling and Analysis

Before morning feeding on day 41 of the total experimental period, blood samples (10 mL) were collected from the carotid arteries of the milk-producing buffalo by a veterinarian. Sodium heparin tubes were chosen to collect the blood samples. Blood with sodium heparin was centrifuged at 1500× g for 20 min at 4 °C, and the obtained serum was then immediately stored at −20 °C for further analysis. Analysis included the measurement of total cholesterol, triglycerides, high-density lipoproteins, and low-density lipoproteins in the serum using an automatic biochemical analyzer (Hitachi 7020, Hitachi Co., Tokyo, Japan). The levels of glutathione peroxidase (GSH-Px, cat. no. A005-1), superoxide dismutase (SOD, cat. no. A001-3), catalase (CAT, cat. no. A007-1), peroxidase (POD, cat. no. A048-2), and malondialdehyde (MDA, cat. no. A003-1) and the total antioxidant capacity (T-AOC, cat. no. A015-1) in the serum were assessed using commercial assay kits (Nanjing Jiancheng Bioengineering Institute, Nanjing, China) according to the manufacturer's instructions.

2.5. Rumen Fluid Sampling and Analysis

Rumen fluids were collected from the experimental milk-producing buffaloes on the last day of the experimental period before 8:30 a.m. feeding and after blood sampling. The rumen fluid was collected through a stomach tube as described by Shen et al. [28]; however, the process was performed with some modifications. The specific process is detailed in our previous study [29]. The collected rumen fluid (500 mL) was divided into two parts; one part (400 mL) was used to determine ruminal pH and NH_3-N, lactic acid, acetate, propionate, butyrate, and total volatile fatty acid contents (these data are shown in Table S2, while the other part (100 mL) was aliquoted into 15 mL polypropylene tubes and frozen at −80 °C for DNA extraction later.

A pH meter (PP-50-PU, Sartorius) was used to measure the ruminal pH; the colorimetric method was used to determine the NH3-N concentration [30]; and a gas chromatograph (Agilent-7890A) equipped with a flame ionization detector and a capillary column (HP-INNOWAX 19091N-133 [30.0 m × 0.25 mm × 0.25 μm]) was used to determine the lactic acid and volatile fatty acid (VFA) concentrations, including acetate, propionate, and butyrate. Briefly, during sample injection, the injection volume was 1 μL, including the internal standard, with nitrogen used as the carrier gas, and the temperature was increased from 60 °C to 200 °C at a rate of 20 °C/min, and then held at 200 °C for 3 min. The inlet and detector temperatures were set at 250 °C and 300 °C, respectively, with a split ratio of 10:1. The flow rate of hydrogen carrier gas was 25 mL/min.

2.6. Rumen Bacterial DNA Extraction and Sequencing

Microbial community DNA was extracted from freeze-dried rumen fluid using previously described methods [31]. The V3–V4 hypervariable region of the bacterial 16S rRNA genes were targeted using the primers 515F (5′-GTTTCGGTGCCAGCMGCCGCGGTAA-3′) and 806R (5′-GCCAATGGACTACHVGGGTWTCTAAT-3′). The 16S rRNA genes were amplified following the protocols described by Magoč and Salzberg [32]. For amplification datasets, paired-end sequencing was conducted using an Illumina MiSeq platform (Shenzhen Huada Gene Technology Co., Ltd., Shenzhen, China). The high-throughput sequence read quality and Good's coverage estimates are shown in Table S3. The raw sequences generated in this study are available from NCBI under the BioProject accession number PRJNA609573 for bacterial (n = 17) community datasets.

The QIIME (v1.80) pipeline was used to filter the raw sequence data as described by Anderson and Walsh [33]. UCLUST was used to conduct operational taxonomic unit

(OTU) clustering at the 97% nucleotide similarity level. Then, the OTUs were assigned taxonomic classifications using the ribosomal database project classifier tool by comparing the bacterial 16S rRNA gene sequences against the Greengenes database. The OTUs were annotated at the phylum, class, order, family, and genus taxonomic levels. Five metrics were used to calculate community alpha diversity, including Chao1 and ACE species richness estimates, the count of unique OTUs (i.e., observed species), and the Simpson and Shannon diversity indices.

2.7. Statistical Analyses

The data for milk fatty acid profiles, blood indicators, and the relative abundances of bacteria were analyzed using SPSS (V19.0; SPSS Inc., Chicago, IL, USA). The statistical model used to analyze the data was as follows, Y =es),$_i$ + T_j + P_k + ε_{ijk}, where Y is the dependent variable; L is the overall mean; A_i is the effect of the weights, DIMs, and milk yields of the milk-producing buffaloes; T_j is the effect of the diet; P_k is the effect of the sampling period; and ε_{ijk} is the residual error term. The model included HSO supplementation as fixed effects, with the weights, DIMs, and milk yields of the milk-producing buffalo considered as a random variable. Considering that the experimental animal groups were unbalanced (6, 5, and 6), statistical differences between the data of dietary treatments were determined using a non-parametric Kruskal–Wallis test. A power test was conducted for each variable to avoid the influence of a low number of experimental units and treatments, which could increase the risk of Type I Error. $p \leq 0.05$ was considered statistically significant, and $p < 0.10$ was considered a trend. All the results are shown as means. SEM represents the standard error of the mean.

3. Results

3.1. DMI and Milk Performance

The effect of dietary HSO on DMI is shown in Table S1. Compared to the group without HSO supplementation, the DMI tended to decrease with dietary HSO supplementation (p = 0.06), with a decrease of 1.44 kg/d in both the H1 and H2 groups, respectively. Meanwhile, the milk production indicators (milk yield, energy-corrected milk (ECM), 4% fat-corrected milk, total milk solids, and non-fat milk solids), as well as the milk content parameters (milk protein, milk fat, total milk solids, non-fat milk solids, and milk lactose) and feed efficiency measures (milk yield/DMI and ECM/DMI) were not changed by dietary HSO supplementation ($p > 0.05$).

3.2. Antioxidant Capacity

As is shown in Table 3, compared to those of the group without HSO supplementation, the T-AOC (p = 0.05), CAT (p <0.01), and GSH-Px (p = 0.02) contents were significantly increased in the blood serum with dietary HSO supplementation. Specifically, the T-AOC, CAT, and GSH-Px contents increased by 0.53 U/mL and 1.16 U/mL, 0.57 U/mL and 1.15 U/mL, and 105.33 U/mL and 134.51 U/mL in the H1 and H2 groups, respectively. Meanwhile, the MDA levels significantly decreased with dietary HSO supplementation, showing decreases of 0.72 nmol/mL and 0.24 nmol/mL in the H1 and H2 groups, respectively. In addition, the POD and SOD were not affected by dietary HSO supplementation ($p > 0.05$).

Table 3. Effect of HSO supplementation on the serum antioxidant capacity in milk-producing buffaloes.

Item [1]	Parameters [2] T-AOC (U/mL)	MDA (nmol/mL)	POD (U/mL)	CAT (U/mL)	SOD (U/mL)	GSH-Px (U/mL)
H0	3.06	2.36	11.11	45.14	41.4	838.81
H1	3.59	1.64	9.28	45.71	38.52	944.14
H2	4.22	2.12	10.39	46.29	44.26	973.32
SEM [3]	0.188	0.122	0.72	0.21	1.221	18.491
p-value [4]	0.05	0.02	0.72	<0.01	0.14	0.02
Power of tests						
H0 vs. H1	0.297	0.882	0.182	0.606	0.222	0.828
H0 vs. H2	0.686	0.131	0.062	0.520	0.140	0.939
H2 vs. H3	0.297	0.318	0.084	0.438	0.372	0.167

[1] Diets supplemented with 0 g/d (H0), 100 g/d (H1), and 200 g/d (H2) of HSO. [2] T-AOC, total antioxidant capacity; MDA, malondialdehyde; POD, peroxidase; CAT, catalase; SOD, superoxide dismutase; GSH-Px, glutathione peroxidase. [3] SEM, standard error of mean. [4] $p \leq 0.05$ is considered statistically significant.

3.3. Blood Lipid Metabolites

The effect of dietary HSO on the blood lipid metabolites in the milk-producing buffaloes is shown in Table 4. Compared to that of the group without HSO supplementation, the HDL-C content tended to increase in blood ($p = 0.09$) with dietary HSO supplementation, and it increased by 0.22 mmol/L and 0.37 mmol/L in the H1 and H2 groups, respectively. However, the total cholesterol, triglyceride, and LDL-C contents were similar among the groups ($p > 0.05$).

Table 4. Effect of HSO supplementation on blood lipid metabolites in milk-producing buffaloes.

Item [1]	Parameters [2] Total Cholesterol (mmol/L)	Triglycerides (mmol/L)	HDL-C (mmol/L)	LDL-C (mmol/L)
H0	3.14	0.33	1.62	0.56
H1	2.58	0.39	1.84	0.61
H2	2.54	0.38	1.99	0.52
SEM [3]	0.363	0.013	0.073	0.039
p-value [4]	0.24	0.19	0.09	0.80
Power of tests				
H0 vs. H1	0.073	0.415	0.246	0.067
H0 vs. H2	0.078	0.475	0.551	0.095
H2 vs. H3	0.054	0.058	0.121	0.114

[1] Diets supplemented with 0 g/d (H0), 100 g/d (H1), and 200 g/d (H2) of HSO. [2] HDL-C, high-density lipoprotein cholesterol; LDL-C, low-density lipoprotein cholesterol. [3] SEM, standard error of mean. [4] $p \leq 0.05$ is considered statistically significant.

3.4. Milk Fatty Acid Composition

As shown in Table 5, compared to the group without HSO supplementation, the proportions of C18:0 ($p = 0.02$), C18:1n9t ($p = 0.02$), C18:2n6c ($p = 0.02$), C18:3n3 ($p < 0.01$), C18:2n9c ($p = 0.04$), and C20:4n6 ($p = 0.05$) were significantly increased with dietary HSO supplementation. Moreover, the proportions of C18:0, C18:1n9t, C18:2n6c, C18:3n3, C18:2n9c, and C20:4n6 increased by 2.69 g/100 g FA and 5.29 g/100 g FA, 0.83 g/100 g FA and 1.81 g/100 g FA, 0.21 g/100 g FA and 0.55 g/100 g FA, 0.03 g/100 g FA and 0.14 g/100 g FA, 0.54 g/100 g FA and 0.75 g/100 g FA, and 0.04 g/100 g FA and 0.01 g/100 g FA in the H1 and H2 groups, respectively. Similarly, the proportions of C18:1n11t ($p = 0.07$) and total CLA

(p = 0.06) tended to increase with the dietary HSO supplementation, and the proportions of C18:1n11t and total CLA increased by 7.03 g/100 g FA and 4.73 g/100 g FA and by 0.26 g/100 g FA and 0.37 g/100 g FA in the H1 and H2 groups, respectively. Meanwhile, the proportion of omega-3 (p = 0.02), omega-6 (p = 0.02), and MUFA (p = 0.02) were significantly increased with dietary HSO supplementation. Specifically, the proportions of omega-3, omega-6, and MUFA increased by 0.04 g/100 g FA and 0.17 g/100 g FA, 0.25 g/100 g FA and 0.56 g/100 g FA, and 0.81 g/100 g FA and 1.48 g/100 g FA in the H1 and H2 groups, respectively. However, the proportion of C20:5n3 was significantly increased in the H2 group (p = 0.04), with an increase of 0.01 g/100 g FA. The omega-3/omega-6 ratio showed an upward tendency in the H2 group (p = 0.06), increasing to 0.21.

Table 5. Effects of HSO supplementation on milk fatty acid profiles in milk-producing buffaloes.

Fatty Acid (g/100 g FA) [1]	Treatment [2]			SEM [3]	p-Value [4]	Power of Tests		
	H0	H1	H2			H0 vs. H1	H0 vs. H2	H2 vs. H3
C4:0	0.17	0.18	0.22	0.009	0.05	0.073	0.599	0.550
C6:0	0.71	0.79	0.75	0.034	0.57	0.234	0.065	0.067
C8:0	0.67	0.78	0.65	0.036	0.35	0.417	0.055	0.215
C10:0	1.45	1.78	1.44	0.093	0.30	0.365	0.050	0.226
C12:0	1.92	2.33	1.97	0.108	0.27	0.392	0.050	0.232
C14:0	9.18	10.69	9.46	0.440	0.16	0.319	0.055	0.180
C14:1n5	1.77	1.30	1.05	0.213	0.14	0.103	0.178	0.440
C16:0	26.74	30.71	28.27	1.010	0.28	0.345	0.081	0.167
C16:1n7	3.89	1.95	1.45	0.739	0.16	0.125	0.171	0.581
C18:0	9.85	12.54	15.14	0.814	0.02	0.368	0.746	0.338
C18:1n9t	2.35	3.18	4.16	0.269	0.02	0.261	0.834	0.555
C18:1n11t	19.80	26.83	24.55	1.296	0.07	0.699	0.274	0.118
C18:1n9c	2.30	0.85	1.44	0.472	0.04	0.153	0.085	0.910
C18:2n6c	1.89	2.10	2.44	0.088	0.02	0.440	0.627	0.356
C18:3n6	0.04	0.04	0.05	0.002	0.36	0.130	0.347	0.347
C20:0	0.18	0.12	0.05	0.049	0.06	0.066	0.131	0.222
C18:3n3	0.32	0.35	0.46	0.018	<0.01	0.218	0.989	0.919
C18:2n9c	1.03	1.57	1.78	0.135	0.04	0.462	0.711	0.091
C18:2n10t	0.25	0.24	0.24	0.023	0.93	0.052	0.052	0.050
C22:0	0.05	0.07	0.06	0.005	0.31	0.876	0.095	0.095
C20:4n6	0.13	0.17	0.14	0.008	0.05	0.797	0.073	0.364
C20:5n3	0.02	0.01	0.03	0.004	0.04	0.053	0.168	0.508
C22:6n3	0.03	0.03	0.04	0.005	0.48	0.050	0.095	0.168
Total CLA	0.64	0.90	1.01	0.070	0.06	0.404	0.661	0.088
SFA	50.93	60.00	58.01	2.111	0.08	0.483	0.199	0.068
USFA	33.82	38.62	37.84	1.462	0.24	0.260	0.142	0.055
MUFA	3.70	4.51	5.18	0.232	0.02	0.618	0.740	0.203
PUFA	30.12	34.11	32.65	1.374	0.26	0.190	0.089	0.075
SFA:USFA	1.59	1.57	1.62	0.104	0.72	0.051	0.051	0.053
Omega-3	0.36	0.40	0.53	0.023	0.01	0.377	0.918	0.803
Omega-6	2.02	2.27	2.58	0.091	0.02	0.489	0.606	0.295
Omega-3:Omega-6	0.18	0.17	0.21	0.007	0.06	0.084	0.318	0.976

[1] CLA, conjugated linoleic acid; SFA, saturated fatty acid; USFA, unsaturated fatty acid; MUFA, monounsaturated fatty acid; PUFA, polyunsaturated fatty acid; any fatty acid(s) not detected were removed from this table. [2] Diets supplemented with 0 g/d (H0), 100 g/d (H1), and 200 g/d (H2) of HSO. [3] SEM, standard error of mean. [4] $p \leq 0.05$ is considered statistically significant.

3.5. Bacterial Diversity

Five indices were used to measure the alpha diversity of the rumen bacteria, including community richness (observed species, Chao1, and ACE) and community diversity (Shannon and Simpson). It can be seen from Table 6 that among the groups, the observed species, the Chao1 estimator, the ACE estimator, the Shannon index, and the Simpson index values do not significantly differ ($p > 0.05$).

Table 6. Statistical α-diversity values for bacterial communities in milk-producing buffalo ruminal fluids after exposure to three dietary treatments.

Items [1]	Parameters [2]				
	Observed Species	**Shannon**	**Simpson**	**Chao1**	**ACE**
H0	947.5	7.62	0.98	1384.13	1429.81
H1	906.8	7.49	0.98	1268.37	1334.93
H2	947	7.63	0.98	1344.55	1412.68
SEM [3]	14.263	0.051	0.001	26.505	26.703
p-value [4]	0.24	0.52	0.38	0.22	0.28
Power of tests					
H0 vs. H1	0.140	0.119	0.050	0.346	0.210
H0 vs. H2	0.050	0.051	0.050	0.082	0.056
H2 vs. H3	0.468	0.284	0.207	0.307	0.352

[1] ACE, abundance-based coverage estimator. [2] Diets supplemented with 0 g/d (H0), 100 g/d (H1), and 200 g/d (H2) of HSO. [3] SEM, standard error of mean. [4] $p \leq 0.05$ is considered statistically significant.

3.6. Bacterial Composition

The relative abundances of bacteria are significant, and two milk-producing buffalo samples were chosen for further analysis, as shown in Table 7 and Figure 1a,b. The three most dominant bacterial phyla were Bacteroidetes, Firmicutes, and Proteobacteria across all the experimental groups, and the three most dominant bacterial genera were *Prevotella*, *Acinetobacter*, and *Butyrivibrio*. Meanwhile, the relative abundance of the genus *Paludibacter* ($p = 0.07$) tended to decrease with HSO supplementation compared to that of the group without HSO supplementation, and it decreased by 0.34% and 0.35% in the H1 and H2 groups, respectively. In contrast, with dietary HSO supplementation, the relative abundances of *Acetobacter* was significantly decreased in the H1 group, while it significantly increased in the H2 group ($p = 0.03$). Specifically, the relative abundance of *Acetobacter* decreased by 0.55% in the H1 group, while it increased by 0.73% in the H2 group.

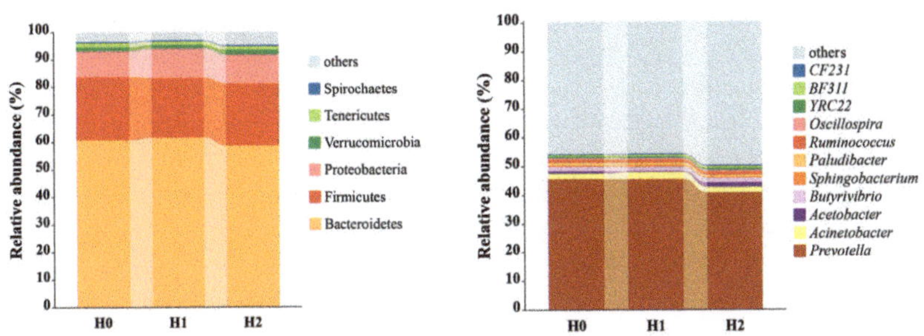

(a) (b)

Figure 1. Relative abundances of dominant bacterial populations at phylum (**a**) and genus (**b**) levels after exposure to three dietary treatments. H0, without HSO supplementation group; H1, group with 100 g/d HSO supplementation; H2, group with 200 g/d HSO supplementation.

Table 7. Relative abundances of dominant bacterial populations (≥0.10% abundance across all 17 milk-producing buffalo samples) at genus and phylum levels after exposure to three dietary treatments.

Taxonomy	Relative Abundance (%) [1]			SEM [2]	*p*-Value [3]	Power of Tests		
	H0	H1	H2			H0 vs. H1	H0 vs. H2	H2 vs. H3
Phyla								
Bacteroidetes	60.60	61.52	58.54	1.797	0.71	0.053	0.091	0.077
Firmicutes	23.04	21.80	22.70	0.779	0.84	0.074	0.054	0.068
Proteobacteria	9.44	10.65	10.47	1.416	0.80	0.057	0.072	0.050
Verrucomicrobia	1.38	1.31	1.74	0.199	0.38	0.052	0.088	0.158
Tenericutes	1.55	1.21	1.17	0.087	0.23	0.285	0.361	0.057
Spirochaetes	0.48	0.52	0.59	0.034	0.32	0.067	0.278	0.110
Genera								
Prevotella	45.54	45.42	40.77	1.938	0.54	0.050	0.154	0.157
Acinetobacter	1.95	2.65	1.95	0.402	0.91	0.078	0.050	0.074
Acetobacter	0.82	0.27	1.55	0.218	0.03	0.278	0.243	0.761
Butyrivibrio	1.48	1.34	1.51	0.102	0.76	0.075	0.052	0.083
Sphingobacterium	0.46	0.83	0.47	0.143	0.95	0.116	0.050	0.102
Paludibacter	0.94	0.60	0.59	0.066	0.07	0.481	0.613	0.051
Ruminococcus	1.09	1.07	1.08	0.059	0.62	0.055	0.050	0.050
Oscillospira	0.61	0.63	0.67	0.033	0.99	0.055	0.103	0.069
YRC22	0.61	0.62	0.77	0.050	0.51	0.051	0.252	0.149
BF311	0.39	0.51	0.55	0.042	0.55	0.199	0.405	0.059
CF231	0.39	0.45	0.53	0.045	0.53	0.112	0.197	0.085

[1] Diets supplemented with 0 g/d (H0), 100 g/d (H1), and 200 g/d (H2) of HSO. [2] SEM, standard error of mean. [3] $p \leq 0.05$ is considered statistically significant.

4. Discussion

4.1. DMI and Milk Performance

In this study, the DMI decreased with HSO supplementation. A decrease in DMI also occurred in ruminants that were fed diets supplemented with linseed oil [34], camelina oil [35], corn oil [36], rapeseed oil [37,38], or sunflower oil [39]. Our results are likely influenced by the following factors: (1) the FAs in HSO may have a direct inhibitory effect on autonomic feeding by inhibiting gastric smooth muscle contraction, thus inhibiting continuous rumen motility [40]; (2) HSO is rich in unsaturated FAs, which may cause disturbances in rumen function by damaging the cell membrane structure of microbial communities, leading to toxicity [41]; and (3) the energy in feed increases after adding HSO, triggering the animal's instinct to stop eating once it has obtained sufficient energy [42]. Perhaps as a result of the differences in the FA composition and the doses of vegetable oil in diet composition or in animal species, Baluchi lambs fed diets supplemented with grapeseed oil (2% grapeseed oil in a hay-based diet) increased their DMI, but not significantly [43]. This finding is consistent with the results of this study. Barletta et al. [44] also reported a decline in milk yield with soybean oil supplementation, though not a significant one. This could be due to fat supplementation decreasing de novo synthesis in the mammary gland [45]. Another reason may be that vegetable oil supplementation affects the rumen microbial structure, which, in turn, leads to changes in rumen fermentation [46]. However, in this study, no significant changes in milk composition, feed efficiency, or fermentation parameters (Tables S1 and S2) were found with HSO supplementation. Therefore, dietary HSO from the "longevity village" of Bama did not negatively affect the rumen fermentation or the performance of milk-producing buffaloes, but like other vegetable oils, it did decrease the DMI in the milk-producing buffaloes.

4.2. Antioxidant Capacity

Animals have an endogenous antioxidant protection system that can help scavenge reactive oxygen species (ROS) generated by lipid oxidative stress to prevent damage to cells and tissues. T-AOC is the cumulative effect of all antioxidants in sera. In the present study, dietary HSO increased the T-AOC content, and this was the highest in the H2 group. Signor et al. also reported a similar increase in the T-AOC content in sera of rats fed a diet supplemented with adlay seed oil [47]. Furthermore, in this study, the CAT and GSH-Px contents also increased with HSO addition, and it was the highest in the H2 group. GSH-Px and CAT have good synergistic antioxidant effects, and both can catalyze the decomposition of H_2O_2 into H_2O and O_2, thereby scavenging H_2O_2 [48,49]. Additionally, MDA acts as a marker of the degree of lipid peroxidation in the body and indirectly reflects the extent of cellular damage [50]. In this study, the MDA content increased with HSO addition and was the lowest in the H1 group. Similarly, rats fed a diet with adlay seed oil (1.25 mL/kg and 5 mL/kg) showed decreased MDA levels in the sera [4], while chickens at 21 days of age fed a diet with palm oil also exhibited a quadratic reduction in the serum MDA content, with the lowest observed in 2% palm oil treatments [5]. These results may be attributed to the following factors: (1) The addition of HSO increases the PUFA content, which, in turn, can lead to the increased production of free radicals, thereby increasing the possibility of spontaneous lipid oxidation [51]. For this reason, when animals are fed a diet high in PUFAs, the body induces an increase in the activity of endogenous antioxidant enzymes to cope with the resulting oxidative stress [52]. (2) The CLA in HSO is considered an antioxidant that inhibits the formation of peroxides from PUFAs [53]. (3) The HSO used in this study was obtained by cold-pressing and may contain a large amount of antioxidant compounds, such as phytosterols and lower-molecular-weight phenolic acids [54]. This indicated that similar to other vegetable oils, dietary HSO from Bama, known as the "longevity village", could notably improve the antioxidant capacity of milk-producing buffaloes, with some indicators of antioxidant activity being affected by its dosage.

4.3. Blood Lipid Metabolites

In this study, dietary HSO increased the HDL-C content in the blood of the milk-producing buffaloes. Chen et al. also reported an increase in HDL-C levels in the sera of mice fed a high-fat diet supplemented with blended oils (including Zanthoxylum bungeanum seed oil, walnut oil, camellia seed oil, and perilla seed oil) [55], with similar results observed in mice fed a diet supplemented with coconut oil [6]. HDL-C plays an important role in transporting cholesterol from extrahepatic tissues to the liver for metabolism, thereby preventing various diseases caused by excess cholesterol [56]. The observed increase in HDL-C levels in the blood may be attributed to the fact that HSO increases the proportion of saturated fatty acids in the diet, especially myristic acid [57]. Elevated levels of HDL-C in the blood can help prevent the occurrence of fatty liver in milk-producing buffaloes supplemented with HSO from the village of Bama.

4.4. Milk Fatty Acid Composition

Other studies have confirmed that the fatty acid profile of milk can be extensively modified by supplementing diets with vegetable oils [58,59]. In particular, the content of some long-chain fatty acids (≥C18:0) in milk increased with the addition of vegetable oils to diets [34,60]. In the present study, the proportions of long-chain fatty acids (e.g., C18:0, C18:1n9t, C18:2n6c, C18:3n3, C18:2n9c, and C20:4n6) increased with dietary HSO supplementation. Meanwhile, the content of other long-chain fatty acids, such as C18:1n11t, showed an upward tendency. Similarly, supplementation with many other vegetable oils, such as soybean oil (Gómez-Cortés et al., who studied goats fed alfalfa hay-based diets supplemented with 6% soybean oil) [61], sunflower oil (Marín et al., who studied lactating dairy goats fed diets supplemented with 48 g/d, and Ollier et al., who studied dairy goats fed low-forage diets supplemented with 130 g/d) [11,62], and linseed oil (Marín et al., who studied lactating dairy goats fed diets supplemented with 48 g/d or 66 g/d), increased the proportion of long-chain (≥C18:0) fatty acids in milk fat [12,13]. The reason for this is that HSO from the village of Bama is rich in long-chain unsaturated fatty acids, which results in increased absorption and the secretion of dietary or rumen-derived FAs [63].

In this study, dietary HSO increased the proportions of C18:3n3. Some studies by Mare et al. found a significant increase in the proportion of C18:3n3 in goat milk fat with high dosages of linseed oil (48 g/d or 66 g/d) fed to goats [12,13]. However, Assaf ewes fed diets supplemented with sunflower oil (2% DM, alfalfa hay-based diets) had decreased milk C18:3n3 levels [64]. The reasons for this may be attributed to diet composition, oil composition, the dosage, or the animal species. At the same time, the proportion of C18:2n6c in milk was also highest in the H2 group. ALA and LA are essential fatty acids that can only be obtained by eating food, and high doses of HSO are rich in these two polyunsaturated fatty acids, as shown in Table 1. In the rumen, C18:0 is a saturated fatty acid formed after the hydrogenation of LA or ALA, while the biohydrogenation intermediate 18:2n9c is only formed via the rumen hydrogenation of LA, and then absorbed through the mammary glands [20]. Therefore, the contents in milk C18:0 and C18:2n9c were the highest in the H2 group. Moreover, in this study, all of the CLA was found in milk C18:2n9c and C18:2n10t, and the content showed an increasing trend after the addition of HSO. Similarly, studies using high levels of soybean oil (3% or 5% soybean oil in the DM of diets) [65,66] and sunflower oil (5.1% sunflower oil in the DM of diets) also reported similar findings [67]. This is possibly due to the isomerization of LA and ALA to CLA in the rumen via the biohydrogenation pathway caused by HSO from the "longevity village" of Bama, which is then absorbed into milk via the mammary glands [68]. Another possibly reason is that CLA is synthesized in the mammary glands of lactating ruminants using oleic acid (C18:1) as a precursor and delta 9-desaturase enzymes [20], and the addition of HSO from Bama provides more C18:1 (Table 1). There are also studies showing that the CLA in diets is directly absorbed through the mammary glands and into milk [69].

In addition, the dietary supplementation of HSO increased the proportions of omega-3 and omega-6. Vargas-Bello-P also reported that the supplementation (3% DM) of diets with soybean oil increases the n-3 and n-6 polyunsaturated fatty acid contents [65]. This is due to the biosynthesis of omega-3 and omega-6 fatty acids through an enzymatic desaturation and elongation process using ALA and LA provided by high doses of HSO from the village of Bama [70]. Both omega-3 and omega-6 are important functional fatty acids for human health; however, they must be consumed in their proper ratio (1:4–1:1) to achieve the desired results, which include improving immunity and promoting cardiovascular health. In the present study, the omega-3/omega-6 ratio increased from 0.18 in the H0 group to 0.21 in the H2 group. A similar change was also found in goats fed diets with linseed oil [12]. This is possibly due to an increased n-3/n-6 ratio, which may also help convert alpha-linolenic acid into long-chain n-3 PUFAs (e.g., DHA), albeit with reduced competition for elongase [70,71]. At the same time, this explains why the proportion of DHA was highest in the H2 group.

Considering the composition of MUFAs, these results also explained why the proportions of MUFA increased after dietary supplementation with HSO. A similar result was also reported by Bernard et al., who found that high doses of sunflower seed oil or linseed oil (130 g/d) added to diets, whether grass hay or maize silage, increased the MUFA contents in goat milk fatty acids [9]. Alfalfa hay-based diets supplemented with high doses of high oleic sunflower oil, regular sunflower oil, or linseed oil (48 g/d) showed consistent results [12].

4.5. Bacterial Diversity and Composition

Consistent with our results, Tapio et al. observed that the bacterial diversity in the rumens of dairy cows did not significantly differ with sunflower oil supplementation [72]. Similar results were also reported for lambs fed diets supplemented with linseed oil [73] and Holstein male calves fed diets supplemented with olive oil, sunflower oil, coconut oil, and other vegetable oils [21–23]. In contrast, high doses of tucumã oil (1%, v/v) supplemented in diets using the nylon bag in situ method resulted in a decrease in the Shannon index of the ruminal bacteria [74]. This variation may be attributed to the differences in diet composition, oil composition, doses, or sampling methods.

A 16S rRNA gene relative abundance $\geq$ 0.10% can be considered the threshold for identifying active core rumen bacterial populations [75]. Thus, the active core bacterial community consisted of six phyla and 11 genera in this study. Similar to our study, whether or not the diets were amended with vegetable oil, Firmicutes and Bacteroidetes have been found to be the top two dominant taxa in the rumens of Holstein dairy cows [21–23], followed by Proteobacteria and other phyla [22,23]. Accordingly, Cancino-Padilla et al., Hu et al., and Liu et al. also observed that *Prevotella* was dominant in dairy cow rumens regardless whether the dairy cows were fed diets with or without vegetable oil supplementation [21–23]. Notably, the genera *Prevotella*, *Acinetobacter*, and *Butyrivibrio* belong to the Bacteroidetes, Firmicutes, and Proteobacteria phyla, respectively. The results indicate that the addition of HSO from Bama did not notably affect the dominant rumen populations at the phylum level in the milk-producing buffaloes.

A previous study documented a significant difference in the bacterial community composition in cattle and lamb rumens after their diets were supplemented with vegetable oil compared to that without vegetable oil [73,76]. However, the bacteria exhibited extensive genetic diversity within higher-level taxonomic levels (e.g., at the phylum and family levels). Thus, to understand the functions of bacteria in rumens after dietary changes, it is necessary to evaluate compositions at a finer taxonomic resolution, such as the genus level [77]. Furthermore, it is particularly important to investigate the active core rumen bacteria that may play important roles, but exhibit lower abundances [78]. Considering the less abundant taxa in this study, the relative abundances of the genus *Paludibacter* decreased with dietary HSO supplementation. This decrease was likely due to the antibacterial effects of vegetable oils from plant seeds with high concentrations of unsaturated fatty acid [46,79]. *Paludibacter* is associated with cellulolysis and hydrogenotrophic methanogens [80]. The inhibition of methane production by long-chain unsaturated fatty acids primarily alters the metabolic use of hydrogen and carbon dioxide by acting as a receptor for hydrogen, or via direct toxic effects on rumen microorganisms [16]. In addition, the relative abundances of *Acetobacter* were highest in the H2 group. Li et al. observed that cellulase activity was efficiently synthesized extracellularly by certain strains of *Acetobacter* [81]. The role of *Acetobacter* in rumen metabolism is not well understood, although their production of acetate, a precursor substance for the synthesis of milk fat, mainly occurs due to certain organic acids (e.g., lactic acid) and ethanol [82]. Notably, Khanna et al. pointed out that acetic acid bacteria (such as *Acetobacter* species) can use glycerol [83], a product of the rumen microbial hydrolysis of exogenous lipids [84], to synthesize DHA.

5. Conclusions

This preliminary study investigated the changes in the milk production performance, the milk fatty acid profiles, the blood parameters, rumen fermentation, and the rumen bacterial communities in milk-producing buffaloes fed diets supplemented with HSO, a native vegetable oil from the "longevity village" of Bama (Guangxi, China). These data suggest that dietary supplementation with HSO improves antioxidant capacity by increasing the T-AOC, CAT, and GSH-Px levels and decreasing the MDA levels in serum, regulate blood lipid metabolism by increasing the HDL-C content, and increases the proportion of functional fatty acids in milk, including C18 saturated and unsaturated fatty acids, omega-6, and omega-3. Meanwhile, only a few less-abundant bacteria (e.g., *Acetobacter*) showed significant changes at the genus level. Although the addition of HSO decreased the DMI, it did not affect feed efficiency, the milk production performance, or rumen fermentation. Moreover, some of the observed parameters in this study were affected by the dosage of HSO, indicating that further research is needed to determine the optimal dosage for milk-producing buffaloes.

Supplementary Materials: The following supporting information can be downloaded at https://www.mdpi.com/article/10.3390/ani15040514/s1: Table S1: Effects of HSO supplementation on DMI and average milk yield and composition in dairy water buffalo; Table S2: Effects of HSO supplementation on rumen fermentation parameters in dairy water buffalo; Table S3: High-throughput sequencing quality statistics and Good's coverage for bacterial communities.

Author Contributions: Conceptualization, Q.G. and D.W.; methodology, Q.G. and D.W.; software, Q.G. and B.L.; validation, Q.G., B.L. and X.C.; formal analysis, Q.G. and Z.K.; investigation, D.W., Z.K., Q.T. and Q.Y.; resources, B.L. and G.Q.; data curation, Q.T., Q.Y., X.C. and H.D.; writing—original draft preparation, Q.G.; writing—review and editing, Q.G., G.Q. and C.Z.; visualization, Q.G.; supervision, G.Q. and C.Z.; project administration, G.Q. and C.Z.; funding acquisition, G.Q. and C.Z. All authors have read and agreed to the published version of the manuscript.

Funding: This research was funded by the Guangxi Natural Science Foundation (Grant No. 2021GXNSFAA196034), the Guangxi Science and Technology Major Program (Grant No. GUIKE AA22068099-5), and the Innovation Project of Guangxi Graduate Education (Grant No. YCBZ2023050).

Institutional Review Board Statement: The experimental procedures were performed in accordance with the protocols approved by the Institutional Animal Care and Use Committee of Guangxi University (No. GXU-2022-179).

Informed Consent Statement: Not applicable.

Data Availability Statement: The data in this article are available upon request from the corresponding author.

Acknowledgments: Thanks to all the researchers in our laboratories for their dedication and hard work.

Conflicts of Interest: The authors declare that they have no conflicts of interest.

References

1. Hua, K.Q.; Yi, S.Z.; Sheng, Q.C.; Fu, W.Y.; Xuan, Z.; Xia, S.X.; Yan, C.X.; Dan, Y.D.; Dong, J.W.; Mei, Y.H.; et al. The Study Progress on Breeding of Flax and Industrial Hemp in China. *J. Nat. Fibers* **2024**, *21*, 2389159. [CrossRef]
2. Chen, C.; Shi, H.; Mo, X.; Qiu, S. Life-Cycle Water Footprint Analysis of Bama's Biomass Fuel in Guangxi Combined with Environment and Economy Assessment. *Sci. Total Environ.* **2024**, *954*, 176345. [CrossRef] [PubMed]
3. Montserrat-de La Paz, S.; Marín-Aguilar, F.; García-Giménez, M.D.; Fernández-Arche, M.A. Hemp (*Cannabis Sativa* L.) Seed Oil: Analytical and Phytochemical Characterization of the Unsaponifiable Fraction. *J. Agric. Food Chem.* **2014**, *62*, 1105–1110. [CrossRef] [PubMed]
4. Yu, F.; Gao, J.; Zeng, Y.; Liu, C.-X. Effects of Adlay Seed Oil on Blood Lipids and Antioxidant Capacity in Hyperlipidemic Rats. *J. Sci. Food Agric.* **2011**, *91*, 1843–1848. [CrossRef]
5. Long, G.; Hao, W.; Bao, L.; Li, J.; Zhang, Y.; Li, G. Effects of Dietary Inclusion Levels of Palm Oil on Growth Performance, Antioxidative Status and Serum Cytokines of Broiler Chickens. *J. Anim. Physiol. Anim. Nutr.* **2019**, *103*, 1116–1124. [CrossRef]

. Santana, L.F.; Cordeiro, K.W.; Soares, F.L.P.; Freitas, K.C. Coconut Oil Increases HDL-c and Decreases Triglycerides in Wistar Rats. *Acta Sci-Health Sci.* **2016**, *38*, 185–190. [CrossRef]

. Ramadan, M.F.; Moersel, J.T. Screening of the Antiradical Action of Vegetable Oils. *J. Food Compos. Anal.* **2006**, *19*, 838–842. [CrossRef]

. Fotschki, B.; Opyd, P.; Juśkiewicz, J.; Wiczkowski, W.; Jurgoński, A. Comparative Effects of Dietary Hemp and Poppy Seed Oil on Lipid Metabolism and the Antioxidant Status in Lean and Obese Zucker Rats. *Molecules* **2020**, *25*, 2921. [CrossRef] [PubMed]

. Bernard, L.; Shingfield, K.J.; Rouel, J.; Ferlay, A.; Chilliard, Y. Effect of Plant Oils in the Diet on Performance and Milk Fatty Acid Composition in Goats Fed Diets Based on Grass Hay or Maize Silage. *Br. J. Nutr.* **2008**, *101*, 213–224. [CrossRef]

0. Vilela, G.; Parente, H.; Bessa, R.; Alves, S.; Cavalcante, H.; Zanine, A.d.M.; Bezerra, L.; Ferreira, D.; de Oliveira, G.; Dos Santos, F.; et al. Effect of Replacing Babassu Oil with Sunflower Oil on Nutrient Utilization, Rumen Fermentation, and Biohydrogenation Estimate in Sheep. *Small Rumin. Res.* **2024**, *237*, 107321. [CrossRef]

1. Renna, M.; Lussiana, C.; Cornale, P.; Battaglini, L.M.; Fortina, R.; Mimosi, A. Effect of Ruminally Unprotected Echium Oil on Milk Yield, Composition and Fatty Acid Profile in Mid-Lactation Goats. *J. Dairy Res.* **2016**, *83*, 28–34. [CrossRef] [PubMed]

2. Marín, A.M.; Gómez-Cortés, P.; Castro, G.G.; Juárez, M.; Alba, L.P.; Hernández, M.P.; De la Fuente, M.A. Effects of Feeding Increasing Dietary Levels of High Oleic or Regular Sunflower or Linseed Oil on Fatty Acid Profile of Goat Milk. *J. Dairy Sci.* **2012**, *95*, 1942–1955. [CrossRef] [PubMed]

3. Marín, A.M.; Gómez-Cortés, P.; Castro, A.G.; Juárez, M.; Alba, L.P.; Hernández, M.P.; De la Fuente, M.A. Animal Performance and Milk Fatty Acid Profile of Dairy Goats Fed Diets with Different Unsaturated Plant Oils. *J. Dairy Sci.* **2011**, *94*, 5359–5368. [CrossRef]

4. Cozma, A.; Andrei, S.; Pintea, A.; Miere, D.; Filip, L.; Loghin, F.; Ferlay, A. Effect of Hemp Seed Oil Supplementation on Plasma Lipid Profile, Liver Function, Milk Fatty Acid, Cholesterol, and Vitamin A Concentrations in Carpathian Goats. *Czech J. Anim. Sci.* **2015**, *60*, 289–301. [CrossRef]

5. Delavaud, C.; Fougère, H.; Bertrand-Michel, J.; Bernard, L. Milk Fat Depression and Plasma Lipids in Dairy Cows and Goats. *Animal* **2022**, *16*, 100635. [CrossRef] [PubMed]

6. Kemp, P.; Lander, D.J.; Gunstone, F.D. The Hydrogenation of Some Cis-and Trans-Octadecenoic Acids to Stearic Acid by a Rumen *Fusocillus* sp. *Br. J. Nutr.* **1984**, *52*, 165–170. [CrossRef] [PubMed]

7. Kemp, P.; Lander, D.J. Hydrogenation in Vitro of -Linolenic Acid to Stearic Acid by Mixed Cultures of Pure Strains of Rumen Bacteria. *Microbiology* **1984**, *130*, 527–533. [CrossRef]

8. Huws, S.A.; Kim, E.J.; Lee, M.R.F.; Scott, M.B.; Tweed, J.K.S.; Pinloche, E.; Wallace, R.J.; Scollan, N.D. As yet Uncultured Bacteria Phylogenetically Classified as *Prevotella*, *Lachnospiraceae Incertae Sedis* and Unclassified Bacteroidales, Clostridiales and Ruminococcaceae May Play a Predominant Role in Ruminal Biohydrogenation. *Environ. Microbiol.* **2011**, *13*, 1500–1512. [CrossRef] [PubMed]

9. Buitenhuis, B.; Lassen, J.; Noel, S.J.; Plichta, D.R.; Sørensen, P.; Difford, G.F.; Poulsen, N.A. Impact of the Rumen Microbiome on Milk Fatty Acid Composition of Holstein Cattle. *Genet. Sel. Evol.* **2019**, *51*, 23. [CrossRef]

0. Dewanckele, L.; Toral, P.G.; Vlaeminck, B.; Fievez, V. Invited Review: Role of Rumen Biohydrogenation Intermediates and Rumen Microbes in Diet-Induced Milk Fat Depression: An Update. *J. Dairy Sci.* **2020**, *103*, 7655–7681. [CrossRef]

1. Cancino-Padilla, N.; Catalán, N.; Siu-Ting, K.; Creevey, C.J.; Huws, S.A.; Romero, J.; Vargas-Bello-Pérez, E. Long-Term Effects of Dietary Supplementation with Olive Oil and Hydrogenated Vegetable Oil on the Rumen Microbiome of Dairy Cows. *Microorganisms* **2021**, *9*, 1121. [CrossRef]

2. Liu, S.; Wei, Z.; Deng, M.; Xian, Z.; Liu, D.; Liu, G.; Li, Y.; Sun, B.; Guo, Y. Effect of a High-Starch or a High-Fat Diet on the Milk Performance, Apparent Nutrient Digestibility, Hindgut Fermentation Parameters and Microbiota of Lactating Cows. *Animals* **2023**, *13*, 2508. [CrossRef] [PubMed]

3. Hu, F.; Piao, M.; Yang, C.; Diao, Q.; Tu, Y. Effects of Coconut Oil and Palm Oil on Growth, Rumen Microbiota, and Fatty Acid Profile of Suckling Calves. *Microorganisms* **2023**, *11*, 655. [CrossRef]

4. Borghese, A. (Ed.) Buffalo meat and meat industry. In *Buffalo Production and Research*; Food and Agriculture Organization of the United Nations: Rome, Italy, 2005; pp. 197–217.

5. Association of Officiating Analytical Chemists (AOAC). *Official Method of Analysis*, 18th ed.; Method 920.39, 934.01, 942.05, 976.06, 978.10, 982.30 E(a,b,c) chp. 45.3.05, and 993.14; AOAC: Washington, DC, USA, 2006.

6. Tyrrell, H.F.; Reid, J.T. Prediction of the Energy Value of Cow's Milk. *J. Dairy Sci.* **1965**, *48*, 1215–1223. [CrossRef]

7. Di Palo, R. Produzione Lattea Nella Bufala con Diete Tradizionali e con L'impiego di Acidi Grassi. Ph.D. Thesis, University of Naples, Naples, Italy, 1992.

8. Shen, J.S.; Chai, Z.; Song, L.J.; Liu, J.X.; Wu, Y.M. Insertion Depth of Oral Stomach Tubes May Affect the Fermentation Parameters of Ruminal Fluid Collected in Dairy Cows. *J. Dairy Sci.* **2012**, *95*, 5978–5984. [CrossRef]

29. Zou, C.; Gu, Q.; Zhou, X.; Xia, Z.; Muhammad, W.I.; Tang, Q.; Liang, M.; Lin, B.; Qin, G. Ruminal Microbiota Composition Associated with Ruminal Fermentation Parameters and Milk Yield in Lactating Buffalo in Guangxi, China—A Preliminary Study. *J. Anim. Physiol. Anim. Nutr.* **2019**, *103*, 1374–1379. [CrossRef]
30. Broderick, G.A.; Kang, J.H. Automated Simultaneous Determination of Ammonia and Total Amino Acids in Ruminal Fluid and in Vitro Media. *J. Dairy Sci.* **1980**, *63*, 64–75. [CrossRef]
31. Lin, B.; Henderson, G.; Zou, C.; Cox, F.; Liang, X.; Janssen, P.H.; Attwood, G.T. Characterization of the Rumen Microbial Community Composition of Buffalo Breeds Consuming Diets Typical of Dairy Production Systems in Southern China. *Anim. Feed Sci. Technol.* **2015**, *207*, 75–84. [CrossRef]
32. Magoč, T.; Salzberg, S.L. FLASH: Fast Length Adjustment of Short Reads to Improve Genome Assemblies. *Bioinformatics* **2011**, *27*, 2957–2963. [CrossRef] [PubMed]
33. Anderson, M.J.; Walsh, D.C.I. PERMANOVA, ANOSIM, and the Mantel Test in the Face of Heterogeneous Dispersions: What Null Hypothesis Are You Testing? *Ecol. Monogr.* **2013**, *83*, 557–574. [CrossRef]
34. Hassanat, F.; Benchaar, C. Corn Silage-Based Diet Supplemented with Increasing Amounts of Linseed Oil: Effects on Methane Production, Rumen Fermentation, Nutrient Digestibility, Nitrogen Utilization, and Milk Production of Dairy Cows. *J. Dairy Sci.* **2021**, *104*, 5375–5390. [CrossRef] [PubMed]
35. Zhang, X.M.; Smith, M.L.; Gruninger, R.J.; Kung Jr, L.; Vyas, D.; McGinn, S.M.; Kindermann, M.; Wang, M.; Tan, Z.L.; Beauchemin, K.A. Combined Effects of 3-Nitrooxypropanol and Canola Oil Supplementation on Methane Emissions, Rumen Fermentation and Biohydrogenation, and Total Tract Digestibility in Beef Cattle. *J. Anim. Sci.* **2021**, *99*, skab081. [CrossRef] [PubMed]
36. Martin, C.; Coppa, M.; Fougère, H.; Bougouin, A.; Baumont, R.; Eugène, M.; Bernard, L. Diets Supplemented with Corn Oil and Wheat Starch, Marine Algae, or Hydrogenated Palm Oil Modulate Methane Emissions Similarly in Dairy Goats and Cows, but Not Feeding Behavior. *Anim. Feed Sci. Technol.* **2021**, *272*, 114783. [CrossRef]
37. Ramin, M.; Chagas, J.C.; Smidt, H.; Exposito, R.G.; Krizsan, S.J. Enteric and Fecal Methane Emissions from Dairy Cows Fed Grass or Corn Silage Diets Supplemented with Rapeseed Oil. *Animals* **2021**, *11*, 1322. [CrossRef]
38. Razzaghi, A.; Leskinen, H.; Ahvenjärvi, S.; Aro, H.; Bayat, A.R. Energy Utilization and Milk Fat Responses to Rapeseed Oil When Fed to Lactating Dairy Cows Receiving Different Dietary Forage to Concentrate Ratio. *Anim. Feed Sci. Technol.* **2022**, *293*, 115454. [CrossRef]
39. Saran Netto, A.; Silva, T.H.; Martins, M.M.; Vidal, A.M.C.; Salles, M.S.; Roma Júnior, L.C.; Zanetti, M.A. Inclusion of Sunflower Oil, Organic Selenium, and Vitamin E on Milk Production and Composition, and Blood Parameters of Lactating Cows. *Animals* **2022**, *12*, 1968. [CrossRef] [PubMed]
40. Chilliard, Y. Dietary Fat and Adipose Tissue Metabolism in Ruminants, Pigs, and Rodents: A Review. *J. Dairy Sci.* **1993**, *76*, 3897–3931. [CrossRef]
41. Zheng, H.C.; Liu, J.X.; Yao, J.H.; Yuan, Q.; Ye, H.W.; Ye, J.A.; Wu, Y.M. Effects of Dietary Sources of Vegetable Oils on Performance of High-Yielding Lactating Cows and Conjugated Linoleic Acids in Milk. *J. Dairy Sci.* **2005**, *88*, 2037–2042. [CrossRef] [PubMed]
42. Raubenheimer, D.; Simpson, S.J. (Eds.) Why Not Just Eat More Protein. In *Eat like the Animals: What Nature Teaches Us about the Science of Healthy Eating*; Houghton Mifflin Harcourt Publishing Company: New York, NY, USA, 2020; pp. 59–73.
43. Sharifi, M.; Bashtani, M.; Naserian, A.A.; Farhangfar, H.; Emami, A. The Effect of Grapeseed Oil on Performance, Rumen Fermentation, Antioxidant Status and Subcutaneous Adipose Fatty Acid Profile in Lambs. *J. Anim. Physiol. Anim. Nutr.* **2018**, *102*, 157–165. [CrossRef]
44. Barletta, R.V.; Gandra, J.R.; Bettero, V.P.; Araújo, C.E.; Del Valle, T.A.; de Almeida, G.F.; de Jesus, E.F.; Mingoti, R.D.; Benevento, B.C.; de Freitas Júnior, J.E. Ruminal Biohydrogenation and Abomasal Flow of Fatty Acids in Lactating Cows: Oilseed Provides Ruminal Protection for Fatty Acids. *Anim. Feed Sci. Technol.* **2016**, *219*, 111–121. [CrossRef]
45. Chilliard, Y.; Glasser, F.; Ferlay, A.; Roy, A.; Rouel, J.; Bernard, L.; Leroux, C.; Doreau, M. Diets, Rumen Biohydrogenation and Nutritional Quality of Cow and Goat Milk Fat. *Eur. J. Lipid Sci. Tech.* **2007**, *109*, 828–855. [CrossRef]
46. Vargas, J.E.; Andrés, S.; López-Ferreras, L.; Snelling, T.J.; Yáñez-Ruíz, D.R.; García-Estrada, C.; López, S. Dietary Supplemental Plant Oils Reduce Methanogenesis from Anaerobic Microbial Fermentation in the Rumen. *Sci. Rep.* **2020**, *10*, 1613. [CrossRef] [PubMed]
47. Signor, M.H.; De Freitas Dos Santos, A.L.; De Vitt, M.G.; Nora, L.; Lago, R.V.P.; Wolschick, G.J.; Correa, N.G.; Klein, B.; Xavier, A.C.H.; Wagner, R.; et al. Grape Seed Oil in the Diet of Primiparous Jersey Cows before and after Parturition: Effects on Performance, Health, Rumen Environment, and Milk Quality. *Trop. Anim. Health Prod.* **2024**, *56*, 202. [CrossRef]
48. Michiels, C.; Raes, M.; Toussaint, O.; Remacle, J. Importance of Se-Glutathione Peroxidase, Catalase, and Cu/Zn-SOD for Cell Survival against Oxidative Stress. *Free Radic. Biol. Med.* **1994**, *17*, 235–248. [CrossRef]
49. Winston, G.W. Oxidants and Antioxidants in Aquatic Animals. *Comp. Biochem. Physiol. C* **1991**, *100*, 173–176. [CrossRef] [PubMed]
50. Peng, S.; Chen, L.; Qin, J.G.; Hou, J.; Yu, N.; Long, Z.; Ye, J.; Sun, X. Effects of Replacement of Dietary Fish Oil by Soybean Oil on Growth Performance and Liver Biochemical Composition in Juvenile Black Seabream, *Acanthopagrus schlegeli*. *Aquaculture* **2008**, *276*, 154–161. [CrossRef]

51. Prisacaru, A.E. Effect of Antioxidants on Polyunsaturated Fatty Acids–Review. *Acta Sci. Pol. Technol. Aliment.* **2016**, *15*, 121–129. [CrossRef] [PubMed]
52. Pamplona, R.; Costantini, D. Molecular and Structural Antioxidant Defenses against Oxidative Stress in Animals. *Am. J. Physiol.-Regul. Integr. Comp. Physiol.* **2011**, *301*, R843–R863. [CrossRef] [PubMed]
53. Ha, Y.L.; Grimm, N.K.; Pariza, M.W. Anticarcinogens from Fried Ground Beef: Heat-Altered Derivatives of Linoleic Acid. *Carcinogenesis* **1987**, *8*, 1881–1887. [CrossRef]
54. Calzolari, D.; Rocchetti, G.; Lucini, L.; Amaducci, S. The Variety, Terroir, and Harvest Types Affect the Yield and the Phenolic and Sterolic Profiles of Hemp Seed Oil. *Food Res. Int.* **2021**, *142*, 110212. [CrossRef] [PubMed]
55. Chen, X.; Ran, J.; Mazhar, M.; Zhu, Y.; Lin, Y.; Qin, L.; Miao, S. The Balanced Unsaturated Fatty Acid Supplement Constituted by Woody Edible Oils Improved Lipid Metabolism and Gut Microbiota in High-Fat Diet Mice. *Front. Nutr.* **2023**, *10*, 1203932. [CrossRef]
56. Lewis, G.F.; Rader, D.J. New Insights Into the Regulation of HDL Metabolism and Reverse Cholesterol Transport. *Circ. Res.* **2005**, *96*, 1221–1232. [CrossRef] [PubMed]
57. Müller, H.; Lindman, A.S.; Brantsæter, A.L.; Pedersen, J.I. The Serum LDL/HDL Cholesterol Ratio Is Influenced More Favorably by Exchanging Saturated with Unsaturated Fat than by Reducing Saturated Fat in the Diet of Women. *J. Nutr.* **2003**, *133*, 78–83. [CrossRef] [PubMed]
58. Markiewicz-Kęszycka, M.; Czyżak-Runowska, G.; Lipińska, P.; Wójtowski, J. Fatty Acid Profile of Milk—A Review. *Bull. Vet. Inst. Pulawy* **2013**, *57*, 135–139. [CrossRef]
59. Manso, T.; Bodas, R.; Vieira, C.; Mantecón, A.R.; Castro, T. Feeding Vegetable Oils to Lactating Ewes Modifies the Fatty Acid Profile of Suckling Lambs. *Animal* **2011**, *5*, 1659–1667. [CrossRef]
60. Benchaar, C.; Romero-Pérez, G.A.; Chouinard, P.Y.; Hassanat, F.; Eugene, M.; Petit, H.V.; Côrtes, C. Supplementation of Increasing Amounts of Linseed Oil to Dairy Cows Fed Total Mixed Rations: Effects on Digestion, Ruminal Fermentation Characteristics, Protozoal Populations, and Milk Fatty Acid Composition. *J. Dairy Sci.* **2012**, *95*, 4578–4590. [CrossRef] [PubMed]
61. Gómez-Cortés, P.; Toral, P.G.; Frutos, P.; Juárez, M.; De La Fuente, M.A.; Hervás, G. Effect of the Supplementation of Dairy Sheep Diet with Incremental Amounts of Sunflower Oil on Animal Performance and Milk Fatty Acid Profile. *Food Chem.* **2011**, *125*, 644–651. [CrossRef]
62. Ollier, S.; Leroux, C.; de La Foye, A.; Bernard, L.; Rouel, J.; Chilliard, Y. Whole Intact Rapeseeds or Sunflower Oil in High-Forage or High-Concentrate Diets Affects Milk Yield, Milk Composition, and Mammary Gene Expression Profile in Goats. *J. Dairy Sci.* **2009**, *92*, 5544–5560. [CrossRef]
63. Palmquist, D.L.; Beaulieu, A.D.; Barbano, D.M. Feed and Animal Factors Influencing Milk Fat Composition. *J. Dairy Sci.* **1993**, *76*, 1753–1771. [CrossRef] [PubMed]
64. Toral, P.G.; Frutos, P.; Hervás, G.; Gómez-Cortés, P.; Juárez, M.; De la Fuente, M.A. Changes in Milk Fatty Acid Profile and Animal Performance in Response to Fish Oil Supplementation, Alone or in Combination with Sunflower Oil, in Dairy Ewes. *J. Dairy Sci.* **2010**, *93*, 1604–1615. [CrossRef]
65. Vargas-Bello-Pérez, E.; Cancino-Padilla, N.; Geldsetzer-Mendoza, C.; Vyhmeister, S.; Morales, M.S.; Leskinen, H.; Romero, J.; Garnsworthy, P.C.; Ibáñez, R.A. Effect of Feeding Cows with Unsaturated Fatty Acid Sources on Milk Production, Milk Composition, Milk Fatty Acid Profile, and Physicochemical and Sensory Characteristics of Ice Cream. *Animals* **2019**, *9*, 568. [CrossRef] [PubMed]
66. Titi, H.H.; Al-Fataftah, A.R. Effect of Supplementation with Vegetable Oil on Performance of Lactating Awassi ewes, Growth of Their Lambs and on Fatty Acid Profile of Milk and Blood of Lambs. *Archiv Fur Tierzucht.* **2013**, *56*, 467–479. [CrossRef]
67. Gómez-Cortés, P.; Frutos, P.; Mantecón, A.R.; Juárez, M.; De la Fuente, M.A.; Hervás, G. Milk Production, Conjugated Linoleic Acid Content, and In Vitro Ruminal Fermentation in Response to High Levels of Soybean Oil in Dairy Ewe Diet. *J. Dairy Sci.* **2008**, *91*, 1560–1569. [CrossRef] [PubMed]
68. Griinari, J.M.; Bauman, D.E. Biosynthesis of conjugated linoleic acid and its incorporation into meat and milk in ruminants. In *Advances in Conjugated Linoleic Acid Research*; Yurawecz, M., Mossoba, M., Kramer, J., Pariza, M., Nelson, G., Eds.; AOCS Press: Champaign, IL, USA, 1999; Volume 1, pp. 180–199.
69. Acosta Balcazar, I.C.; Granados Rivera, L.D.; Salinas Chavira, J.; Estrada Drouaillet, B.; Albarrán, M.R.; Bautista Martínez, Y. Relationship between the Composition of Lipids in Forages and the Concentration of Conjugated Linoleic Acid in Cow's Milk: A Review. *Animals* **2022**, *12*, 1621. [CrossRef]
70. Ponnampalam, E.N.; Sinclair, A.J.; Holman, B.W.B. The Sources, Synthesis and Biological Actions of Omega-3 and Omega-6 Fatty Acids in Red Meat: An Overview. *Foods* **2021**, *10*, 1358. [CrossRef]
71. Sprecher, H. Metabolism of Highly Unsaturated N-3 and n-6 Fatty Acids. Bba-Mol. *Cell Biol L.* **2000**, *1486*, 219–231. [CrossRef]
72. Tapio, I.; Fischer, D.; Blasco, L.; Tapio, M.; Wallace, R.J.; Bayat, A.R.; Ventto, L.; Kahala, M.; Negussie, E.; Shingfield, K.J.; et al. Taxon Abundance, Diversity, Co-Occurrence and Network Analysis of the Ruminal Microbiota in Response to Dietary Changes in Dairy Cows. *PLoS ONE* **2017**, *12*, e0180260. [CrossRef]

73. Lyons, T.; Boland, T.M.; Storey, S.; Doyle, E. Linseed Oil Supplementation of Lambs' Diet in Early Life Leads to Persistent Changes in Rumen Microbiome Structure. *Front. Microbiol.* **2017**, *8*, 1656. [CrossRef]
74. Ramos, A.F.; Terry, S.A.; Holman, D.B.; Breves, G.; Pereira, L.G.; Silva, A.G.; Chaves, A.V. Tucumã Oil Shifted Ruminal Fermentation, Reducing Methane Production and Altering the Microbiome but Decreased Substrate Digestibility within a RUSITEC Fed a Mixed Hay–Concentrate Diet. *Front. Microbiol.* **2018**, *9*, 1647. [CrossRef] [PubMed]
75. Li, X.Z.; Park, B.K.; Shin, J.S.; Choi, S.H.; Smith, S.B.; Yan, C.G. Effects of dietary linseed oil and propionate precursors on ruminal microbial community, composition, and diversity in Yanbian yellow cattle. *PLoS ONE* **2015**, *10*, e0126473. [CrossRef] [PubMed]
76. Lillis, L.; Boots, B.; Kenny, D. The effect of dietary concentrate and soya oil inclusion on microbial diversity in the rumen of cattle. *J. Appl. Microbiol.* **2011**, *111*, 1426–1435. [CrossRef]
77. Li, F.; Guan, L.L. Metatranscriptomic profiling reveals linkages between the active rumen microbiome and feed efficiency in beef cattle. *Appl. Environ. Microbiol.* **2017**, *83*, e0061-17. [CrossRef]
78. Xue, M.Y.; Sun, H.Z.; Wu, X.H.; Guan, L.L.; Liu, J.X. Assessment of rumen bacteria in dairy cows with varied milk protein yield. *J. Dairy Sci.* **2019**, *102*, 5031–5041. [CrossRef] [PubMed]
79. Petropoulos, S.A.; Fernandes, Â.; Calhelha, R.C.; Rouphael, Y.; Petrović, J.; Soković, M.; Ferreira, I.C.F.R.; Barros, L. Antimicrobial Properties, Cytotoxic Effects, and Fatty Acids Composition of Vegetable Oils from Purslane, Linseed, Luffa, and Pumpkin Seeds. *Appl. Sci.* **2021**, *11*, 5738. [CrossRef]
80. Wang, J.; Liu, M.; Wu, Y.; Wang, L.; Liu, J.; Jiang, L.; Yu, Z. Medicinal Herbs as a Potential Strategy to Decrease Methane Production by Rumen Microbiota: A Systematic Evaluation with a Focus on Perilla Frutescens Seed Extract. *Appl. Microbiol. Biotechnol.* **2016**, *100*, 9757–9771. [CrossRef]
81. Li, G.; Wang, L.; Deng, Y.; Wei, Q. Research Progress of the Biosynthetic Strains and Pathways of Bacterial Cellulose. *J. Ind. Microbiol. Biotechnol.* **2022**, *49*, kuab071. [CrossRef]
82. Sengun, I.Y. Acetic Acid Bacteria in Food Fermentations. In *Fermented Foods, Part I: Biochemistry and Biotechnology*; Montet, D., Ray, R.C., Eds.; CRC Press: Boca Raton, FL, USA, 2016; pp. 75–96.
83. Khanna, S.; Goyal, A.; Moholkar, V.S. Microbial Conversion of Glycerol: Present Status and Future Prospects. *Crit. Rev. Biotechnol.* **2012**, *32*, 235–262. [CrossRef]
84. Garton, G.A.; Lough, A.K.; Vioque, E. Glyceride Hydrolysis and Glycerol Fermentation by Sheep Rumen Contents. *J. Gen. Microbiol.* **1961**, *25*, 215–225. [CrossRef]

animals

MDPI

Article

Influence of Dietary Forage Neutral Detergent Fiber on Ruminal Fermentation, Chewing Activity, Nutrient Digestion, and Ruminal Microbiota of Hu Sheep

Zhian Zhang [1], Fei Li [1,*], Fadi Li [1], Zongli Wang [1], Long Guo [1], Xiuxiu Weng [1], Xuchun Sun [2], Zhenhu He [3], Xianyu Meng [4], Zhaoqing Liang [5] and Xiong Li [6]

1 State Key Laboratory of Grassland Agro-Ecosystems, Key Laboratory of Grassland Livestock Industry Innovation, Ministry of Agriculture and Rural Affairs, Engineering Research Center of Grassland Industry, Ministry of Education, College of Pastoral Agriculture Science and Technology, Lanzhou University, Lanzhou 730020, China
2 Linxia Animal Husbandry Technical Extension Station, Linxia 731199, China
3 Linxia County Animal Husbandry Development Centre, Linxia 731804, China
4 Dongxiang County Animal Husbandry Development Centre, Linxia 731199, China
5 Linxia State Xiaokangcun Feed Co., Linxia 731801, China
6 Hezheng County Animal Husbandry Development Centre, Linxia 731299, China
* Correspondence: lfei@lzu.edu.cn

Simple Summary: The role of dietary forage neutral detergent fiber in maintaining rumen function is critical for ruminants, yet it remains unclear what an optimal proportion of forage neutral detergent fiber that is suitable for sheep is in the pelleted total mixed ration feeding model. This investigation aimed to explore the influence of dietary forage neutral detergent fiber on ruminal pH, fermentation characteristics, chewing activity, nutrient digestibility, and the ruminal bacterial community of Hu sheep fed a pelleted total mixed ration. The results showed that increasing dietary forage neutral detergent fiber content shifted the rumen fermentation pattern towards acetate production, reduced lactate concentration in the rumen, and stimulated both chewing and ruminating, thereby mitigating the severity of subacute rumen acidosis in sheep. Additionally, increasing dietary forage neutral detergent fiber content increased the proliferation of cellulolytic bacteria in the rumen, and it promoted fiber utilization. A ratio of forage neutral detergent fiber (12.48%) to rumen-degradable starch (15.57%) with 0.8 was found to be optimal for maintaining stable ruminal function.

Abstract: As the key components of dietary carbohydrates, ensuring a balance between forage-neutral detergent fiber (FNDF) and rumen-degradable starch (RDS) is essential for ruminant health. Eight male Hu sheep equipped with rumen cannulas were randomly divided into four groups based on dietary FNDF content: low FNDF (L-FNDF, 6.08%), middle low FNDF (ML-FNDF, 9.47%), middle high FNDF (MH-FNDF, 12.48%), and high FNDF (H-FNDF, 15.68%), while the RDS levels (15.65% of DM on average) were similar among the four groups. A replicated 4 × 4 Latin square design was employed in this study. The results indicated that mean and minimum ruminal pH increased linearly with increasing dietary FNDF content, while the duration and area of pH under 5.8 and 5.6, along with the acidosis index, reduced linearly ($p \leq 0.002$). There were no differences between the MH-FNDF group and the H-FNDF group in these indicators ($p > 0.05$). The molar proportions of acetate, butyrate, isobutyrate, and isovalerate, as well as the acetate-to-propionate ratio, increased linearly, while propionate and valerate molar proportions and lactate concentration displayed a linear decrease with increasing FNDF content in the diet ($p < 0.001$). Increasing dietary FNDF content extended ruminating and chewing time while enhancing ruminal microbial diversity, promoting the proliferation of Fibrobacterota

 https://doi.org/10.3390/ani15030314

and *Butyrivibrio* in the rumen, and improving fiber degradability ($p < 0.05$). When the dietary FNDF content exceeded 12.48%, no effects of FNDF on acetate to propionate ratio and fiber utilization were observed ($p > 0.05$). The results suggest that augmenting FNDF content in the PTMR can reshape ruminal fermentation towards acetate production and promote rumination to enhance ruminal pH, thereby alleviating the risk of ruminal acidosis. When the RDS content in the PTMR was 15.57%, an FNDF content of 12.48% was optimal for maintaining stable ruminal function in sheep, and the recommended ratio of FNDF to RDS was 0.8.

Keywords: chewing activity; forage neutral detergent fiber; fermentation; ruminal bacterial community; pelleted total mixed ration; sheep

1. Introduction

In regions with limited access to high-quality forages, such as China, the prevalent dietary strategy for high-yielding ruminants involves feeding an excess of rapidly fermentable carbohydrates to meet energy requirements [1,2]. However, due to deficient physically effective fiber in the diet, this approach leads to an unbalanced diet structure and induces subacute rumen acidosis (SARA) in ruminants, which thereby adversely affects ruminal function and fiber digestion [3]. Subacute ruminal acidosis is generally believed to occur when the duration of ruminal pH under 5.6 is above 3 h/day. Physically effective neutral detergent fiber (peNDF) or forage neutral detergent fiber (FNDF) and rumen degradable starch (RDS) are pivotal dietary constituents for ruminants [4,5]. Maintaining a balanced ratio of peNDF to RDS or FNDF to RDS in the diet can enhance feed efficiency, mitigate SARA risk, uphold optimal ruminal metabolism, and ensure the overall health of ruminants.

Given that the dietary carbohydrate balance predominantly relies on the levels of peNDF or FNDF and RDS, it is essential to evaluate the effects of their respective dietary contributions on rumen health in ruminants when controlling for another variable is consistent. In our earlier study [5], maintaining a constant proportion of FNDF at 9.6% across four experimental groups, we found that increasing the proportion of RDS content (9.70%, 12.00%, 16.00%, 23.66% of DM) in the diets of sheep amplified the risk of SARA, altered fermentation dynamics to favor propionate generation, and diminished the utilization of dietary fibers. In addition, it also indicated that the RDS content of 12.00% was optimal to maintain rumen health when FNDF content was 9.6% [5]. On the contrary, previous studies reported that increasing peNDF content proved advantageous in boosting ruminal pH by enhancing chewing and ruminating to stimulate salivation in lactating dairy cows fed a high-grain diet [1]. Salivary buffering serves as a crucial mechanism for acid neutralization, contributing to about 37% of the total ruminal buffering capacity [6]. In some cases, cattle receiving diets rich in concentrate deliberately opt for longer forage particles to stimulate salivation, as a behavioral adaptation believed to respond to low ruminal pH [7,8]. But it is well established that the application of pelleted total mixed rations (PTMR) in the feeding regimen of fattening ruminants effectively curtails their sorting behavior. This effect arises from the uniform chopping of roughage in PTMR to a similar length and making pellets [9], which restricts the ability of animals to select longer particle sizes to mitigate the reduction in ruminal pH. Consequently, ensuring sufficient FNDF content in the PTMR would avert SARA occurrences and promote stable ruminal function. Nevertheless, data on balancing dietary carbohydrate profile by regulating dietary FNDF content in ruminants under the PTMR feeding mode remains scant; further investigation is required.

Therefore, we hypothesized that increasing the dietary FNDF content would enhance ruminating behavior, improve ruminal fermentation, and subsequently elevate ruminal pH, thereby mitigating the risk of ruminal acidosis in finishing ruminants consuming a PTMR. This investigation was designed to determine the impacts of varying FNDF content in PTMR on ruminal pH, fermentation characteristics, nutrient digestibility, chewing activity, and the ruminal bacterial community in Hu sheep. Furthermore, we sought to identify the optimal FNDF to RDS ratio that is suitable for sheep in the PTMR feeding model.

2. Materials and Methods

This experiment was conducted according to established guidelines of the Animal Care Committee of Lanzhou University (approval number: LZU 201803002).

2.1. Animals and Experimental Design

The present investigation was a continuation of our earlier study, focusing on the modulation of dietary FNDF in the reduction of ruminal pH in sheep consuming a PTMR characterized by high levels of RDS. A comprehensive description of the study design has previously been delineated [5]. In brief, we selected eight male Hu sheep (body weight of 59.18 ± 4.94 kg, mean ± SD), each equipped with a rumen cannula, and employed a replicated 4 × 4 Latin square design for this investigation. The PTMR for all sheep was featured with similar CP content (average: 13.92% of DM), RDS levels (average: 15.65% of DM), and metabolizable energy (ME) content (average: 10.04 MJ/kg DM). However, the content of dietary FNDF differed among the four treatments, including the low FNDF group (L-FNDF, 6.08%), the middle low FNDF group (ML-FNDF, 9.47%), the middle high FNDF group (MH-FNDF, 12.48%), and the high FNDF group (H-FNDF, 15.68%), which was achieved by varying roughage corn stover proportion (as shown in Table 1). Each experiment period lasted for 31 d, comprising a 5-d washout period (where all sheep received a washout diet to negate carryover effects, as shown in Supplementary Materials "Section S1.1"), a 15-d adaptation period (during which sheep were fed their respective designated diets), and an 11-d sampling period. The sheep were accommodated in individual cages (size: 140 × 75 × 150 cm). Diets were provided twice daily at 08:00 and 18:00, with clean water available ad libitum throughout the trial. The amount of feed offered and orts for each animal were recorded to determine the average daily feed intake (DMI).

Table 1. Composition and nutrient composition of the experimental diet.

Items	Diet [1]			
	L-FNDF	ML-FNDF	MH-FNDF	H-FNDF
Corn stover	9.50	14.80	19.50	24.50
Corn bran	24.00	16.00	13.80	5.30
Corn	35.00	37.00	39.00	41.00
Molasses	4.00	4.00	4.00	4.00
Corn germ meal	11.80	11.50	4.10	4.75
Cottonseed meal	7.00	7.00	8.00	8.50
Soybean meal	6.50	7.50	9.40	9.75
Limestone	1.20	1.20	1.20	1.20
NaCl	0.50	0.50	0.50	0.50
Premix [2]	0.50	0.50	0.50	0.50
Total	100.00	100.00	100.00	100.00
Nutrient content, % of DM				
CP	13.92	13.91	13.89	13.96
Ca	0.79	0.81	0.84	0.86
P	0.28	0.29	0.29	0.31

Table 1. *Cont.*

Items	Diet [1]			
	L-FNDF	ML-FNDF	MH-FNDF	H-FNDF
Starch	25.66	25.66	25.77	25.81
NDF	33.96	32.04	31.13	29.17
ADF	13.94	14.51	15.26	15.79
ME, MJ/kg	10.13	10.08	9.99	9.95
RDS	15.70	15.70	15.57	15.61
FNDF	6.08	9.47	12.48	15.68
FNDF/RDS	0.4	0.6	0.8	1.0

[1] L-FNDF, low forage neutral detergent fiber; ML-FNDF, middle low forage neutral detergent fiber; MH-FNDF, middle high forage neutral detergent fiber; H-FNDF, high forage neutral detergent fiber; CP, crude protein; NDF, neutral detergent fiber; ADF, acid detergent fiber; RDS, rumen degradable starch; FNDF, forage neutral detergent fiber. [2] The premix provided the following per kg of diets: Fe 25 mg; Mn 40 mg; Zn 40 mg; Cu 8 mg; I 0.3 mg; Se 0.2 mg; Co 0.1 mg; VA 1500 IU; VD3 280 IU; VE 20 IU.

2.2. Rumen Sampling and Analysis

During each experimental period, rumen digesta (approximately 100 mL) was collected through rumen cannula every two hours from 08:00 to 18:00 on d 21 and d 22 and passed through four layers of cheesecloth for filtration. Five milliliters of filtrate were blended with 1 milliliter of 25% HPO_3 to prevent volatilization and stored at −20 °C for volatile fatty acid (VFA) measurement, following the method described by Zhang et al. [5]. Simultaneously, another 15 mL of filtrate of ruminal fluid was collected and stored at −20 °C for the lactate determination, which was performed in accordance with the instructions of the lactate assay kit (A019-2-1, Nanjing Jiancheng Bioengineering Institute, Nanjing, China). In addition, 15 mL of filtrate collected at 0 h before morning feeding was preserved at −80 °C for subsequent bacterial DNA extraction.

2.3. Ruminal pH and Chewing Activity

Dynamic fluctuations in ruminal pH were monitored in each sheep from day 23 (08:00 h) to d 25 (08:00 h) of each trial period using a ruminal pH monitoring system, as described by Ma et al. [10]. This system captured pH values at 30 s intervals over a span of 48 h. Ruminal pH data were summarized by calculating mean, maximum, and minimum pH, as well as the duration and area with ruminal pH under 5.8 and 5.6. The acidosis index was calculated as the ratio of the area with pH below 5.8 to dry matter (DM) consumption [11]. Concurrently, chewing behavior of individual sheep was continuously monitored for 48 h using video surveillance from day 23 (08:00 h) to d 25 (08:00 h) of each trial period. Through the viewing of the recorded video by a trained person, the time spent on eating and ruminating in 5 min intervals was recorded, assuming that any activity lasted at least 5 min [12,13]. The total chewing time was determined by adding the durations of eating and ruminating, with data reported on a daily (24 h) basis [1].

2.4. In Situ Degradability of Alfalfa

The nylon bag technique was used to determine the degradability of DM, NDF, and ADF of alfalfa hay on day 26 of each period. This methodology was used to assess the influence of FNDF content in the diet on fiber degradation. Initially, alfalfa hay samples underwent grinding and were sieved through a 2 mm sieve. Subsequently, three grams of the ground samples were accurately weighed and placed in nylon bags with a pore size of 50 μm, with three replicates of alfalfa hay samples for each sheep. Following sealing, bags were incubated in the ventral sac of the rumen for 24 h. After incubation, all nylon bags were removed, promptly washed six times with cold water to halt fermentation, and dried at 65 °C for 48 h. The degradability of DM, NDF, and ADF was subsequently analyzed following the methodology outlined by Li et al. [14].

2.5. Nutrient Digestion

The measurement of nutrient digestibility was performed on 27 to 31 days of each period, during which feed samples were collected daily before morning feeding. Fecal samples were obtained from the rectum of each sheep at 8:00 h, 16:00 h, and 24:00 h daily and were promptly stored at −20 °C. At the end of each period, the collected feed and fecal samples were mixed according to each individual. For each sheep, 150 g of fecal samples were transferred into a brown vial and mixed with 15 mL of 10% H_2SO_4 for subsequent determination of CP. Moreover, an extra 200 g of feces were continuously dried at 65 °C for 72 h. Both feed and dried fecal samples were ground and passed through a 1 mm sieve for subsequent analysis. The samples were assessed for DM, organic matter (OM), and CP according to AOAC protocols [15]. Neutral detergent fiber (NDF) and acid detergent fiber (ADF) were determined with the aid of heat-stable α-amylase and sodium sulfite, including residual ash in the measurements. Acid-insoluble ash was assayed using the method described by Allen [16]. The calculation of nutrient digestibility was referred to Zhang et al. [5].

2.6. Rumen Microbial DNA Extraction and Sequencing

The total bacterial DNA extraction from the rumen fluid samples was conducted following the YM + SB procedure outlined by Ma et al. [17]. Following the evaluation of the quality and quantity of the DNA, the V3-V4 regions of the 16S rRNA gene were amplified with primers 341F and 806R [18]. The PCR products were purified utilizing the Qiagen Gel Extraction Kit (Qiagen, Hilden, Germany) and sequenced with the Illumina NovaSeq platform. More details are provided in Supplementary Material (Section S1.2).

2.7. Statistical Analysis

The results of ruminal pH, chewing activity, fiber degradability, and nutrient digestibility were analyzed using the MIXED program of SAS 9.4 (SAS Institute Inc., Cary, NC, USA). The fixed effects were treatment_i (i = 1–4) and period_j (j = 1–4), and the random effects were block_k (k = 1–2) and sheep_l (l = 1–8). In the assessment of ruminal fermentation parameters, sampling times were incorporated as repeated measurements. Microbial composition data were analyzed via a generalized linear model with gamma distribution as the response variable in SPSS 21.0 software (SPSS Inc., Chicago, IL, USA). Detailed procedures were provided in Supplementary material (Section S1.3). The results were presented as means accompanied by their standard errors. Statistical significance was defined as $p \leq 0.05$, with a trend towards significance indicated for $0.05 < p < 0.10$.

3. Results

3.1. Ruminal pH

The area of pH < 5.80 ($p = 0.096$) and the acidosis index ($p = 0.081$) exhibited a quadratic response trend as dietary FNDF content increased (Table 2), while the area of pH < 5.60 showed a quadratic response ($p = 0.042$). No effect of dietary FNDF content on DMI was observed ($p = 0.992$), but the FNDF intake displayed a linear increase with augmenting FNDF content in the diet ($p < 0.001$). The ruminal mean pH ($p = 0.002$) and minimum pH ($p < 0.001$) increased linearly with increasing dietary FNDF content. In contrast, the duration and area of ruminal pH < 5.8 and 5.6, and the acidosis index demonstrated a linear decrease ($p < 0.05$). The ruminal maximum pH remained similar across the four groups ($p > 0.05$).

Table 2. Ruminal pH of Hu sheep among the four groups.

Items [1]	L-FNDF	ML-FNDF	MH-FNDF	H-FNDF	SEM	p Value		
						Treatment	Linear	Quadratic
DMI, kg/d	1.62	1.61	1.63	1.59	0.049	0.992	0.914	0.986
FNDFI, kg/d	0.10 ^{d}	0.15 ^{c}	0.20 ^{b}	0.25 ^{a}	0.008	<0.001	<0.001	0.654
Mean pH	5.74 ^{c}	5.81 ^{bc}	5.93 ^{ab}	5.95 ^{a}	0.042	0.014	0.002	0.596
Maximum pH	6.89	6.86	6.84	6.85	0.045	0.908	0.552	0.696
Minimum pH	4.86 ^{b}	4.96 ^{b}	5.18 ^{a}	5.13 ^{a}	0.046	0.001	<0.001	0.123
Duration pH < 5.8, min/d	815 ^{a}	809 ^{a}	599 ^{ab}	503 ^{b}	75.2	0.059	0.013	0.815
Duration pH < 5.6, min/d	645 ^{a}	523 ^{ab}	278 ^{bc}	260 ^{c}	70.8	0.009	0.002	0.281
Area pH < 5.8, pH × min/d	371 ^{a}	272 ^{a}	136 ^{b}	133 ^{b}	33.8	0.001	<0.001	0.096
Area pH < 5.8, pH × min/d	226.64 ^{a}	141.01 ^{b}	54.37 ^{c}	74.35 ^{bc}	23.151	0.001	<0.001	0.042
Acidosis index	235.94 ^{a}	167.44 ^{b}	84.83 ^{c}	97.59 ^{c}	21.296	0.001	<0.001	0.081

[1] L-FNDF, low forage neutral detergent fiber; ML-FNDF, middle low forage neutral detergent fiber; MH-FNDF, middle high forage neutral detergent fiber; H-FNDF, high forage neutral detergent fiber; DMI, dry matter intake; FNDFI, forage neutral detergent fiber intake. $^{a–d}$ Values within a row with uncommon letters differ among the four groups ($p < 0.05$).

3.2. Volatile Fatty Acids in the Rumen

As shown in Table 3, increasing dietary FNDF content resulted in a quadratic response for the molar ratios of acetate ($p = 0.015$), propionate ($p = 0.009$), valerate ($p = 0.007$), and acetate to propionate ratio ($p = 0.018$), and a trend for a quadratic response for the molar ratio of butyrate ($p = 0.072$) in the rumen. With an increase in dietary FNDF, there was a linear increase in the molar proportions of acetate ($p < 0.001$), isobutyrate ($p < 0.001$), butyrate ($p < 0.001$), and isovalerate ($p < 0.001$), as well as the acetate to propionate ratio ($p < 0.001$) in the rumen. Conversely, a linear decrease was observed for propionate ($p < 0.001$) valerate molar proportions ($p < 0.001$) and lactate concentration ($p = 0.002$). The total volatile fatty acid (TVFA) concentration in the rumen remained unaffected by dietary FNDF content ($p > 0.05$).

Table 3. Ruminal fermentation of Hu sheep among the four groups.

Items [1]	L-FNDF	ML-FNDF	MH-FNDF	H-FNDF	SEM	p Value				
						Treatment	Time	T × T [2]	Linear	Quadrati
TVFA, mmol/L	68.48	69.94	69.34	72.66	2.158	0.552	<0.001	0.634	0.217	0.667
Acetate, %	51.66 ^{c}	54.36 ^{b}	57.29 ^{a}	56.64 ^{a}	0.679	<0.001	<0.001	0.754	<0.001	0.015
Propionate, %	32.71 ^{a}	29.19 ^{b}	23.39 ^{c}	24.40 ^{c}	0.864	<0.001	0.001	0.924	<0.001	0.009
Isobutyrate, %	0.63 ^{b}	0.70 ^{b}	0.68 ^{b}	0.85 ^{a}	0.031	<0.001	<0.001	0.615	<0.001	0.112
Butyrate, %	11.66 ^{c}	12.72 ^{b}	14.93 ^{a}	14.62 ^{a}	0.378	<0.001	0.019	0.747	<0.001	0.072
Isovalerate, %	0.95 ^{c}	1.23 ^{b}	1.30 ^{b}	1.64 ^{a}	0.067	<0.001	<0.001	0.945	<0.001	0.721
Valerate, %	2.39 ^{a}	1.87 ^{b}	1.67 ^{b}	1.77 ^{b}	0.111	<0.001	0.182	0.900	<0.001	0.007
Acetate: propionate	1.86 ^{c}	2.16 ^{b}	2.68 ^{a}	2.55 ^{a}	0.090	<0.001	<0.001	0.750	<0.001	0.018
Lactate, mmol/L	1.47 ^{a}	0.86 ^{b}	0.76 ^{b}	0.65 ^{b}	0.15	0.010	-	-	0.002	0.128

[1] L-FNDF, low forage neutral detergent fiber; ML-FNDF, middle low forage neutral detergent fiber; MH-FNDF, middle high forage neutral detergent fiber; H-FNDF, high forage neutral detergent fiber; TVFA, total volatile fatty acid. [2] T×T, the interaction of treatment and time. $^{a–c}$ Values within a row with uncommon letters differ among the four groups ($p < 0.05$).

3.3. Chewing Activity

There was no observed quadratic response in the chewing activity of sheep as the dietary FNDF content increased ($p > 0.05$; Table 4). Nevertheless, a linear increase was noted in the time spent on ruminating per day ($p = 0.041$), along with a trend towards a linear increase in the total chewing time per day ($p = 0.052$). Moreover, both ruminating time ($p = 0.019$) and total chewing time ($p = 0.027$) spent on per kilogram of dry matter intake exhibited a linear increase with the increasing dietary FNDF content. Alterations in dietary FNDF content did not impact the eating time per day or each kilogram of dry matter intake in sheep ($p > 0.05$).

Table 4. Chewing activity of Hu sheep among the four groups.

Items [1]	L-FNDF	ML-FNDF	MH-FNDF	H-FNDF	SEM	*p* Value		
						Treatment	Linear	Quadratic
Eating time, min/d	86.56	103.13	96.56	97.19	9.286	0.660	0.554	0.408
Ruminating time, min/d	180	203	234	253	24.5	0.209	0.041	0.942
Total chewing time, min/d	267	306	331	351	28.6	0.244	0.052	0.738
Eating time, min/kg DM	54.90	67.41	60.46	62.64	6.300	0.582	0.574	0.428
Ruminating time, min/kg DM	109	127	141	161	14.0	0.116	0.019	0.980
Total chewing time, min/kg DM	164	195	202	223	16.5	0.138	0.027	0.776

[1] L-FNDF, low forage neutral detergent fiber; ML-FNDF, middle low forage neutral detergent fiber; MH-FNDF, middle high forage neutral detergent fiber; H-FNDF, high forage neutral detergent fiber.

3.4. Nutrient Digestibility and Fiber Degradability

Dietary FNDF content had an effect on the apparent digestibility of DM, organic matter (OM), CP, NDF, and ADF ($p < 0.05$; Table 5). However, there was no difference in starch digestibility among the groups ($p = 0.407$). Specifically, the apparent digestibility of DM in the ML-FNDF group was higher than that in the L-FNDF group ($p = 0.002$), but not different from that in the MH-FNDF and H-FNDF groups ($p > 0.05$). The apparent digestibility of CP and OM in the L-FNDF group was lower than that in the MH-FNDF and H-FNDF groups ($p < 0.05$). The apparent digestibility of NDF and ADF in the ML-FNDF group was higher than that in the MH-FNDF and H-FNDF groups ($p < 0.05$). A quadratic response trend in the apparent digestibility of CP emerged with the increasing dietary FNDF content ($p = 0.061$). As the dietary FNDF content increased, the apparent digestibility of DM ($p = 0.022$), OM ($p = 0.002$), and CP ($p < 0.001$) increased linearly, while the apparent digestibility of NDF ($p = 0.029$) and ADF ($p = 0.028$) exhibited a linear decrease. As can be seen in Table 6, dietary FNDF content had an effect on DM degradability of alfalfa hay ($p = 0.022$), with a tendency to affect the degradability of NDF ($p = 0.099$) and ADF ($p = 0.057$). The DM degradability was higher in the MH-FNDF group than in the ML-FNDF and L-FNDF groups ($p < 0.05$), but no distinction was observed when compared to the H-FNDF group ($p > 0.05$). Furthermore, the degradability of DM ($p = 0.008$), NDF ($p = 0.035$), and ADF ($p = 0.027$) increased linearly with increasing dietary FNDF content.

Table 5. Nutrient apparent digestibility of Hu sheep among the four groups.

Digestibility [1]	L-FNDF	ML-FNDF	MH-FNDF	H-FNDF	SEM	*p* Value		
						Treatment	Linear	Quadratic
DM, %	71.59 ^{b}	73.40 ^{a}	73.22 ^{ab}	73.12 ^{ab}	0.712	0.002	0.022	0.535
OM, %	73.46 ^{b}	75.46 ^{a}	75.52 ^{a}	75.81 ^{a}	0.681	0.001	0.002	0.403
CP, %	80.56 ^{b}	81.64 ^{ab}	81.99 ^{a}	82.67 ^{a}	0.548	0.001	<0.001	0.061
NDF, %	52.07 ^{ab}	54.59 ^{a}	50.00 ^{b}	49.23 ^{b}	1.444	0.000	0.029	0.742
ADF, %	47.38 ^{ab}	49.73 ^{a}	46.22 ^{b}	45.08 ^{b}	1.289	0.000	0.028	0.398
Starch, %	98.63	98.74	98.64	98.65	0.051	0.407	0.904	0.319

[1] L-FNDF, low forage neutral detergent fiber; ML-FNDF, middle low forage neutral detergent fiber; MH-FNDF, middle high forage neutral detergent fiber; H-FNDF, high forage neutral detergent fiber; DM, dry matter; OM, organic matter; CP, crude protein; NDF, neutral detergent fiber; ADF, acid detergent fiber. ^{a,b} Values within a row with uncommon letters differ among the four groups ($p < 0.05$).

Table 6. The 24 h in situ degradability of alfalfa hay.

Degradability [1]	L-FNDF	ML-FNDF	MH-FNDF	H-FNDF	SEM	*p* Value		
						Treatment	Linear	Quadratic
DM, %	50.16 [c]	52.25 [bc]	56.30 [a]	54.67 [ab]	1.243	0.022	0.008	0.160
NDF, %	45.65	45.23	47.91	47.84	0.949	0.099	0.035	0.858
ADF, %	38.82	37.88	41.18	41.26	1.034	0.057	0.027	0.621

[1] L-FNDF, low forage neutral detergent fiber; ML-FNDF, middle low forage neutral detergent fiber; MH-FNDF, middle high forage neutral detergent fiber; H-FNDF, high forage neutral detergent fiber; DM, dry matter; NDF, neutral detergent fiber; ADF, acid detergent fiber. [a–c] Values within a row with uncommon letters differ among the four groups ($p < 0.05$).

3.5. *Rumen Bacterial Communities*

The ZOTUs (p = 0.098), Chao1 (p = 0.061), and Ace index (p = 0.083) exhibited a quadratic response trend with increasing dietary FNDF content (Table 7). The Shannon index of the ruminal microbial community displayed a linear increase with increasing dietary FNDF content (p = 0.028), while the Simpson index displayed a trend towards linear decrease (p = 0.056).

Table 7. Microbial alpha-diversity in rumen of Hu sheep among the four groups.

Items [1]	L-FNDF	ML-FNDF	MH-FNDF	H-FNDF	SEM	*p* Value		
						Treatment	Linear	Quadratic
ZOTUs	2143	2388	2384	2331	82.9	0.178	0.156	0.098
Chao1	2727	2970	2968	2872	82.2	0.176	0.261	0.061
Ace	2730	2967	2958	2889	80.9	0.193	0.217	0.083
Shannon	5.22	5.36	5.60	5.62	0.128	0.136	0.028	0.639
Simpson	0.03	0.03	0.02	0.02	0.004	0.237	0.056	0.923

[1] L-FNDF, low forage neutral detergent fiber; ML-FNDF, middle low forage neutral detergent fiber; MH-FNDF, middle high forage neutral detergent fiber; H-FNDF, high forage neutral detergent fiber.

The predominant phyla present in the rumen ecosystem were identified as Bacteroidetes and Firmicutes (Figure 1A). As shown in Figure 1B, increasing the dietary FNDF content resulted in a quadratic response in the relative abundance of Bacteroidetes ($p < 0.001$), Fibrobacteres (p = 0.004), and Proteobacteria ($p < 0.001$). Additionally, the relative abundance of Actinobacteria ($p < 0.001$), Bacteroidetes (p = 0.001), Fibrobacteres ($p < 0.001$), and Proteobacteria ($p < 0.001$) in the rumen increased linearly with increasing dietary FNDF content, while the relative abundance of Synergistetes decreased linearly (p = 0.009).

At the genus level (Table 8), when dietary FNDF content increased, the relative abundances of Bacteria_unclassified (p = 0.027), Erysipelotrichaceae_unclassified (p = 0.001), Bacteroides ($p < 0.001$), S24_7_unclassified ($p < 0.001$), Butyrivibrio (p = 0.004), and RF16_unclassified (p = 0.002) in the rumen had a quadratic response. The relative abundances of Lachnospiraceae_unclassified (p = 0.033), Prevotellaceae_unclassified (p = 0.037), Coriobacterineae (p = 0.026), RF16_unclassified ($p < 0.001$), and Syntrophococcus (p = 0.049) in the rumen decreased linearly as the dietary FNDF content increased, while the relative abundances of Prevotella (p = 0.072) and Firmicutes_unclassified (p = 0.080) tended to decrease linearly. The relative abundances of Incertae_Sedis ($p < 0.001$), Succinivibrio ($p < 0.001$), Erysipelotrichaceae_unclassified (p = 0.006), Bacteroides ($p < 0.001$), and Butyrivibrio ($p < 0.001$) increased linearly with increasing FNDF content in the diet.

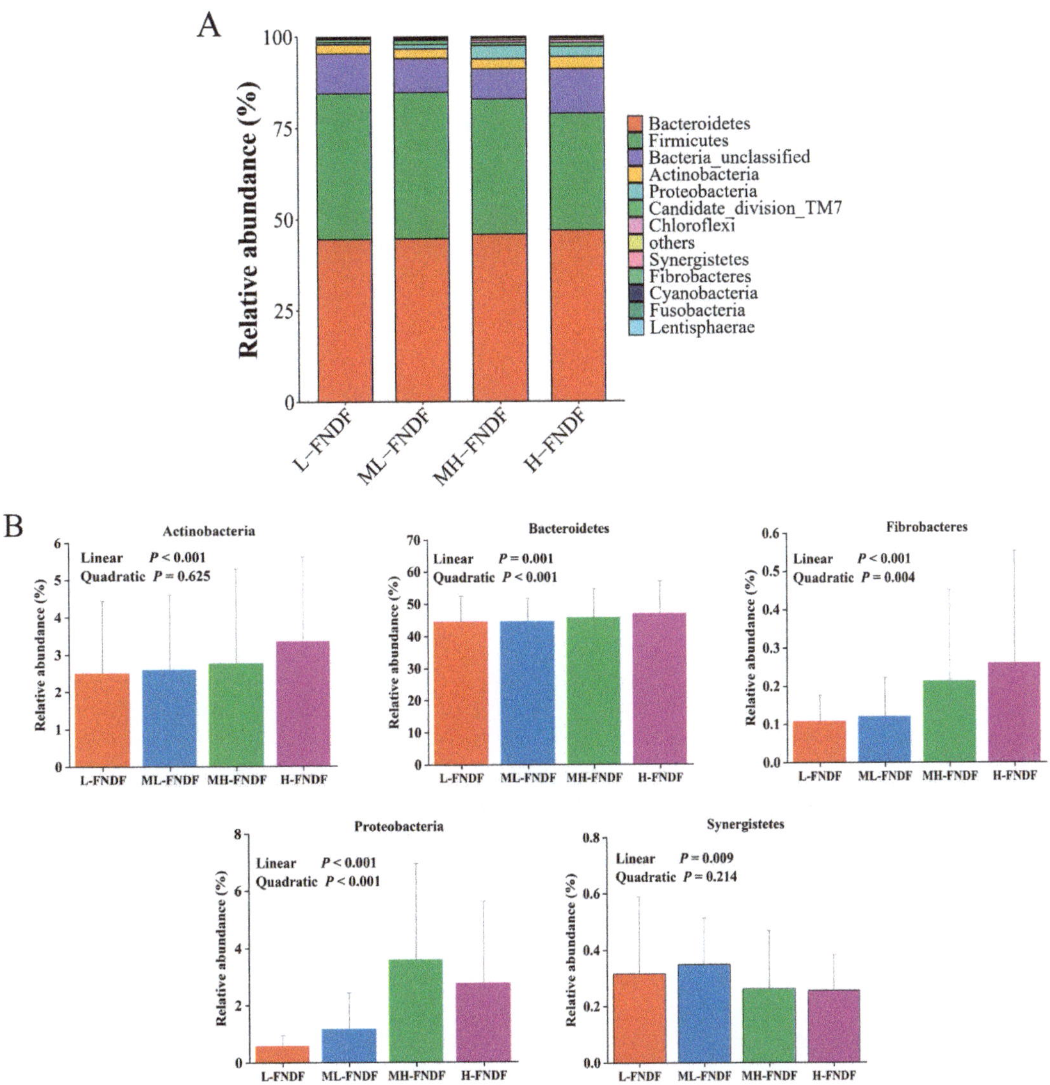

Figure 1. Ruminal microbiota composition at phylum level (**A**) and differential bacteria (**B**) of Hu sheep among the treatments. L-FNDF, low forage neutral detergent fiber; ML-FNDF, middle low forage neutral detergent fiber; MH-FNDF, middle high forage neutral detergent fiber; H-FNDF, high forage neutral detergent fiber.

Table 8. The relative abundance (top 25) of ruminal microbiota at genus level of Hu sheep among the four groups.

Items [1]	L-FNDF	ML-FNDF	MH-FNDF	H-FNDF	SEM	*p* Value	
						Linear	Quadratic
Prevotella	25.97	25.55	21.39	22.56	1.761	0.072	0.569
Lachnospiraceae_unclassified	12.68	15.19	11.74	9.85	1.208	0.033	0.156
Bacteria_unclassified	10.90	9.27	8.25	12.15	1.263	0.548	0.027
Ruminococcaceae_unclassified	5.98	5.72	6.24	6.17	0.402	0.521	0.751

Table 8. *Cont.*

Items [1]	L-FNDF	ML-FNDF	MH-FNDF	H-FNDF	SEM	p Value	
						Linear	Quadratic
Prevotellaceae_unclassified	8.20	7.79	6.11	6.69	0.720	0.037	0.146
uncultured	5.71	7.09	7.07	6.67	0.496	0.519	0.051
RC9	4.75	4.96	5.33	5.69	0.489	0.284	0.740
Bacteroidales_unclassified	3.69	2.77	4.21	3.95	0.320	0.144	0.287
Bacteroidetes_unclassified	3.62	3.16	3.12	1.94	0.365	0.320	0.897
Clostridiales_unclassified	2.51	1.96	3.04	2.07	0.180	0.954	0.290
BS11_unclassified	2.22	1.63	2.81	2.67	0.344	0.656	0.444
Incertae_Sedis	1.23	1.01	1.97	1.59	0.115	<0.001	0.694
Ruminococcus	1.20	1.79	1.99	2.25	0.249	0.473	0.215
Coriobacterineae	1.35	0.89	1.30	0.82	0.121	0.026	0.692
Bifidobacteriaceae	0.96	1.16	0.68	1.40	0.159	0.367	0.094
Succinivibrio	0.41	0.55	2.03	1.07	0.118	<0.001	0.057
Erysipelotrichaceae_unclassified	0.36	0.71	2.44	0.63	0.061	0.006	0.001
Candidate_division_TM7_unclassified	0.85	0.77	0.80	1.05	0.120	0.220	0.103
Firmicutes_unclassified	0.97	0.95	0.89	0.73	0.101	0.080	0.615
Bacteroides	0.55	0.60	0.88	0.76	0.105	<0.001	<0.001
S24_7_unclassified	0.65	0.47	0.76	0.52	0.138	<0.001	<0.001
Butyrivibrio	0.41	0.41	0.43	0.65	0.042	<0.001	0.004
RF16_unclassified	0.62	0.33	0.27	0.40	0.044	<0.001	0.002
Syntrophococcus	0.67	0.28	0.24	0.27	0.056	0.049	0.204
Blautia	0.28	0.32	0.39	0.29	0.032	0.959	0.629

[1] L-FNDF, low forage neutral detergent fiber; ML-FNDF, middle low forage neutral detergent fiber; MH-FNDF, middle high forage neutral detergent fiber; H-FNDF, high forage neutral detergent fiber.

4. Discussion

In the current study, as expected, the FNDF intake exhibited a linear increase, but DMI remained similar across the four groups when the dietary FNDF content increased from 6.08 to 15.68%. According to previous literature, a linear decrease in DMI in goats was observed with increased dietary peNDF$_{8.0}$ concentrations (from 1.9 to 11.7%) [19]. As revealed in a meta-analysis by Ferraretto et al. [20], a decreased DMI was observed when increasing FNDF content in the diet, which was attributed to the filling effect of forage. Caetano et al. [21] found that DMI of cattle fed a high-concentrate diet has a quadratic response with the increasing FNDF content. The lack of DMI difference in this study would be due to the PTMR feeding, which avoided the sorting behavior of sheep [5]. In addition, roughage in the diet was the standard chopped length, which would not result in a serious filling effect of forage. The FNDF and RDS are key nutrients in ruminant diets, with FNDF promoting saliva production and RDS indicating the contribution of grain-based diets to acid accumulation in the rumen [4,5]. In our earlier study, Zhang et al. [5] found that increasing RDS (9.70~23.66% of DM) content in the diet that controlled for the potential confounding effects of FNDF among treatments increased the risk of SARA in sheep. Nevertheless, increasing peNDF concentration could reduce the risk of SARA in dairy cows fed high-concentrate diets [1]. In the current study, increasing FNDF content indeed resulted in increased ruminal mean pH and minimum pH, accompanied by a concomitant decrease in the diurnal duration of pH < 5.8 and 5.6, the area of pH < 5.8 and 5.6, as well as the acidosis index. These findings suggested a reduction in the severity of rumen acidosis with increasing the dietary FNDF content, consistent with a previous study [1]. However, according to the definition where subacute ruminal acidosis (SARA) is identified by ruminal pH dropping below 5.6 for over three hours daily [22], the sheep across all four treatments exhibited signs of SARA based on this criterion in the present study. Notably, a pH < 5.6 sustained for approximately 217 min/d can be regarded as a mild SARA, which does not negatively affect the populations of ruminal fibrolytic bacteria [23]. In this study, the duration of pH < 5.6 in the MH-FNDF and H-FNDF was 278 and 260 min/d, respectively, which could be considered as mild SARA. These results indicated that increasing dietary FNDF content reduced the severity of SARA in sheep; 12.48% of FNDF in the high-RDS (15.65%) diet was the minimum FNDF level allowed to avoid the negative effects of rumen acidosis on ruminal function.

In the present investigation, the increased acetate molar proportion accompanied by a decline in propionate and valerate molar proportions as dietary FNDF content increased was consistent with the findings by Beauchemin and Yang [24], which was attributed to the increased fiber content in the diet. Fiber degradation promotes acetate production in the rumen, while starch degradation favors propionate production, revealing a positive correlation between acetate-to-propionate ratio and FNDF content but a negative correlation with nonstructural carbohydrate content [1,25]. The increased butyrate molar proportion with increasing dietary FNDF content would be attributed to the increased relative abundance of butyrate-producing bacteria, Butyrivibrio, in the rumen [26]. Notably, the concentration of TVFA in the rumen remained similar across the four groups, while the lactate concentration displayed a linear decrease with increasing FNDF content in the diet in this study. The accumulation of organic acids, including volatile fatty acids and lactate, serves as a prominent driver of subacute ruminal acidosis (SARA) in ruminants [27], which suggested that lactate was the main factor to affect the risk of SARA in sheep in the present study.

No difference in the eating time (min/d or min/kg DM) was observed with increasing dietary FNDF content in this study, which was consistent with the observed similar DMI. This finding was consistent with the report by Cao et al. [1], which indicated that ruminating, rather than eating behavior, was more responsive to variations in $peNDF_{8.0}$ concentration in high-concentrate diets. Ruminating and chewing activities are important to regulate the ruminal function of ruminants, which neutralizes approximately 30 to 40% of organic acids produced in the rumen through stimulating saliva secretion [6,16,28]. As expected, increasing FNDF content in the diet led to an increase in the average time spent on ruminating and total chewing per day or per kilogram of DMI, which underscored the mitigation of SARA severity achieved by stimulating salivary flow and finally enhancing ruminal buffering.

Increasing dietary FNDF content would enhance ruminal NDF and ADF digestibility, aligning with the observed increase in the degradability of 24 h in situ NDF and ADF of alfalfa hay in the current study. This phenomenon might be attributed to the higher ruminal pH and the increased substrate that was beneficial for the proliferation of ruminal cellulolytic bacteria with increasing the dietary FNDF content [5,29], which was supported by the increased abundance of Fibrobacterota and Butyrivibrio in the rumen. However, it was interesting that the apparent total-tract digestibility of NDF and ADF exhibited a decline with increasing dietary FNDF content in the present study, while the total-tract starch digestibility remained similar among the four groups. The decreased apparent total-tract digestibility of NDF and ADF was attributed to the increased flow of NDF and ADF to the hindgut (especially the large intestine) with the increase in FNDF content in the diet, where the NDF and ADF digested inefficiently. Increasing dietary FNDF content would not be beneficial for ruminal starch digestibility; however, the lower ruminal starch digestibility with increasing FNDF content in the diet was compensated by higher intestinal starch digestibility [30], thus no differences were observed across the four groups in the current study. Increased digestibility of DM, OM, and CP with increasing FNDF content was due to the increased ruminal pH, which improved the microbial diversity in the rumen, as shown by the increased Shannon index in the present study. In addition, ZOTUs, Chao1, and Ace indexes had a quadratic response, which indicated that an appropriate FNDF content could increase ruminal microbial richness, and 12.48% of FNDF was recommended in the present study.

In the present study, Bacteroidetes and Firmicutes were the predominant bacterial phyla in the rumen, which was consistent with previous reports [5,31,32]. Increasing the FNDF content in the diet increased the relative abundance of Fibrobacterota, Actinobacteria, and Bacteroidetes in the rumen. Bacteroidetes are gram-negative bacteria and sensitive to ruminal pH [33]. The ruminal pH exhibited an increase with increasing the dietary FNDF, which facilitated the proliferation of Bacteroidetes in the rumen. Fibrobacterota and Actinobacteria contribute to the decomposition of plant cellulose [34,35], increasing the FNDF content in the diet provided adequate substrate for these bacteria in the current study. Reduced risk of SARA would increase the relative abundance of Proteobacteria [36]. Increasing FNDF content in the diet reduced the risk of SARA in this study, which promoted Proteobacteria proliferation. Butyrivibrio and Ruminococcus are known to be cellulolytic bacteria, whereas Prevotella is known for the utilization of starch and protein [31,37,38]. In the present study, the relative abundance of Butyrivibrio increased with increasing FNDF content in the diet, while the Ruminococcus increased numerically, which was attributed to the increased nutrient substrates provided by FNDF in the diet. Additionally, the proliferation of cellulolytic bacteria in the rumen requires branched-chain fatty acids (isobutyrate, isovalerate), and the increased molar ratio of isobutyrate and isovalerate in the rumen promoted the multiplication of cellulolytic bacteria in the rumen [39]. Lachnospiraceae_unclassified ferments glucose and fermentable carbohydrates to produce lactate [40]; the decreased relative abundance of this bacteria was consistent with the declined concentration of lactate in the rumen.

5. Conclusions

The current study indicated that increasing FNDF content in the PTMR rich in RDS reduced the severity of SARA in sheep and improved the ruminal fermentation, microbial diversity, and fiber degradability. When the rumen-degradable starch content was 15.57% in the PTMR, the FNDF content of 12.48% was optimal to uphold stable ruminal function, which suggested that the ratio of FNDF to RDS with 0.8 was recommended for finishing sheep. Of course, further study is needed to investigate the effects of FNDF and RDS at different levels on rumen microbial function and host gene expression by metagenomics and transcriptomics techniques, which will achieve more accurate nutrition supply.

Supplementary Materials: The following supporting information can be downloaded at: https://www.mdpi.com/article/10.3390/ani15030314/s1, Section S1.1: Composition and nutrient composition of the experimental diet during the transition period; Section S1.2: DNA extraction and 16S rRNA sequencing; Section S1.3: Statistical analysis.

Author Contributions: Conceptualization, F.L. (Fei Li) and Z.Z.; data curation, Z.Z.; formal analysis, Z.Z.; funding acquisition, F.L. (Fei Li); investigation, Z.Z., X.S. and Z.H.; methodology, F.L. (Fei Li), Z.Z., F.L. (Fadi Li), Z.W., L.G. and X.W.; project administration, F.L. (Fei Li), X.M., Z.L. and X.L.; resources, F.L. (Fei Li), F.L. (Fadi Li), Z.W., L.G. and X.W.; supervision, F.L. (Fei Li), F.L. (Fadi Li), Z.W., X.S. and Z.H.; validation, F.L. (Fei Li) and Z.Z.; visualization, Z.Z.; writing—original draft preparation, Z.Z.; writing—review and editing, F.L. (Fei Li) and Z.Z. All authors have read and agreed to the published version of the manuscript.

Funding: This research is supported by the Cattle and Sheep Industry Development Research Institute of Linxia (KJJC-LX-2023-06), the Tianchi Talent Introduction Program of Xinjiang Uygur Autonomous Regions, and the Science and Technology Program of Gansu Province (24CXNA032).

Institutional Review Board Statement: This experiment was conducted according to established guidelines of the Animal Care Committee of Lanzhou University (approval number: LZU 201803002).

Informed Consent Statement: Not applicable.

Data Availability Statement: All raw sequences were deposited in the NCBI Sequence Read Archive (SRA) database and can be accessed via accession number PRJNA1145563.

Conflicts of Interest: The co-author (Z.L.) is an employee of a company (Linxia State Xiaokangcun Feed Co.) that has contributions on investigation and project administration in this manuscript and declares no conflict of interest. The other authors also declare no conflicts of interest.

Abbreviations

The following abbreviations are used in this manuscript:

ADF	acid detergent fiber
CP	crude protein
DMI	dry matter intake
FNDF	forage neutral detergent fiber
ME	metabolic energy
NDF	neutral detergent fiber
OM	organic matter
peNDF	physically effective neutral detergent fiber
PTMR	pelleted total mixed ration
RDS	rumen degradable starch
SARA	subacute ruminal acidosis
VFA	volatile fatty acid

References

1. Cao, Y.; Wang, D.; Wang, L.; Wei, X.; Li, X.; Cai, C.; Lei, X.; Yao, J. Physically effective neutral detergent fiber improves chewing activity, rumen fermentation, plasma metabolites, and milk production in lactating dairy cows fed a high-concentrate diet. *J. Dairy Sci.* **2021**, *104*, 5631–5642. [CrossRef] [PubMed]
2. Dong, H.; Wang, S.; Jia, Y.; Ni, Y.; Zhang, Y.; Zhuang, S.; Shen, X.; Zhao, R. Long-term effects of subacute ruminal acidosis (SARA) on milk quality and hepatic gene expression in lactating goats fed a high-concentrate diet. *PLoS ONE* **2013**, *8*, e82850. [CrossRef]
3. Mulakala, B.K.; Smith, K.M.; Snider, M.A.; Ayers, A.; Honan, M.C.; Greenwood, S.L. Influence of dietary carbohydrate profile on the dairy cow rumen meta-proteome. *J. Dairy Sci.* **2022**, *105*, 8485–8496. [CrossRef] [PubMed]
4. Li, F.; Li, Z.; Li, S.; Ferguson, J.D.; Cao, Y.C.; Yao, J.H.; Sun, F.F.; Wang, X.; Yang, T. Effect of dietary physically effective fiber on ruminal fermentation and the fatty acid profile of milk in dairy goats. *J. Dairy Sci.* **2014**, *97*, 2281–2290. [CrossRef] [PubMed]
5. Zhang, Z.; Wang, L.; Li, Q.; Li, F.; Ma, Z.; Li, F.; Wang, Z.; Chen, L.; Yang, X.; Wang, X.; et al. Effects of dietary forage neutral detergent fiber and rumen degradable starch ratios on chewing activity, ruminal fermentation, ruminal microbes and nutrient digestibility of Hu sheep fed a pelleted total mixed ration. *J. Anim. Sci.* **2024**, *102*, skae100. [CrossRef] [PubMed]
6. Chibisa, G.E.; Beauchemin, K.A.; Penner, G.B. Relative contribution of ruminal buffering systems to pH regulation in feedlot cattle fed either low- or high-forage diets. *Animal* **2016**, *10*, 1164–1172. [CrossRef]
7. Ran, T.; Saleem, A.M.; Beauchemin, K.A.; Penner, G.B.; Yang, W. Processing index of barley grain and dietary undigested neutral detergent fiber concentration affected chewing behavior, ruminal pH, and total tract nutrient digestibility of heifers fed a high-grain diet. *J. Anim. Sci.* **2021**, *99*, skab011. [CrossRef]
8. Dykier, K.C.; Oltjen, J.W.; Robinson, P.H.; Sainz, R.D. Effects of finishing diet sorting and digestibility on performance and feed efficiency in beef steers. *Animal* **2020**, *14*, 59–65. [CrossRef]
9. Xue, Y.; Yin, Y.; Trabi, E.B.; Xie, F.; Lin, L.; Mao, S. Transcriptome analysis reveals the effect of high-grain pelleted and non-pelleted diets on ruminal epithelium of Hu-lamb. *Animal* **2021**, *15*, 100278. [CrossRef] [PubMed]
10. Ma, X.; Zhou, W.; Guo, T.; Li, F.; Li, F.; Ran, T.; Zhang, Z.; Guo, L. Effects of dietary barley starch contents on the performance, nutrient digestion, rumen fermentation, and bacterial community of fattening Hu sheep. *Front. Nutr.* **2021**, *8*, 797801. [CrossRef] [PubMed]
11. Penner, G.B.; Aschenbach, J.R.; Gabel, G.; Rackwitz, R.; Oba, M. Epithelial capacity for apical uptake of short chain fatty acids is a key determinant for intraruminal pH and the susceptibility to subacute ruminal acidosis in sheep. *J. Nutr.* **2009**, *139*, 1714–1720. [CrossRef] [PubMed]
12. Colenbrander, V.F.; Noller, C.H.; Grant, R.J. Effect of fiber content and particle size of alfalfa silage on performance and chewing behavior. *J. Dairy Sci.* **1991**, *74*, 2681–2690. [CrossRef]
13. Beauchemin, K.A.; Yang, W.Z.; Rode, L.M. Effects of particle size of alfalfa-based dairy cow diets on chewing activity, ruminal fermentation, and milk production. *J. Dairy Sci.* **2003**, *86*, 630–643. [CrossRef]

14. Li, F.; Cao, Y.; Liu, N.; Yang, X.; Yao, J.; Yan, D. Subacute ruminal acidosis challenge changed in situ degradability of feedstuffs in dairy goats. *J. Dairy Sci.* **2014**, *97*, 5101–5109. [CrossRef] [PubMed]
15. AOAC. *Official Methods of Analysis*, 18th ed.; Association of Official Analytical Chemists International: Gaithersburg, MD, USA, 2006.
16. Allen, M.S. Relationship between fermentation acid production in the rumen and the requirement for physically effective fiber. *J. Dairy Sci.* **1997**, *80*, 1447–1462. [CrossRef] [PubMed]
17. Ma, Z.; Zhang, X.; Wang, R.; Wang, M.; Liu, T.; Tan, Z. Effects of Chemical and mechanical lysis on microbial dna yield, integrity, and downstream amplicon sequencing of rumen Bacteria and Protozoa. *Front. Microbiol.* **2020**, *11*, 2812. [CrossRef] [PubMed]
18. Zhang, Z.; Li, F.; Ma, Z.; Li, F.; Wang, Z.; Li, S.; Wang, X.; Li, K. Variability in chewing, ruminal fermentation, digestibility and bacterial communities between subacute ruminal acidosis-susceptible and acidosis-tolerant sheep. *Animal* **2023**, *17*, 100902. [CrossRef] [PubMed]
19. Zhao, X.; Zhang, T.; Xu, M.; Yao, J. Effects of physically effective fiber on chewing activity, ruminal fermentation, and digestibility in goats. *J. Anim. Sci.* **2011**, *89*, 501–509. [CrossRef]
20. Ferraretto, L.F.; Crump, P.M.; Shaver, R.D. Effect of cereal grain type and corn grain harvesting and processing methods on intake, digestion, and milk production by dairy cows through a meta-analysis. *J. Dairy Sci.* **2013**, *96*, 533–550. [CrossRef]
21. Caetano, M.; Goulart, R.S.; Silva, S.L.; Drouillard, J.S.; Leme, P.R.; Lanna, D.P. Effect of flint corn processing method and roughage level on finishing performance of Nellore-based cattle. *J. Anim. Sci.* **2015**, *93*, 4023–4033. [CrossRef]
22. Gozho, G.N.; Plaizier, J.C.; Krause, D.O.; Kennedy, A.D.; Wittenberg, K.M. Subacute ruminal acidosis induces ruminal lipopolysaccharide endotoxin release and triggers an inflammatory response. *J. Dairy Sci.* **2005**, *88*, 1399–1403. [CrossRef] [PubMed]
23. Khafipour, E.; Li, S.; Plaizier, J.C.; Krause, D.O. Rumen microbiome composition determined using two nutritional models of subacute ruminal acidosis. *Appl. Environ. Microbiol.* **2009**, *75*, 7115–7124. [CrossRef] [PubMed]
24. Beauchemin, K.A.; Yang, W.Z. Effects of physically effective fiber on intake, chewing activity, and ruminal acidosis for dairy cows fed diets based on corn silage. *J. Dairy Sci.* **2005**, *88*, 2117–2129. [CrossRef] [PubMed]
25. Firkins, J.L.; Eastridge, M.L.; St-Pierre, N.R.; Noftsger, S.M. Effects of grain variability and processing on starch utilization by lactating dairy cattle. *J. Anim. Sci.* **2001**, *79*, E218–E238. [CrossRef]
26. Koike, S.; Ueno, M.; Miura, H.; Saegusa, A.; Inouchi, K.; Inabu, Y.; Sugino, T.; Guan, L.L.; Oba, M.; Kobayashi, Y. Rumen microbiota and its relation to fermentation in lactose-fed calves. *J. Dairy Sci.* **2021**, *104*, 10744–10752. [CrossRef]
27. Herdt, T.H. Ruminant adaptation to negative energy balance. Influences on the etiology of ketosis and fatty liver. *Vet. Clin. N. Am. Food Anim. Pract.* **2000**, *16*, 215–230. [CrossRef] [PubMed]
28. Lechartier, C.; Peyraud, J.L. The effects of forage proportion and rapidly degradable dry matter from concentrate on ruminal digestion in dairy cows fed corn silage-based diets with fixed neutral detergent fiber and starch contents. *J. Dairy Sci.* **2010**, *93*, 666–681. [CrossRef] [PubMed]
29. Weimer, P.J. Degradation of cellulose and hemicellulose by ruminal microorganisms. *Microorganisms* **2022**, *10*, 2345. [CrossRef] [PubMed]
30. Li, C.; Beauchemin, K.A.; Yang, W. Feeding diets varying in forage proportion and particle length to lactating dairy cows: I. Effects on ruminal pH and fermentation, microbial protein synthesis, digestibility, and milk production. *J. Dairy Sci.* **2020**, *103*, 4340–4354. [CrossRef] [PubMed]
31. Zhang, J.; Shi, H.; Wang, Y.; Li, S.; Cao, Z.; Ji, S.; He, Y.; Zhang, H. Effect of dietary forage to concentrate ratios on dynamic profile changes and interactions of ruminal microbiota and metabolites in Holstein heifers. *Front. Microbiol.* **2017**, *8*, 2206. [CrossRef] [PubMed]
32. Freetly, H.C.; Lindholm-Perry, A.K. Rumen and cecum bacteria of beef cattle that differ in feed efficiency fed a forage diet. *J. Anim. Sci.* **2023**, *101*, skad292. [CrossRef]
33. Kampmann, K.; Ratering, S.; Kramer, I.; Schmidt, M.; Zerr, W.; Schnell, S. Unexpected stability of bacteroidetes and firmicutes communities in laboratory biogas reactors fed with different defined substrates. *Appl. Environ. Microbiol.* **2012**, *78*, 2106–2119. [CrossRef]
34. Sulak, M.; Sikorova, L.; Jankuvova, J.; Javorsky, P.; Pristas, P. Variability of Actinobacteria, a minor component of rumen microflora. *Folia. Microbiol.* **2012**, *57*, 351–353. [CrossRef] [PubMed]
35. Jewell, K.A.; McCormick, C.A.; Odt, C.L.; Weimer, P.J.; Suen, G. Ruminal bacterial community composition in dairy cows is dynamic over the course of two lactations and correlates with feed efficiency. *Appl. Environ. Microbiol.* **2015**, *81*, 4697–4710. [CrossRef]
36. Li, F.; Wang, Z.; Dong, C.; Li, F.; Wang, W.; Yuan, Z.; Mo, F.; Weng, X. Rumen bacteria communities and performances of fattening lambs with a lower or greater subacute ruminal acidosis risk. *Front. Microbiol.* **2017**, *8*, 2506. [CrossRef] [PubMed]
37. Fernando, S.C.; Purvis, H.T.; Najar, F.Z.; Sukharnikov, L.O.; Krehbiel, C.R.; Nagaraja, T.G.; Roe, B.A.; Desilva, U. Rumen microbial population dynamics during adaptation to a high-grain diet. *Appl. Environ. Microbiol.* **2010**, *76*, 7482–7490. [CrossRef] [PubMed]

38. Betancur-Murillo, C.L.; Aguilar-Marin, S.B.; Jovel, J. Prevotella: A key player in ruminal metabolism. *Microorganisms* **2022**, *11*, 1. [CrossRef]
39. Allison, M.J.; Bryant, M.P.; Doetsch, R.N. Volatile fatty acid growth factor for cellulolytic cocci of bovine rumen. *Science* **1958**, *128*, 474–475. [CrossRef]
40. Han, L.; Xue, W.; Cao, H.; Chen, X.; Qi, F.; Ma, T.; Tu, Y.; Diao, Q.; Zhang, C.; Cui, K. Comparison of rumen fermentation parameters and microbiota of yaks from different altitude regions in Tibet, China. *Front. Microbiol.* **2021**, *12*, 807512. [CrossRef]

animals

MDPI

Review

Exploration of Multi-Source Lignocellulose-Degrading Microbial Resources and Bioaugmentation Strategies: Implications for Rumen Efficiency

Xiaokang Lv [1,†], Zhanhong Qiao [1,†], Chao Chen [1], Jinling Hua [1,*] and Chuanshe Zhou [2,*]

1 College of Animal Science, Anhui Science and Technology University, Fengyang 233100, China; lvxk@ahstu.edu.cn (X.L.); 18731310109@163.com (Z.Q.); chenchao@ahstu.edu.cn (C.C.)

2 Hunan Provincial Key Laboratory of Animal Nutritional Physiology and Metabolic Process, National Engineering Laboratory for Pollution Control and Waste Utilization in Livestock and Poultry Production, Institute of Subtropical Agriculture, Chinese Academy of Sciences, Changsha 410125, China

* Correspondence: huajl@ahstu.edu.cn (J.H.); zcs@isa.ac.cn (C.Z.)

† These authors contributed equally to this work.

Simple Summary: This review explores the diverse world of microbes (bacteria and fungi) capable of breaking down lignocellulose, found in environments from cow rumens to insect guts and forest soils. We cover how researchers identify these microbes and, crucially, the shift towards using advanced multi-omics techniques to pinpoint highly efficient straw-degrading enzymes, alongside genome-editing tools to engineer improved microbes. Our aim is to guide the discovery of new microbial resources and superior enzymes to enhance the utilization of roughage like straw, especially within ruminant feeding systems.

Abstract: Utilizing straw feed is an effective strategy to optimize straw resource utilization by incorporating microbial degradation agents to expedite lignocellulose breakdown and enhance feed efficiency. Lignocellulose-degrading species and microbial communities are present in various Earth ecosystems, including the rumen of ruminants, insect digestive tracts, forest soil, and microbial populations in papermaking processes. The rumen of ruminants harbors a diverse range of microbial species, making it a promising source of lignocellulose-degrading microorganisms. Exploring alternative systems like insect intestines and forest soil is essential for future research. Current studies primarily rely on traditional microbial isolation techniques to identify lignocellulose-degrading strains, underscoring the necessity to transition to utilizing microbial culturomics and genome-editing technologies for discovering and manipulating cellulose-degrading microbes. This review provides an overview of lignocellulose-degrading microbial communities from diverse environments, encompassing bacterial and fungal populations. It also delves into the use of metagenomic, metatranscriptomic, and metaproteomic approaches to pinpoint highly efficient cellulase genes, along with the application of genome-editing tools for engineering lignocellulose-degrading microorganisms. The primary objective of this review is to offer insights for further exploration of potential lignocellulose-degrading microbial resources and high-performance cellulase genes to enhance roughage utilization in ruminant rumen ecosystems.

Keywords: lignocellulose degrading bacteria; rumen microorganisms; metagenomics; cellulase; gene editing technology

1. Introduction

The global availability of straw resources is substantial, with an annual production exceeding 7 billion tons worldwide, of which China contributes 25% [1]. Currently, crop straw is primarily utilized through various methods, including direct field incorporation (3–5%), composting for field application (7–9%), livestock feed (8–11%), organic fertilizer production (2–5%), edible fungi cultivation (4–5%), inefficient combustion (approximately 15%), among others [2]. Although straw combustion is widespread, it results in significant resource loss and environmental harm [3]. Given the growing global demand for cattle and sheep feed, converting straw into animal feed offers a sustainable and productive alternative [4]. Lignocellulose, a ubiquitous substance on Earth, is the primary component of straw, constituting a significant portion of plant cell walls in various plant tissues. The lignocellulose content in straw varies with plant species. Wheat straw contains 32–45% cellulose, 20–45% hemicellulose, and 11–26% lignin [5]. Corn stalk contains cellulose 36.89%, hemicellulose 20.42%, and lignin 17.38% [6]. Cellulose and hemicellulose of straw are 50–80% of dry matter, and lignin is 10–15% of dry matter [7]. The complex structure of lignocellulose makes it inherently resistant to natural degradation. Consequently, there is a continuous research focus on improving the efficient breakdown of lignocellulose. Microorganisms capable of cellulose degradation produce cellulases, which facilitate the conversion of lignocellulose into mono- or oligosaccharides, thereby enhancing the nutritional value of straw for animal feed [8].

The recent rapid progress in biotechnology has sparked interest in identifying effective lignocellulose-degrading microorganisms for straw feed processing [9–11]. Significant headway has been achieved in isolating and characterizing highly efficient lignocellulose-degrading microorganisms [12]. Techniques such as metagenomics and microbial culturomics have successfully identified potent cellulose-degrading microorganisms from various environments, including insect guts [13], ruminant gastrointestinal tracts [14], and forest soils [15]. Moreover, the integration of metagenomic [9], metatranscriptomic [10], and metaproteomic methodologies [11] has substantially broadened the pool of cellulase genes. The isolation of these potent cellulose-degrading microorganisms and the elucidation of cellulase genes serve as valuable microbial assets for straw feed processing. The genetic manipulation of cellulose-degrading microorganisms is expected to be a crucial technical approach for enhancing cellulose utilization in the future. This review provides a comprehensive summary of recent progress in isolating cellulose-degrading microorganisms, identifying cellulase genes, and genetically modifying cellulose-degrading strains. Utilizing multi-omics technology and gene editing is essential for enhancing the cellulose degradation capacity of cellulose-degrading bacteria. This review broadens the scope of cellulose-degrading bacterial sources, emphasizing the investigation of the extreme environmental adaptability of these bacteria as a crucial area for future research. Such efforts have the potential to significantly enhance the global utilization of agricultural waste and straw feed.

2. Lignocellulose Degradation Mechanisms

Lignocellulose is primarily composed of cellulose (45–50%), hemicellulose (20–30%), and lignin (20–30%) [16].

2.1. Degradation Mechanism of Lignin

Lignin is a complex and heterogeneous polyphenolic polymer. Enzymes responsible for lignin degradation primarily include lignin peroxidase (LiP), manganese peroxidase (MnP), and laccase [17]. LiP, containing a heme group, aids in substrate oxidation by utilizing hydrogen peroxide or similar peroxides as co-substrates. This peroxidase catalysis leads to the electron loss of aromatic rings, resulting in the formation of aryl cation radicals that subsequently degrade into smaller molecules in the presence of molecular oxygen [18]. MnP, an oxidoreductase containing heme and Mn^{2+} in its catalytic center, initiates its catalytic cycle through the oxidation of Mn^{2+} to Mn^{3+} by hydrogen peroxide during the reaction. The Mn^{3+} chelate acts as an oxidant, directly targeting phenolic structures within lignin. Laccase, a copper-containing polyphenol oxidase, utilizes molecular oxygen as an electron acceptor to oxidize both phenolic and non-phenolic compounds, producing free radicals while converting oxygen to water. With a wide substrate range, laccase catalyzes the oxidation of various compounds such as phenols, aromatic amines, and aliphatic amines [19]. These enzymes exhibit substrate specificity: laccase and MnP preferentially degrade phenolic structures, whereas LiP targets non-phenolic lignin components.

2.2. Degradation Mechanism of Cellulose

Cellulose, a linear macromolecular polymer comprising β-1,4-glycosidic linkages of D-glucose units, is insoluble in water and common organic solvents [20]. It is the predominant component of plant cell walls, where cellulose chains are connected by hydrogen bonds [21]. Cellulose accounts for over 50% of the carbon content in plants, making it the most abundant organic material in nature. Its strong structural integrity and stability make it highly resistant to digestion and absorption by most organisms [22].

Cellulases are predominantly synthesized by microorganisms, encompassing bacteria, fungi, and protozoa. These enzymes operate within a system comprising three key elements: endo-β-1,4-glucanase (EG), cellobiohydrolase (CBH), and β-glucosidase (BGL). Endoglucanases catalyze the breakdown of β-1,4-glycosidic bonds present in cellulose chains, disrupting the elongated chain structure and yielding free chain ends. Exoglucanases, exemplified by CBH, target these liberated chain ends, cleaving glycosidic bonds to release cellobiose units. Subsequently, β-Glucosidases further degrade cellobiose and short-chain oligosaccharides into glucose monomers. Extensive investigation has underscored the indispensable nature of the cooperative action of these three enzymes for thorough cellulose degradation [23]. Cellulose, hemicellulose, and other non-starch polysaccharides (NSPs) constitute principal anti-nutritional components in plant-based feed. Glycoside hydrolases, including cellulases, glucanases, xylanases, and mannanases, possess the capability to cleave the glycosidic bonds within NSPs, thereby mitigating their anti-nutritional properties and yielding low-molecular-weight products such as oligosaccharides or monosaccharides [24].

2.3. Degradation Mechanism of Hemicellulose

Hemicellulose undergoes biodegradation into monosaccharides and acetic acid. Xylan, a significant constituent of hemicellulose, displays intricate structural diversity, necessitating a complex enzymatic system for its breakdown. Efficient degradation of xylan-based hemicellulose requires the coordinated action of multiple enzymes. The enzymatic system responsible for hemicellulose degradation includes endo-β-1,4-xylanase, β-xylosidase, and α-L-arabinofuranosidase, collectively targeting most naturally occurring xylan-type hemicelluloses [25]. Specifically, endo-β-1,4-xylanase cleaves β-1,4-glycosidic bonds in the xylan backbone, producing xylooligosaccharides. Subsequently, β-xylosidase acts on the non-reducing ends of these oligosaccharides to liberate xylose residues [22].

3. Sources and Systems of Cellulose-Degrading Microorganisms

3.1. Insect Gut

The insect gut acts as a natural bioreactor, containing a rich reservoir of microbial communities for cellulose degradation. Various studies have highlighted the impressive lignocellulose-degrading capacities of different insect gut systems (Figure 1). For instance, Dar et al. [26] isolated *Bacillus tequilensis G9*, a cellulase-producing strain with high yield, from the hemolymph of African giant snails (Table 1). By combining 16S rRNA sequencing with enzyme activity assays (DNS method), the strain exhibited cellulase activity of 956.9 IU/mL on day 8 in a sugarcane bagasse-induced medium. This discovery not only expands the range of cellulolytic microorganisms but also provides new insights into identifying high-activity strains. Luo et al. [13] conducted a comprehensive assessment of bamboo weevils (*Cyrtotrachelus buqueti*) in degrading bamboo lignocellulose through both in vivo and in vitro experiments. The results demonstrated in vivo degradation efficiencies of 61.82%, 87.65%, and 69.05% for cellulose, hemicellulose, and lignin, respectively. In contrast, in vitro efficiencies were lower at 24.98%, 37.52%, and 26.67%. This significant difference emphasizes the crucial regulatory function of the insect gut microenvironment in cellulose degradation mechanisms.

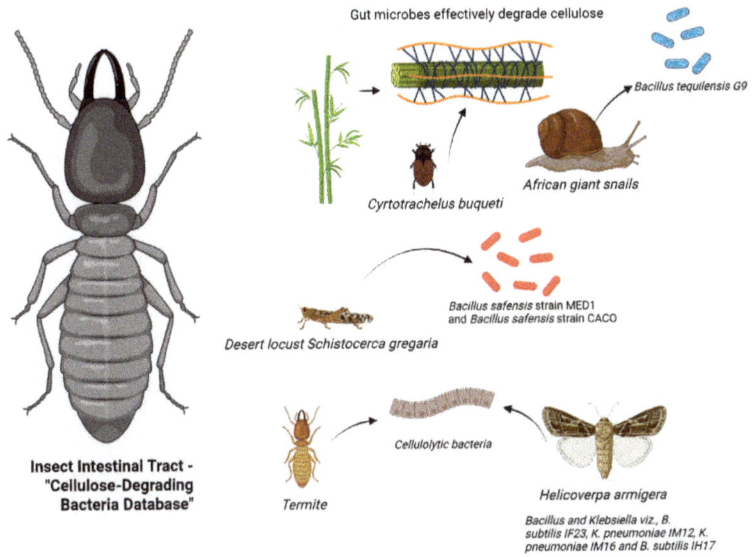

Figure 1. Database of insect gut cellulose-degrading bacteria (generated by bioRender).

Nelson et al. isolated two highly efficient cellulose-degrading bacterial strains from the gut of desert locusts. The cellulolytic activity index (ICA) of *Bacillus safensis MED1* and *Bacillus* sp. *CACO* was determined as 1.146 ± 0.109 and 0.8442 ± 0.09203, respectively, using the agar plate clearance zone method. Oktiarni et al. [27] screened 24 cellulose-degrading strains from the gut of peatland termites (*Macrotermes* sp.), with strains 37 (*Staphylococcus* sp.), 39 (*Microbacterium* sp.), 40 (*Bacillus* sp.), and 62 (*Brevibacterium* sp.) exhibiting cellulolytic indices of 3, 4, 2.5, and 2.5, respectively. Biochemical characterization confirmed these strains as catalase-positive and citrate-positive but negative for indole and urease tests. Dar et al. [28] demonstrated that the foregut of cotton bollworms (Helicoverpa armigera) is the primary enrichment site for cellulose-degrading microorganisms. Isolation and cultivation yielded 71 bacterial strains, with 54% isolated from the foregut. Sodium carboxymethyl cellulose (CMC-Na) plate assays revealed significant cellulose-degrading activity in 25 strains belonging to the phyla *Firmicutes* and *Proteobacteria*.

Table 1. Isolation of cellulose-degrading bacteria from the intestinal tract of insects.

Year	Host Source	Cellulose-Degrading Bacteria	Microbe Phylum, Genus	References
2018	*Achatina fulica*	*B. tequilensis G9*	*Firmicutes*	[26]
2019	*Bamboo weevils*	No specific strain was identified	—	[13]
2021	*Desert Locust*	*Bacillus safensis MED1*, *Bacillus* sp. *CACO*	*Firmicutes*	[29]
2021	*Indralaya*	*Staphylococcus* sp. *Microbacterium* sp. *Bacillus* sp. *Brevibacterium* sp.	*Staphylococcus*, *Microorganisms*, *Bacillus*, *Brevibacterium*	[27]
2021	*Cotton bollworm*	25 strains (*Firmicutes*, *Proteobacteria*)	*Firmicutes and Proteobacteria*	[28]

Bacterial strains exhibiting cellulolytic activity, identified from studies on bamboo weevils (*Cyrtotrachelus buqueti*), desert locusts, peatland termites (*Macrotermes* sp.), Giant African land snail, and cotton bollworms (*Helicoverpa armigera*).

3.2. Ruminant Gastrointestinal Tract

Ruminants predominantly consume cellulose-rich forage, with their rumen serving as a natural anaerobic fermentation system housing abundant cellulose-degrading microbial populations. Research indicates that the rumen microbiota demonstrates a substantial biomass, consisting of bacteria (10^{10}–10^{11} cells/mL), fungi (10^3–10^6 cells/mL), protozoa (10^4–10^7 cells/mL), phages (10^7–10^9 particles/mL), and archaea (10^6 cells/mL) [30]. These microorganisms collaboratively break down cellulose into glucose for the host's benefit. Notably, the *phyla Bacteroidetes* and *Firmicutes* are prevalent in the core rumen microbiota, collectively representing over 80% [31]. The rumen microbial genome encodes a wide array of carbohydrate-active enzymes (CAZymes), establishing it as a crucial repository for novel cellulases (Figure 2).

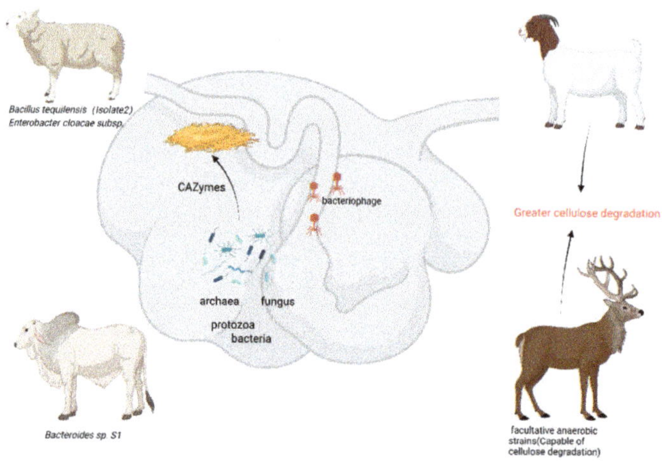

Figure 2. Cellulose-degrading bacteria from rumen of ruminants—potential resources for the future (generated by bioRender).

3.2.1. Cellulose-Degrading Microorganisms from Cattle Rumen

Sari et al. [14] isolated a highly cellulolytic strain, *S1*, from the rumen fluid of Aceh cattle in Indonesia. This strain demonstrated notable cellulolytic activity on BHM-CMC agar medium (cellulolytic index > 1, clearance zone diameter 2.5 cm). Identified as a Gram-negative bacillus, it exhibited optimal growth conditions at pH 6.0–6.5 and 36–39 °C (Table 2).

Pang et al. [32] isolated a cellulolytic strain, *ZH4*, from the rumen of Inner Mongolia cattle. Through morphological, physiological, and biochemical characterization, as well as 16S rDNA gene sequencing, the strain was identified as *Escherichia coli*. Analysis of extracellular enzyme activity revealed that the strain exhibited cellulase production. Specifically, at pH 6.8, the exoglucanase activity was measured at 9.13 IU, endoglucanase activity at 5.31 IU, and β-glucosidase activity at 7.27 IU. Furthermore, when utilizing corn stalk as a substrate, the strain produced 0.36 g/L ethanol and 4.71 mL/g hydrogen. The degradation rates of cellulose and hemicellulose were determined to be 14.30% and 11.39%, respectively [32].

Zhao et al. [33] utilized metagenomics to demonstrate that the yak rumen microbiota not only significantly increases the relative abundance of cellulolytic bacteria but also enhances the expression of cellulase and xylanase-encoding genes. Furthermore, the rumen microbial community enriches *succinate-metabolizing bacteria*, facilitating pyruvate fermentation to propionate and enhancing energy utilization efficiency.

3.2.2. Cellulose-Degrading Microorganisms from Sheep Rumen

Guder et al. [34] evaluated 20 cellulose-degrading strains isolated from the sheep rumen, with 8 strains displaying significant cellulose degradation capability, notably Enterobacter cloacae subsp. *strain KLCD08* exhibiting the highest enzymatic activity. Ben Ghalib et al. [35] identified five highly cellulolytic strains from the sheep gastrointestinal tract using MALDI-TOF MS and 16S rRNA sequencing. These strains were classified under *Bacillus, Lysinibacillus*, and *Priestia genera*, with *Bacillus tequilensis (Isolate 2)* demonstrating the most pronounced cellulase activity.

Yang et al. [36] investigated rumen microorganisms in Wujimu sheep, identifying 22 isolates through 16S rDNA or ITS sequencing. Phylogenetic analysis revealed that these isolates were classified into four phyla: *Actinomycota, Firmicutes, Proteobacteria*, and *Ascomycota*. Notably, *Bacillus subtilis RE251* exhibited remarkable enzyme activities (including endoglucanase, exoglucanase, and β-glucosidase) during the degradation of alfalfa hay. After 10 days, this strain significantly disrupted the fiber structure, demonstrating promising practical application.

3.2.3. Cellulose-Degrading Microorganisms from Other Ruminants

Tulsani et al. [37] successfully isolated *Aspergillus sydowii C6d* from the rumen of camels. Their analysis of CAZymes demonstrated that *C6d* expressed a total of 543 polysaccharide-degrading CAZymes, with 148 of these enzymes potentially participating in lignocellulose breakdown. Comparative genomic analysis between *C6d* and 30 widely studied lignocellulolytic fungi, including *white rot fungi, brown rot fungi*, and *soft rot fungi*, revealed a notable enrichment of cellulolytic and pectinolytic CAZymes in the *C6d* genome when compared to other fungal species.

Pramartaa et al. [38] identified five facultative anaerobic strains capable of degrading cellulose from the feces of Southeast Asian banteng, Javan banteng, muntjac, and Timor deer. The cellulolytic indices indicated high cellulolytic activity (index $\approx$ 1.2) in isolates from Southeast Asian and Javan banteng feces, suggesting their promise as cellulose-degrading candidates.

Representative bacteria isolated from the rumen of diverse ruminants (cattle, sheep, yak, camel, deer, banteng) demonstrating significant cellulolytic potential. Highlighted strains exhibit high cellulolytic indices, enzymatic activities (endoglucanase, exoglucanase, β-glucosidase), and lignocellulose degradation capabilities, positioning the rumen microbiome as a key resource for future biotechnological applications (especially for ruminant animals such as goats, yaks, and deer that are tolerant to roughage).

Table 2. Isolation of cellulose-degrading bacteria from rumen in ruminants.

Year	Host Source	Isolated Cellulose-Degrading Bacteria	Microbes Phylum, Genus	References
2017	Indonesian Aceh bovine rumen	*Bacteroides* sp. *S1*	Probably a new species of Enterobacteriaceae.	[14]
2017	Cattle rumen	*ZH-4*	*Proteobacteria*	[32]
2022	Rumen microbial community in yak	Uncultured microbiota	—	[33]
2019	Sheep rumen	*Enterobacter cloacae* subsp. *KLCD08*	*Genus Bacillus*	[34]
2024	Sheep digestive tract	*Bacillus tequilensis Isolate2*	*Bacillus*, *Lysinibacillus*, *Priestia*	[35]
2022	Wujimu Sheep	*22 isolates*	*Actinomycetales*, *Firmicutes*, *Proteobacteria* and *Ascomycetes*	[36]
2022	Camel rumen	*Aspergillus sydowii C6d*	*Ascomycota*	[37]
2024	The feces of Southeast Asian banteng, Javan banteng, muntjac, and Timor deer.	*Five facultative anaerobic strains*	—	[38]

3.3. *Forest Soil-Derived Cellulose-Degrading Microorganisms*

Forest soil is a crucial habitat for cellulose-degrading microorganisms due to the abundance of cellulose substrates in the thick layer of decaying plant litter. Several research groups have isolated and characterized highly efficient cellulose-degrading strains from diverse forest soils. For instance, Mokale Kognou et al. [39] utilized a carboxymethyl cellulose (CMC)-Congo red screening method to isolate six cellulose-degrading strains from mixed soil samples collected on a Canadian university campus (Table 3). Through 16S rRNA sequencing, *Bacillus* sp. *MKAL6* exhibited the highest cellulase activity. Optimization studies using Box–Behnken design revealed that the addition of 10 mM sucrose significantly enhanced cellulase activity. SDS-PAGE analysis indicated that the cellulases produced by this strain had molecular weights concentrated around 25 kDa, suggesting a potential link to the enzyme's efficient catalytic structure. In a separate study, Ma et al. [40] isolated 55 cellulolytic bacterial strains from decaying wood in the Qinling Mountains, with *Methylobacillus* sp. 1EJ7 demonstrating optimal degradation efficiency for wheat straw, corn stover, and switchgrass. Naresh et al. [41] isolated seven thermophilic cellulose-degrading strains from Malaysian tropical forest soil, with *Anoxybacillus* sp. *UniMAP-KB06* exhibiting the highest hydrolytic activity. Additionally, Khosravi et al. [42] identified *Brevibacillus borstelensis strains US5* and *US7* from forest soil in Kerman, Iran, which maintained high cellulase activity even at 65 °C, indicating their potential for industrial applications.

Table 3. Isolation of cellulose-degrading bacteria from forest ecosystems.

Year	Source of Flora	Isolated Cellulose-Degrading Bacteria	Microbe Phylum, Genus	References
2018	Canadian Campus Ginkgo Grove	*Bacillus* sp. MKAL6	*Firmicutes*	[39]
2020	Qinling Mountain deadwood	*Bacillus methylotrophicus 1EJ7*	*Proteobacteria*	[40]
2019	Tropical Forest of Malaysia	*Anoxybacillus* sp. *UniMAP-KB06*	*Firmicutes*	[41]
2021	Kerman Forest, Iran	*Brevibacillus borstelensis* US5	*Firmicutes*	[42]
2021	Rice rhizosphere soil	*Massilia* sp. *NEAU-DD11T*	*Proteobacteria Massilia*	[43]
2024	Soil	*Rhodococcus wratislaviensis YZ02*, *Pseudomonas anthosomatis YZ03*	*Actinobacteria* *Proteobacteria*	[15]
2022	Mangrove soil	*Robertkochia solimangrovi*	*Bacteroidetes*	[44]

Du et al. [43] isolated *Massilia* sp. *NEAU-DD11T*, a novel cellulose-degrading strain, from rice rhizosphere soil using polyphasic taxonomic approaches. This Gram-negative bacillus was characterized through various methods, including 16S rRNA gene sequence analysis (>98.4% similarity to Massilia species), whole-genome phylogeny, fatty acid profiling (predominantly hexadecanoic acid (C16:), cycloheptadecanoic acid (C17:-cyclo), and hexadecenoic acid (C16:1ω7c)), and polar lipid composition (e.g., diphosphatidylglycerol, phosphatidylglycerol). Ma et al. [15] identified seven potential cellulolytic strains from soil using 16S rRNA sequencing. Through DNS-based enzyme activity screening, two highly efficient strains, *Rhodococcus wratislaviensis YZ02* and *Pseudomonas anthosomatis YZ03*, were selected, with genome sizes of 8.51 Mb (8466 genes) and 6.66 Mb (5745 genes), respectively. Analysis of CAZyme revealed the presence of glycoside hydrolase (GH) family genes and other cellulose degradation-related genes in both strains.

Lam et al. [44] isolated *Robertkochia solimangrovi* from mangrove soil and conducted genomic analysis, revealing 87 lignocellulose-degrading enzymes. These enzymes include cellulases (GH3, GH5, GH9, and GH30), xylanases (GH5, GH10, GH43, GH51, GH67, and GH115), mannanases (GH2, GH26, GH27, and GH113), and xyloglucanases (GH2, GH5, GH16, GH29, GH31, and GH95). *R. solimangrovi* exhibits the capability to produce lignocellulolytic enzymes for the decomposition of empty fruit bunches (EFB) from oil palm. The integration of genomic analysis with experimental investigations has elucidated the potential of *R. solimangrovi* as a valuable candidate for the seawater biorefinery sector.

Hao et al. [45] isolated cellulose-degrading strains from humus soil in cold regions through a screening process involving sodium carboxymethyl cellulose culture and Congo red staining. Further characterization of cellulase activity included morphological observation, physiological and biochemical tests, and phylogenetic analysis of internal transcribed spacer sequences. A cold-active cellulose-degrading strain (LC-6) belonging to Tausonia pullulans was identified, exhibiting peak activities of 4.93 U/mL for carboxymethyl cellulase, 7.62 U/mL for cellobiohydrolase, 14.17 U/mL for β-glucosidase, and 7.15 U/mL for filter paper enzyme at temperatures ranging from 10 °C to 15 °C. Following fermentation at 15 °C for 15 days, LC-6 achieved degradation rates of 25.01% for hemicellulose, 16.13% for cellulose, and 24.35% for lignin. These findings demonstrate the efficient degradation of straw by strain LC-6 at low temperatures, highlighting its significant potential in lignocellulosic biomass utilization.

3.4. Other Source Systems

Apart from traditional sources, specialized ecological niches harbor microorganisms with highly effective cellulose-degrading capabilities. These strains hold significant promise for applications in treating agricultural waste, purifying industrial wastewater, and converting organic waste (Table 4). Notably, cellulolytic microorganisms naturally decompose cellulose-rich waste tissues (e.g., leaves, pseudostems) from banana plantations [46]. Legodi et al. [47] identified six highly efficient fungal strains, such as *Trichoderma longibrachiatum*, *T. harzianum*, and *Aspergillus fumigatus*, from compost. These strains displayed significant endoglucanase activity on carboxymethyl cellulose (CMC)-Congo red agar plates, with *T. harzianum* exhibiting the highest total cellulase activity at 28.6 U/mL. Sinza et al. [48] identified five bacterial strains with complete cellulase systems out of 17 isolates from paper mill wastewater, including *Bacillus sporothermodurans*, *B. cereus*, and *B. flexus*. Among these, *B. flexus* showed the highest cellobiose degradation efficiency (89% degradation rate at 72 h) with optimal activity observed at pH 6.5–7.0. Demissie et al. [49] compared the distribution of cellulose-degrading microbes in various environments such as forest soil, cattle dung, brewery waste, and industrial refuse. *Bacillus* sp. *CD1* isolated from cattle dung exhibited the highest enzymatic activity, generating a 4.5 mm clearance zone and a cellulolytic index of 3.1. This strain demonstrated peak activity under conditions of 1% CMC, 40 °C, and pH 7, highlighting livestock feces as a rich source for isolation. In contrast, only a limited number of active strains were isolated from brewery waste and industrial refuse, with no efficient degraders found in forest soil samples. These investigations broaden the possibilities for acquiring cellulose-degrading microorganisms: strains derived from agricultural waste are conducive to bio-composting, those from industrial sources demonstrate robust environmental resilience for wastewater treatment, and strains isolated from livestock feces possess comprehensive enzyme systems suitable for industrial-scale cultivation. Subsequent research should emphasize the exceptional environmental adaptability of these distinct strains, the synergistic mechanisms of multi-enzyme systems, and the enhancement of large-scale cultivation techniques (Figure 3).

Table 4. Isolation of cellulose-degrading bacteria from other systems.

Year	Source of Flora	Isolated Cellulose-Degrading Bacteria	Microbe Phylum, Genus	References
2019	Banana plantation compost	*Trichoderma longibrachiatum* *Trichoderma harzianum* *Aspergillus fumigatus*	*Ascomycota, Trichoderma*	[47]
2021	Paper Mill Wastewater Contaminated Soil (Tanzania)	*Bacillus gemmatus* *Bacillus cereus* *Bacillus flexus*	*Firmicutes*	[48]
2024	Cow dung	*Bacillus* sp. *CD1*	*Firmicutes*	[49]

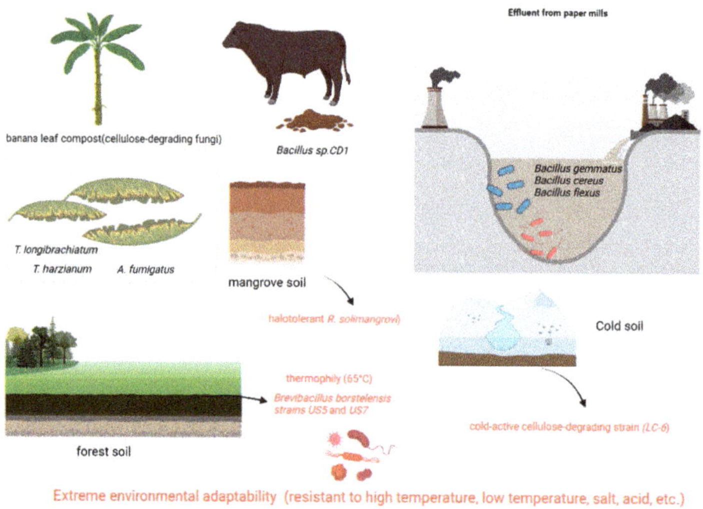

Figure 3. Cellulose-degrading bacteria from different sources (extreme environmental adaptability) (generated by bioRender).

Representative bacteria and fungi sourced from non-traditional environments, including forest soil, decaying wood, cold humus, mangrove soil, compost, industrial wastewater (paper mill), livestock dung, and agricultural waste. Key isolates exhibit exceptional tolerance to extremes: thermophily (65 °C), psychrophily (active $\leq$ 15 °C), halotolerance (mangrove), and varied pH. This inherent adaptability, coupled with significant lignocellulose degradation capabilities (e.g., cold-active *LC-6*, thermostable *Brevibacillus borstelensis US5/US7*, halotolerant *R. solimangrovi*), positions these strains as valuable resources for applications in bio-composting, wastewater treatment, biomass conversion, and biorefineries under challenging conditions.

4. Mining and Separation Techniques for Cellulose-Degrading Bacteria

4.1. Screening Strategies for Aerobic and Anaerobic Bacteria

Isolation techniques are customized based on the physiological characteristics of the target microbial strains. The streak plate method is commonly utilized for aerobic bacteria [50]. This involves enrichment culture, serial dilution, and streaking on selective media (e.g., cellulose-specific media like CMC-Na or filter paper-based media) until pure colonies are obtained. Anaerobic strains necessitate oxygen-free environments [51].

Lanjekar et al. [52] demonstrated a standard anaerobic isolation procedure using rumen samples from Pune cattle. Enrichment in PYGR broth under anaerobic conditions, followed by 16S rRNA sequencing (GenBank alignment via MAFFT) and phylogenetic tree analysis, identified Actinomyces sp. CtC72T as an obligate anaerobe. *Strain CtC72T* displayed a generation time of 2.7 h in PYGR broth under optimal conditions (48 h incubation, no aerobic growth observed).

4.2. Microbial Culturomics

The use of multiple culture conditions combined with high-throughput sequencing technology to identify the isolated strains is defined as culturomics. Compared to traditional microbial isolation and culture methods, the high-throughput and parallel nature of culturomics has earned it the status of "omics". Previously, traditional microbial culture methods were constrained by the complex nutritional requirements and interactions of microorganisms, resulting in more than 99% microorganisms in environmental samples remaining uncultured [53]. Lagier et al. [54] introduced culturomics into the field of environmental microbiology, enabling the visualization and functional analysis of uncultivable microbial resources. Currently, the primary technologies employed in culturomics fall into the following two categories: (1) droplet-based microfluidics; (2) microarray-based technology. Droplet-based microfluidics leverage two-phase fluid systems to encapsulate individual microbial cells in biocompatible oil droplets, enabling high-throughput screening and isolation under controlled conditions. Due to its advantages in high throughput, parallelization, standardization, and dynamic process, this technology is considered a key component in the workflow of single-cell microbial isolation and culture [55]. However, microorganisms that thrive in anaerobic environments, such as the rumen and intestines, encounter additional challenges when using microfluidics for the separation, culture, and analysis of individual microbial cells. Specifically, sample processing, growth detection, and microfluidic sorting must be conducted in anaerobic conditions [56]. In contrast to microfluidic-based culturomics, microarray-based technology offers a larger microbial culture space, enabling the expansion of single cells into larger volumes of high-density microbial colonies. Compared to microfluidics, separation of selected microcolonies from discrete microcompartments is generally more straightforward. At present, culturomics is infrequently employed in rumen microbial research; however, existing studies have already demonstrated its enormous potential in reducing the unknown range of rumen microbial species and exploring rumen microbial functions. Zehavi et al. [57] identified that various factors, including culture medium type, sample dilution, and phylogenetic relationships, can affect the cultivability of rumen microorganisms, with sample dilution notably increases the possibility of isolating 'uncultured' microorganisms. Additionally, microbial abundance and phylogenetic relationships are the primary determinants of the cultivability of rumen microorganisms. Rumen microorganisms are intricately associated with essential functions such as polysaccharide degradation, protein hydrolysis and synthesis, cellulose degradation, and rumen detoxification. Consequently, rumen microbial purification and cultivation based on culturomics can isolate certain key functional strains, thereby providing new insights into the rumen microbiome [53].

Culturomics not only expands microbial genetic resources but also modifies the composition and abundance of microbial communities in specific environments. The physiological health of hosts and the occurrence of diseases are closely related to changes in the composition of the gut microbiota. By transplanting strains isolated through culturomics, the structure of microbiota can be regulated, thereby playing a crucial role in restoring microbial homeostasis, with the aim of treating clinical diseases of the digestive tract in ruminant animals. Given the potential of microbial culturomics in restoring microbial balance and correcting microbial composition, isolating and culturing key microorganisms associated with diseases may provide insights into microbial communication mechanisms under specific environmental conditions.

Specialized techniques are essential for isolating cellulose-degrading bacteria. The emerging field of culturomics, which merges high-throughput culture and sequencing methods, has revolutionized microbial isolation by facilitating the growth of previously deemed "unculturable" species. Technological advancements such as microfluidics and microarray-based culturomics have bolstered efficiency, although challenges persist in anaerobic cultivation. This strategy is particularly advantageous for investigating the rumen microbiome, discovering novel cellulose-degrading strains, and elucidating their roles in biomass degradation. Future integration with multi-omics approaches will deepen understanding of microbial functions, propelling progress in biofuel production, waste management, and animal nutrition. Continuous advancements in microbial isolation are pivotal for unlocking further biotechnological and medical prospects.

5. Classification of Major Cellulose-Degrading Microorganisms

5.1. *Bacteria*

5.1.1. *Fibrobacter succinogenes*

Succinic acid, also known as butanedioic acid, is a crucial industrial chemical synthesized biologically through pathways involving *Fibrobacter succinogenes*, a strictly anaerobic Gram-negative bacterium originally isolated from the bovine rumen. This bacterium possesses unique capabilities in breaking down complex cellulose substrates such as straw [58]. *F. succinogenes* metabolizes cellulose, cellobiose, and glucose to produce acetate, succinate, and formate via anaerobic fermentation. Of particular note is its succinic acid biosynthesis pathway, which holds considerable industrial significance. Research has shown that *F. succinogenes*, a predominant cellulolytic species in the rumen, effectively degrades cellulose and hemicellulose in straw by secreting a combination of hydrolytic enzymes.

Metagenomic analyses have unveiled that the genome of *F. succinogenes* not only contains cellulase systems but also essential enzymes like pectinases, glucanases, arabinogalactanases, and xylanases. This array of enzymes works synergistically to efficiently break down intricate plant cell wall structures, particularly the interconnected network of cellulose, hemicellulose, and pectin.

P.C. Lee et al. [59] isolated a new bacterium capable of producing succinic acid from the rumen of cattle. This bacterium, named *Mannheimia succiniciproducens MBEL55E*, is a non-motile, non-spore-forming, mesophilic, and capnophilic Gram-negative bacillus or coccobacillus. Through phylogenetic classification based on 16S rRNA sequencing and physiological assessments, it was assigned to the redefined *genus Mannheimia*. Thriving within a pH range of 6–7.5 under conditions of less than 100% CO_2, this bacterium consistently produces succinic acid, acetate, and formate in a fixed ratio of 2:1:1. When cultivated anaerobically with 20 g/L glucose as the carbon source, this strain yielded 13.5 g/L of succinic acid.

Cellulase preparations have proven effective as feed additives, enhancing nutrient availability, fiber digestibility, and animal growth performance. Particularly, the recombinant expression of cellobiohydrolase genes sourced from *F. succinogenes* in lactic acid bacteria notably enhanced neutral detergent fiber degradation rates in corn stover silage.

5.1.2. *Ruminococcus albus*

Ruminococcus albus, a predominant Gram-positive strict anaerobe in the rumen, relies on exogenous vitamins and growth factors for growth and demonstrates significant enzymatic activity in lignocellulose breakdown (Table 5).

Table 5. Fiber-degrading enzyme systems and key applications of filamentous bacteria succinogenes, white ruminococcus, and yellow ruminococcus.

Culture	Key Enzyme Systems	Substrate Specificity	Critical Applications
Fibrobacter succinogenes	Cellulase system (GH family unspecified) pectinase, glucanase, arabinogalactase, xylanase	Cellulose, cellobiose, and glucose preferentially degrade cellulose-hemicellulose-pectin crosslinking network	Industrial production of succinic acid (yield 13.5 g/L) Feed additive (improve the degradation rate of neutral detergent fiber)
Ruminococcus Albus	GH5 family enzymes (62.3% of CAZymes) have prominent cellobiase activity	Cellobiose (38% more available than glucose) pH 6.7–5.5 optimum (less than 6.0 growth inhibition)	Complementarity with *xanthococcus* to maintain rumen microecological balance (proton power 60 mV)
Ruminococcus flavus	GH9 family enzymes (51.8% CAZymes). Complex enzyme system (cellulose, hemicellulose, pectin degrading enzymes)	Alkaline hydrogen peroxide treatment of wheat straw (AHPWS degradation rate 6.1 mg/d) needs phenylacetic acid/phenylacetic acid synergism	Co-culture improves fiber digestibility (e.g., *OS14 +S137 strains*) and promotes acetic acid and propionic acid production

In co-culture experiments, Yeoman et al. [60] observed spatially distinct microcolonies of *R. albus*, *Fibrobacter succinogenes*, and *Ruminococcus flavefaciens* on cellulose substrates. Introduction of *R. flavefaciens* led to a 43.2% decrease in *F. succinogenes*, yet overall cellulose degradation remained unaffected. Metatranscriptomic analysis revealed that *R. albus* predominantly expresses GH5 family enzymes (62.3% of total CAZymes), while *R. flavefaciens* relies on GH9 family enzymes (51.8%), suggesting functional complementarity through distinct enzymatic approaches to maintain efficient ruminal fiber degradation.

5.1.3. *Ruminococcus flavefaciens*

Ruminococcus flavefaciens, a key cellulolytic bacterium in the rumen, plays a crucial ecological role in ruminant digestion by producing a diverse array of enzymes that effectively degrade cellulose, hemicellulose, and pectin. Recent research has shed light on its distinctive degradation mechanisms and potential applications [61]. Boonsaen [62] and colleagues isolated six *R. flavefaciens* strains from the rumen of swamp buffaloes, among which strain *OS14* exhibited the highest fiber digestibility, either in isolation or in co-culture with non-cellulolytic bacteria. Co-culturing *OS14* with strain *S137* resulted in optimal fiber degradation and a significant increase in acetate and propionate production.

Odenyo AA et al. [63] investigated the degradation of wheat straw (WS) and alkaline hydrogen peroxide (AHP)-treated wheat straw (AHPWS) by *Ruminococcus albicans No. 8* and *Flavobacterium pyogenes FD-1* through monitoring bacterial growth (OD600) and determining dry matter disappearance (DMD) of the substrate. In the presence of phenylpropionic acid (PPA) and phenylacetic acid (PAA) in the established medium, the degradation of AHPWS (39.6 $\pm$ 2.6%) was enhanced compared to AHPWS without PPA and PAA (24.9 $\pm$ 7.6%). However, there was no synergistic effect on degradation when both species were co-cultured with WS or AHPWS substrates. Specifically, PPA and PAA did not affect the disappearance of AHPWS by *Flavobacterium pyogenes FD-1*, nor was there any observed impact when both bacteria were cultivated together. Dry matter disappearance analysis in a complex medium revealed that *Aspergillus flavus FD-1* degraded AHPWS more rapidly (6.1 mg/d) than *Aspergillus kawachii 8* (4.2 mg/d).

5.2. *Fungi*

5.2.1. White-Rot Fungi

White-rot fungi are recognized for their significant potential in selectively delignifying biomass due to their distinctive decomposing capabilities. Among *filamentous fungi*, *white-rot fungi* stand out as highly researched organisms, exhibiting exceptional efficacy and specificity in degrading lignocellulose. This unique feature substantially enhances the efficiency of lignocellulose degradation in the rumen [64].

White-rot fungi degrade lignin in roughage by employing enzymes such as lignin peroxidase, manganese peroxidase, multifunctional peroxidase, laccase, and hydrogen peroxide-generating enzyme. Manganese peroxidase and laccase specifically act on phenolic components of lignin [65]. In addition, cellulases, hemicellulases, and esterases are also thought to be extremely important in the degradation of lignocellulosic biomass [66] (Table 6).

Table 6. Fiber-degrading enzyme systems and key applications of white-rot fungi, brown-rot fungi, and yeast.

Culture	Key Enzyme Systems	Substrate Specificity	Critical Applications
White-rot fungi	Lignin-degrading enzymes (lignin peroxidase, manganese peroxidase, laccase); cellulases, hemicellulases, esterases	Preferential degradation of lignin (β-O-4′ aryl ether bond) significantly reduced lignin ($p < 0.05$) in wheat straw treated for 30 days	Feed pretreatment (increase crude protein content, metabolic energy); industrial delignification (selective degradation)
Brown-rot fungi	Simplified enzyme system (GH family hemicellulases, chitinases), hydroxyl radical mediated oxidative depolymerization of cellulose	Rapid depolymerization of cellulose/hemicellulose in Pinus yunnanensis preferential degradation of polysaccharides	Biomass pretreatment (to increase wood porosity) and *white-rot fungi* synergize to enhance digestibility
Yeast	Indirect regulation (promotion of cellulase secretion by rumen microorganisms), metabolites (e.g., short chain fatty acids)	Free sugar promotes fiber decomposition-Digestibility of crude fiber in wheat straw diet increased by 24% ($p < 0.05$)	Feed additive (increased dry matter digestibility 7.3%) improved rumen fermentation (increased TVFA, NH^3-N)

Shrivastava et al. [67] conducted solid-state fermentation of wheat straw employing *white-rot fungi* to improve rumen digestibility and nutrient availability. After a 30-day fungal treatment, a significant reduction ($p < 0.05$) was noted in acid detergent fiber, neutral detergent fiber, hemicellulose, lignin, and cellulose components of the cell wall. This fungal fermentation of wheat straw resulted in a marked elevation in crude protein content, organic matter digestibility, short-chain fatty acid production, and metabolic energy, while concurrently lowering the C/N ratio.

Erven et al. [68] isolated intact lignin to investigate the vulnerability of structural motifs within wheat straw lignin to fungal degradation. Their findings indicate that the susceptibility of β-O-4′ aryl ethers to cleavage is influenced by the electronic density and local steric hindrance of the 4′-O-coupling ring. This sensitivity of β-O-4′ aryl ether substructures aligns with the lignin decomposition mechanism that is triggered by single electron transfer.

5.2.2. Brown-Rot Fungi

Brown-rot fungi, obtained from decaying peach tissues, efficiently break down cellulose and hemicellulose while modifying lignin through demethylation, hydroxylation, and side-chain oxidation [69]. Genomic analyses have revealed a substantial simplification in the lignocellulose degradation system of *brown-rot fungi* during evolution, particularly in comparison to *white-rot fungi*, leading to a significant reduction in the abundance of cellulose-hydrolyzing and lignin-modifying enzymes [70].

Brown-rot basidiomycetes exhibit rapid cellulose depolymerization while preserving altered lignin residues. In a study by Jill Gaskell et al. [71], the analysis of Wolfiporia cocos transcriptomes and secretomes during growth on various substrates such as glucose, crystalline cellulose, aspen, or pine revealed significant gene upregulation on cellulose, aspen, and pine compared to glucose. Specifically, 30, 183, and 207 genes were upregulated by at least 4-fold on cellulose, aspen, and pine, respectively. Mass spectrometry analysis identified 64 glycoside hydrolase (GH) proteins, with 17 associated with upregulated transcripts on woody substrates, predominantly categorized as hemicellulases or chitinases. The high transcription levels of genes related to iron homeostasis, reduction, and extracellular peroxide generation support the cellulose depolymerization driven by reactive oxygen species (ROS), distinguishing it from the *brown-rot fungus* Postia placenta (Table 6).

Qi et al. [72] investigated the enzymatic decomposition and transformation of biomass by *white-rot fungi* (*Trametes versicolor*) and *brown-rot fungi* (*Gloeophyllum trabeum and Rhodonia placenta*). The study revealed that these fungi play a crucial role in breaking down wood components. Specifically, T. versicolor, a *white-rot fungus*, induces simultaneous decay and uniform biotransformation of wood constituents. In contrast, *brown-rot fungi* possess the capability to depolymerize carbohydrates and degrade polysaccharides via distinct biotransformation pathways, leading to diverse wood chemical transformations. Notably, both *white-rot* and *brown-rot fungi* pretreatments enhance wood cell porosity and accessibility, ultimately improving digestibility. Overall, the findings underscore the significant impact of these fungi on biomass degradation and transformation processes.

5.2.3. Yeast

Yeast and its metabolites are effective functional feed additives that enhance livestock production performance and are commonly utilized in the industry. *Yeast* culture improves fiber degradation in the rumen of animals consuming cellulose-containing feed by utilizing free sugars to expedite the decomposition of fiber substances, thereby altering the fermentation process.

Comert et al. [73] investigated the impact of *yeast* supplementation on in situ dry matter degradability in lambs, as well as on rumen fermentation and growth performance. Their findings indicated that the group receiving *yeast* supplementation exhibited elevated rumen total volatile fatty acid (TVFA) levels. Moreover, the combination of ammonia treatment and *yeast* supplementation resulted in enhanced dry matter degradation efficiency, average daily weight gain, and NH^3-N levels. The study concluded that the synergistic effect of ammonia treatment and *yeast* supplementation yielded superior enhancement in the feed value of wheat straw compared to either treatment in isolation. This combined approach was shown to enhance rumen fermentation efficiency and the potential productivity of ruminants.

Ben et al. [74] investigated the impact of Saccharomyces cerevisiae on the feed intake and nutrient digestibility of rams fed a wheat straw-based diet. Their findings revealed that, in comparison to the control group, the inclusion of Saccharomyces cerevisiae led to a significant increase in dry matter digestibility by 7.3% ($p < 0.05$), organic matter by 11.9%, and crude fiber by 24%. These results suggest that Saccharomyces cerevisiae has the potential to enhance nutrient digestibility in the diet of rams.

Yeast enhances fiber digestion and animal performance through various mechanisms, including the regulation of cellulose-degrading bacteria activity and rumen microbial populations. Research findings indicate that *yeast* additives facilitate early colonization of rumen microflora, leading to a notable increase in protozoa, *fungi*, and *lactic acid bacteria* populations [75]. Studies have identified yeast and its cultures as beneficial for improving fiber digestibility (Table 6).

6. Mining of Cellulose-Degrading Enzyme Genes

6.1. Metagenomics

Metagenomics, initially proposed by Handelsman in 1998, entails the extraction of total microbial DNA from environmental samples for subsequent sequencing. This method enables the direct identification of microbial functions, genomic relative abundance, and taxonomic resolution at the species level. By scrutinizing the genomes of pivotal cellulolytic strains, researchers can reconstruct potential functional pathways and modules to investigate microbial functionalities and genetic diversity. Metagenomics also unveils taxonomic profiles and relative abundances while delving into community-wide biological functions.

Bai [76] utilized metagenomics to scrutinize the gut microbiota of Chinese bamboo rats (Rhizomys sinensis), known for their high-fiber diets. The investigation pinpointed abundant lignocellulose-degrading bacterial populations, with crucial functions associated with carbohydrate, amino acid, and nucleic acid metabolism. Carbohydrate esterases (CEs) and glycoside hydrolases (GHs) predominated among the CAZyme families in the cecum, primarily contributed by *Firmicutes*, *Bacteroidetes*, and *Proteobacteria*. A total of 587 CAZyme genes were identified, encompassing 7 CE and 21 GH families. Chai et al. [77] unearthed a novel bifunctional cellulase/hemicellulase (ZFYN184) from a soil metagenomic library. Sequence analysis categorized ZFYN184 into the GH44 family, featuring a single catalytic domain responsible for both endo-β-1,4-glucanase and β-mannanase activities. Meng et al. [9] characterized CelXyn2, a stable bifunctional cellulase-xylanase sourced from buffalo rumen metagenomes. CelXyn2 exhibited hydrolytic activity on rice straw, wheat straw, Leymus chinensis, and beet pulp, showcasing enhanced lignocellulose hydrolysis and in vitro rumen fermentation.

6.2. Metatranscriptomics

Although metagenome can identify potential functional genes, transcriptome can verify their actual expression. Macrotranscriptomics is a method used to extract all RNA from environmental samples collected at a specific time and location. This technique allows for the analysis of genome transcription and regulation at the transcriptional level, providing insights into the growth status and patterns of microorganisms, as well as enabling the investigation of the metabolic potential of microbial communities [78].

Fang et al. [10] investigated the utilization of corncob lignocellulose by Auricularia heimuer at various growth stages, analyzing the expression of CAZymes. They identified 46 differentially expressed CAZyme genes associated with lignocellulose degradation, demonstrating efficient cellulose utilization from corncobs. He et al. [79] studied the rumen microbiota of Hu sheep using metatranscriptomics, identifying *Firmicutes* and *Bacteroidetes* as the dominant phyla contributing to CAZyme expression. Among 14,489 genes annotated to 15 cellulase-containing GH families, GH3, GH5, and GH9 were the most prevalent. Two GH5 proteins (Cel5A-h28 and Cel5A-h11) showed high specific activities (222.2 and 142.8 U/mg) against carboxymethyl cellulose and p-nitrophenyl-β-d-cellobioside. Yu et al. [80] characterized fungal species in termite guts through internal transcribed spacer sequencing and genomic analysis. They identified *Talaromyces funiculosus*, an ascomycete containing an AA9 lytic polysaccharide monooxygenase (TfAA9A). Reverse transcription-PCR confirmed TfAA9A expression in termite guts, and heterologous expression in Pichia pastoris validated its cellulose-oxidizing capability.

6.3. Metaproteomics

Metaproteomics, the comprehensive protein profile of a microbial community in a specific environment at a particular time, represents an emerging field subsequent to metagenomics [81]. While genes and transcripts can predict functionality, protein abundance directly reflects functional activities within specific spatiotemporal contexts [82]. Wang et al. [11] conducted a study on bacterial and fungal diversity in mushroom residue compost using 16S rRNA sequencing and label-free quantitative proteomics. They identified 22,815 proteins, with cellulase-related proteins exhibiting stage-specific expression patterns that correlated with different composting phases. In a separate study, Toyoda et al. [83] employed metaproteomics to isolate cellulose-binding proteins (CBPs) from sheep rumen. Through SDS-PAGE and mass spectrometry analysis, they identified four CBPs from *Fibrobacter succinogenes*, which included a tetratricopeptide repeat domain protein, an

OmpA family protein, a fibronectin domain protein, and cellulase F (EGF), all crucial for cellulose degradation in the rumen (Figure 4).

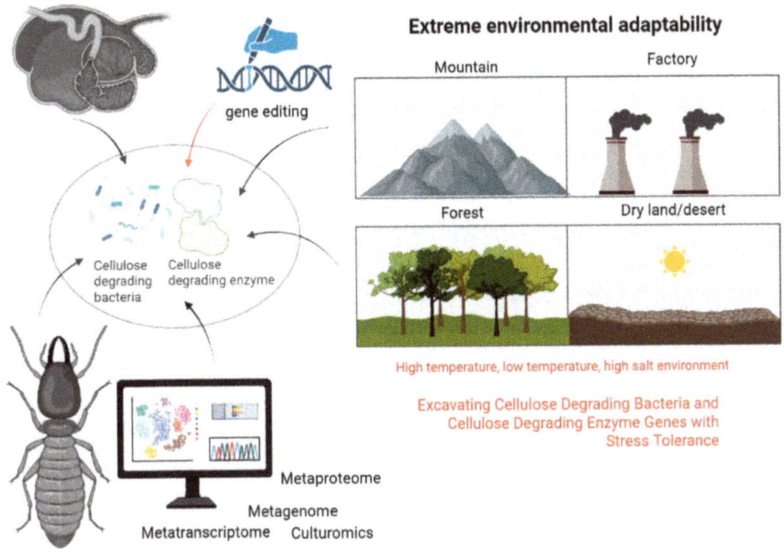

Figure 4. Multi-omics and gene editing methods for mining and modifying cellulose-degrading bacteria and enzymes (generated by bioRender).

6.4. Gene Editing Technology

Recently, CRISPR/Cas9 systems have shown significant promise for genome manipulation in *filamentous fungi* and industrial microorganisms, particularly in coordinating the expression of multiple genes. Several investigations have successfully achieved the targeted evolution and effective production of enzymes that degrade lignocellulose using this method.

Liu et al. [84] have developed an efficient CRISPR/Cas9 system for multiplex genome engineering in *T. thermophila* and *M. heterothallica*. This system enables precise mutation of the amdS gene via NHEJ-mediated events. The cre-1, res-1, gh1-1, and alp-1 genes involved in cellulase production were targeted for editing. By a single transformation with the CRISPR/Cas9 system, up to four sites could be simultaneously disrupted through neomycin selection marker integration. Using this tool, multiple strains were created, showing significantly enhanced cellulase production, with extracellular secretory protein and lignocellulase activity increasing by up to 5-fold and 13-fold, respectively, compared to the wild-type strain.

Liu et al. [85] markedly improved the cellulose degradation capacity of *Bacillus subtilis RLI2019* by employing CRISPR/Cas9-mediated multi-gene integration. Specifically, eglS, cel48S, and bglS genes were successively integrated into the aprE, epr, and amyE loci of *Bacillus subtilis RLI2019*, resulting in the creation of the optimized strain *Bacillus subtilis AEA3*. This engineered strain demonstrates efficient degradation of crop straw, leading to the production of monosaccharides.

Bacillus licheniformis 2709 underwent marker-free genome editing to disrupt the natural lchAC gene linked to foam production and the eps cluster responsible for extracellular mucopolysaccharides. Zhou et al. [86] optimized the expression of the alkaline protease gene (aprE) by evaluating various modular plasmids and genomic loci to identify the most efficient expression system. The study revealed that genomic expression of aprE outperformed plasmid expression. By optimizing the host and integrating the gene near the replication origin on the chromosome, the transcription level of aprE increased by 1.67-fold, leading to a significant 62.19% enhancement in enzyme activity compared to the wild-type strain producing alkaline protease.

7. Conclusions and Perspectives

Cellulose-degrading microorganisms, such as anaerobic bacteria in the rumen of ruminants, symbiotic bacteria in insect intestines, *white-rot fungi*, and *brown-rot fungi*, are prevalent in nature. Recent advancements in multi-omics technologies have facilitated a comprehensive analysis of the degradation mechanisms of these microbial celluloses. Techniques such as macrogenomics, macrotranscriptomics, and macroproteomics have been instrumental in elucidating these processes. Moreover, gene editing tools like CRISPR/Cas9 have been employed to enhance the cellulose-degrading capabilities of bacteria, leading to notable improvements in enzyme activity and substrate utilization efficiency. The utilization of complex microbial consortia in straw feed has yielded significant advancements in this field.

In the future, there is a need to focus on harnessing microbial resources from extreme environments. This includes identifying strains with high salt tolerance and low-temperature resilience, isolating cellulose-degrading bacteria from ruminants' rumen, and addressing challenges related to anaerobic bacteria that hinder their practical application. Understanding the cooperative degradation mechanisms within microbial communities, leveraging multi-component technologies and synthetic biology to engineer efficient bacterial consortia for lignocellulose degradation, and optimizing fermentation processes using machine learning algorithms are crucial steps. Enhancing the catalytic efficiency and stability of cellulose-degrading enzymes through gene editing and directed evolution techniques is essential for advancing industrial applications. Furthermore, evaluating the ecological impact of large-scale microbial applications, developing cost-effective pretreatment methods, and promoting the sustainable utilization of agricultural residues are imperative for the industrialization and ecological sustainability of these technologies (Figure 4).

The technology for microbial cellulose degradation holds significant promise across various applications. Future advancements will necessitate interdisciplinary collaboration to explore microbial resources, enhance degradation techniques, facilitate industrial implementation, and offer technical assistance to achieve the high-value utilization and carbon neutrality of agricultural residues.

The integrated use of metagenomics, metatranscriptomics, and macroproteomics holds promise for identifying new cellulose-degrading microorganisms and enzymes across diverse environments, including the rumen, insect gut, and extreme ecological habitats. By leveraging CRISPR/Cas9 technology for genetic engineering, enhancements in enzyme performance, stability, and microbial consortia efficacy can be achieved, leading to enhanced lignocellulose conversion for applications like straw utilization and biomass valorization.

Author Contributions: Wrote the paper: X.L. and Z.Q. Revised the manuscript: X.L., C.C., C.Z. and J.H. All authors have read and agreed to the published version of the manuscript.

Funding: This work was jointly supported by the High-level Talents Introduction Project of Anhui Science and Technology University (DKYJ202303), Projects of International Cooperation and Exchanges NSFC (32261143467), and the National Key Research and Development Program (2022YFD1300805).

Institutional Review Board Statement: Not applicable.

Informed Consent Statement: Not applicable.

Data Availability Statement: Data sharing is not applicable to this article as no datasets were generated or analyzed during the current study.

Conflicts of Interest: The authors declare no conflicts of interest.

References

1. Zhang, A.; Zhang, X.; Liang, Q.; Sun, M. Co-application of straw incorporation and biochar addition stimulated soil N_2O and NH_3 productions. *PLoS ONE* **2024**, *19*, e0289300. [CrossRef] [PubMed]
2. Sun, M.; Xu, X.; Wang, C.; Bai, Y.; Fu, C.; Zhang, L.; Fu, R.; Wang, Y. Environmental burdens of the comprehensive utilization of straw: Wheat straw utilization from a life-cycle perspective. *J. Clean. Prod.* **2020**, *259*, 120702. [CrossRef]
3. Chen, L.; Hong, F.; Yang, X.-X.; Han, S.-F. Biotransformation of wheat straw to bacterial cellulose and its mechanism. *Bioresour. Technol.* **2013**, *135*, 464–468. [CrossRef] [PubMed]
4. Gong, X.; Zou, H.; Qian, C.; Yu, Y.; Hao, Y.; Li, L.; Wang, Q.; Jiang, Y.; Ma, J. Construction of in situ degradation bacteria of corn straw and analysis of its degradation efficiency. *Ann. Microbiol.* **2020**, *70*, 62. [CrossRef]
5. Trubetskaya, A.; Jensen, P.A.; Jensen, A.D.; Steibel, M.; Spliethoff, H.; Glarborg, P.; Larsen, F.H. Comparison of high temperature chars of wheat straw and rice husk with respect to chemistry, morphology and reactivity. *Biomass Bioenergy* **2016**, *86*, 76–87. [CrossRef]
6. Zhang, Y.; Wang, H.; Sun, X.; Wang, Y.; Liu, Z. Separation and characterization of biomass components (cellulose, hemicellulose, and lignin) from corn stalk. *BioResources* **2021**, *16*, 7205–7219. [CrossRef]
7. Hsu, T.-C.; Guo, G.-L.; Chen, W.-H.; Hwang, W.-S. Effect of dilute acid pretreatment of rice straw on structural properties and enzymatic hydrolysis. *Bioresour. Technol.* **2010**, *101*, 4907–4913. [CrossRef]
8. Qiu, C.; Liu, N.; Diao, X.; He, L.; Zhou, H.; Zhang, W. Effects of Cellulase and Xylanase on Fermentation Characteristics, Chemical Composition and Bacterial Community of the Mixed Silage of King Grass and Rice Straw. *Microorganisms* **2024**, *12*, 561. [CrossRef]
9. Meng, Z.; Ma, J.; Sun, Z.; Yang, C.; Leng, J.; Zhu, W.; Cheng, Y. Characterization of a novel bifunctional enzyme from buffalo rumen metagenome and its effect on in vitro ruminal fermentation and microbial community composition. *Anim. Nutr.* **2023**, *13*, 137–149. [CrossRef]
10. Fang, M.; Sun, X.; Yao, F.; Lu, L.; Ma, X.; Shao, K.; Kaimoyo, E. A Combination of Transcriptome and Enzyme Activity Analysis Unveils Key Genes and Patterns of Corncob Lignocellulose Degradation by Auricularia heimuer under Cultivation Conditions. *J. Fungi* **2024**, *10*, 545. [CrossRef]
11. Wang, C.; Wang, Y.; Ru, H.; He, T.; Sun, N. Study on Microbial Community Succession and Functional Analysis during Biodegradation of Mushroom Residue. *BioMed Res. Int.* **2021**, *2021*, 6620574. [CrossRef] [PubMed]
12. Mei, J.; Shen, X.; Gang, L.; Xu, H.; Wu, F.; Sheng, L. A novel lignin degradation bacteria-Bacillus amyloliquefaciens SL-7 used to degrade straw lignin efficiently. *Bioresour. Technol.* **2020**, *310*, 123445. [CrossRef]
13. Luo, C.; Li, Y.; Chen, Y.; Fu, C.; Nong, X.; Yang, Y. Degradation of bamboo lignocellulose by bamboo snout beetle Cyrtotrachelus buqueti in vivo and vitro: Efficiency and mechanism. *Biotechnol. Biofuels* **2019**, *12*, 75. [CrossRef]
14. Sari, W.N.; Safika; Darmawi; Fahrimal, Y. Isolation and identification of a cellulolytic Enterobacter from rumen of Aceh cattle. *Vet. World* **2017**, *10*, 1515–1520. [CrossRef] [PubMed]
15. Ma, D.; Chen, H.; Liu, D.; Feng, C.; Hua, Y.; Gu, T.; Guo, X.; Zhou, Y.; Wang, H.; Tong, G.; et al. Soil-derived cellulose-degrading bacteria: Screening, identification, the optimization of fermentation conditions, and their whole genome sequencing. *Front. Microbiol.* **2024**, *15*, 1409697. [CrossRef] [PubMed]
16. Chen, J.; Zhang, B.; Luo, L.; Zhang, F.; Yi, Y.; Shan, Y.; Liu, B.; Zhou, Y.; Wang, X.; Lü, X. A review on recycling techniques for bioethanol production from lignocellulosic biomass. *Renew. Sustain. Energy Rev.* **2021**, *149*, 111370. [CrossRef]
17. Zhang, S.; Dong, Z.; Shi, J.; Yang, C.; Fang, Y.; Chen, G.; Chen, H.; Tian, C. Enzymatic hydrolysis of corn stover lignin by laccase, lignin peroxidase, and manganese peroxidase. *Bioresour. Technol.* **2022**, *361*, 127699. [CrossRef] [PubMed]
18. Falade, A.O.; Nwodo, U.U.; Iweriebor, B.C.; Green, E.; Mabinya, L.V.; Okoh, A.I. Lignin peroxidase functionalities and prospective applications. *Microbiol. Open* **2017**, *6*, e00394. [CrossRef]
19. Atiwesh, G.; Parrish, C.C.; Banoub, J.; Le, T.-A.T. Lignin degradation by microorganisms: A review. *Biotechnol. Prog.* **2022**, *38*, e3226. [CrossRef]
20. Faruk, O.; Bledzki, A.K.; Fink, H.-P.; Sain, M. Biocomposites reinforced with natural fibers: 2000–2010. *Prog. Polym. Sci.* **2012**, *37*, 1552–1596. [CrossRef]
21. Wohlert, M.; Benselfelt, T.; Wågberg, L.; Furó, I.; Berglund, L.A.; Wohlert, J. Cellulose and the role of hydrogen bonds: Not in charge of everything. *Cellulose* **2022**, *29*, 1–23. [CrossRef]
22. Weimer, P.J. Degradation of Cellulose and Hemicellulose by Ruminal Microorganisms. *Microorganisms* **2022**, *10*, 2345. [CrossRef] [PubMed]

23. Wang, Z.Y.; Wang, R.X.; Zhou, J.S.; Cheng, J.F.; Li, Y.H. An assessment of the genomics, comparative genomics and cellulose degradation potential of Mucilaginibacter polytrichastri strain RG4-7. *Bioresour. Technol.* **2020**, *297*, 122389. [CrossRef] [PubMed]
24. Ríos-Ríos, K.L.; Dejonghe, W.; Vanbroekhoven, K.; Rakotoarivonina, H.; Rémond, C. Enzymatic Production of Xylo-oligosaccharides from Destarched Wheat Bran and the Impact of Their Degree of Polymerization and Substituents on Their Utilization as a Carbon Source by Probiotic Bacteria. *J. Agric. Food Chem.* **2021**, *69*, 13217–13226. [CrossRef]
25. Tao, J.; Song, S.; Qu, C. Recent Progress on Conversion of Lignocellulosic Biomass by MOF-Immobilized Enzyme. *Polymers* **2024**, *16*, 1010. [CrossRef]
26. Dar, M.A.; Pawar, K.D.; Pandit, R.S. Prospecting the gut fluid of giant African land snail, Achatina fulica for cellulose degrading bacteria. *Int. Biodeterior. Biodegrad.* **2018**, *126*, 103–111. [CrossRef]
27. Oktiarni, D.; Hermansyah; Hasanudin; Miksusanti; Nofyan, E.; Kasmiarti, G. Isolation and Identification Cellulolytic Bacteria from Termite Gut Obtained from Indralaya Peatland area. *IOP Conf. Ser. Earth Environ. Sci.* **2021**, *926*, 012024. [CrossRef]
28. Dar, M.A.; Shaikh, A.F.; Pawar, K.D.; Xie, R.; Sun, J.; Kandasamy, S.; Pandit, R.S. Evaluation of cellulose degrading bacteria isolated from the gut-system of cotton bollworm, Helicoverpa armigera and their potential values in biomass conversion. *PeerJ* **2021**, *9*, e11254. [CrossRef]
29. Nelson, K.; Muge, E.; Wamalwa, B. Cellulolytic Bacillus species isolated from the gut of the desert locust Schistocerca gregaria. *Sci. Afr.* **2021**, *11*, e00665. [CrossRef]
30. Silva, É.B.R.d.; Silva, J.A.R.d.; Silva, W.C.d.; Belo, T.S.; Sousa, C.E.L.; Santos, M.R.P.d.; Neves, K.A.L.; Rodrigues, T.C.G.d.C. Camargo-Júnior, R.N.C.; Lourenço-Júnior, J.d.B. A Review of the Rumen Microbiota and the Different Molecular Techniques Used to Identify Microorganisms Found in the Rumen Fluid of Ruminants. *Animals* **2024**, *14*, 1448. [CrossRef]
31. Morgavi, D.P.; Kelly, W.J.; Janssen, P.H.; Attwood, G.T. Rumen microbial (meta)genomics and its application to ruminant production. *Animal* **2013**, *7*, 184–201. [CrossRef] [PubMed]
32. Pang, J.; Liu, Z.-Y.; Hao, M.; Zhang, Y.-F.; Qi, Q.-S. An isolated cellulolytic Escherichia coli from bovine rumen produces ethanol and hydrogen from corn straw. *Biotechnol. Biofuels* **2017**, *10*, 165. [CrossRef]
33. Zhao, C.; Wang, L.; Ke, S.; Chen, X.; Kenéz, Á.; Xu, W.; Wang, D.; Zhang, F.; Li, Y.; Cui, Z.; et al. Yak rumen microbiome elevates fiber degradation ability and alters rumen fermentation pattern to increase feed efficiency. *Anim. Nutr.* **2022**, *11*, 201–214. [CrossRef] [PubMed]
34. Guder, D.G.; Krishna, M.S.R. Isolation and Characterization of Potential Cellulose Degrading Bacteria from Sheep Rumen. *J. Pure Appl. Microbiol.* **2019**, *13*, 1831–1839. [CrossRef]
35. Ben Ghalib, K.; Chadli, M.; Daştan, S.D.; Elmtili, N. Isolation and molecular identification of cellulose-degrading bacteria from rumen sheep "Ovis aries" and evaluation of their cellulase production. *Sci. Afr.* **2024**, *26*, e02439. [CrossRef]
36. Yang, J.; Zhao, J.; Wang, B.; Yu, Z. Unraveling aerobic cultivable cellulolytic microorganisms within the gastrointestinal tract of sheep (Ovis aries) and their evaluation for cellulose biodegradation. *Can. J. Microbiol.* **2022**, *68*, 237–248. [CrossRef]
37. Tulsani, N.J.; Jakhesara, S.J.; Hinsu, A.T.; Jyotsana, B.; Dafale, N.A.; Patil, N.V.; Purohit, H.J.; Joshi, C.G. Genome analysis and CAZy repertoire of a novel fungus Aspergillus sydowii C6d with lignocellulolytic ability isolated from camel rumen. *Electron. J Biotechnol.* **2022**, *59*, 36–45. [CrossRef]
38. Pramartaa, I.Q.; Wiryawan, K.G.; Suharti, S. Microbial Protein Synthesis by Cellulolytic Bacterial Isolates from Feces of Indonesian Endemic Herbivores. *Indones. J. Appl. Res. (IJAR)* **2024**, *5*, 147–155. [CrossRef]
39. Mokale Kognou, A.L.; Chio, C.; Khatiwada, J.R.; Shrestha, S.; Chen, X.; Han, S.; Li, H.; Jiang, Z.-H.; Xu, C.C.; Qin, W. Characterization of Cellulose-Degrading Bacteria Isolated from Soil and the Optimization of Their Culture Conditions for Cellulase Production. *Appl. Biochem. Biotechnol.* **2022**, *194*, 5060–5082. [CrossRef] [PubMed]
40. Ma, L.; Lu, Y.; Yan, H.; Wang, X.; Yi, Y.; Shan, Y.; Liu, B.; Zhou, Y.; Lü, X. Screening of cellulolytic bacteria from rotten wood of Qinling (China) for biomass degradation and cloning of cellulases from Bacillus methylotrophicus. *BMC Biotechnol.* **2020**, *20*, 2. [CrossRef]
41. Naresh, S.; Kunasundari, B.; Gunny, A.A.N.; Teoh, Y.P.; Shuit, S.H.; Ng, Q.H.; Hoo, P.Y. Isolation and Partial Characterisation of Thermophilic Cellulolytic Bacteria from North Malaysian Tropical Mangrove Soil. *Trop. Life Sci. Res.* **2019**, *30*, 123–147. [CrossRef]
42. Khosravi, F.; Khaleghi, M.; Naghavi, H. Screening and identification of cellulose-degrading bacteria from soil and leaves at Kerman province, Iran. *Arch. Microbiol.* **2021**, *204*, 88. [CrossRef] [PubMed]
43. Du, C.; Li, C.; Cao, P.; Li, T.; Du, D.; Wang, X.; Zhao, J.; Xiang, W. Massilia cellulosiltytica sp. nov., a novel cellulose-degrading bacterium isolated from rhizosphere soil of rice (*Oryza sativa* L.) and its whole genome analysis. *Antonie Van Leeuwenhoek* **2021**, *114*, 1529–1540. [CrossRef]
44. Lam, M.Q.; Oates, N.C.; Leadbeater, D.R.; Goh, K.M.; Yahya, A.; Md Salleh, M.; Ibrahim, Z.; Bruce, N.C.; Chong, C.S. Genomic Analysis to Elucidate the Lignocellulose Degrading Capability of a New Halophile Robertkochia solimangrovi. *Genes* **2022**, *13*, 2135. [CrossRef] [PubMed]
45. Hao, X.; Jianzheng, L.; Jun, X.; Yiyang, F.; Furao, W.; Meng, J. Isolation and characterization of a low-temperature and cellulose-degrading fungus *Tausonia pullulans* LC-6. *Environ. Technol.* **2025**, 1–10. [CrossRef] [PubMed]

46. Li, K.; Fu, S.; Zhan, H.; Zhan, Y.; Lucia, L.A. Analysis of the chemical composition and morphological structure of banana pseudo-stem. *BioResources* **2010**, *5*, 576–585. [CrossRef]
47. Legodi, L.M.; La Grange, D.; van Rensburg, E.L.J.; Ncube, I. Isolation of Cellulose Degrading Fungi from Decaying Banana Pseudostem and Strelitzia alba. *Enzym. Res.* **2019**, *2019*, 1390890. [CrossRef]
48. Sinza, E.; Mwakilili, A.; Mpinda, C.; Lyantagaye, S. Cellulase-producing bacteria isolated from Mufindi Paper Mill industrial effluent, Iringa Tanzania. *Tanzan. J. Sci.* **2021**, *47*, 204–213.
49. Demissie, M.S.; Legesse, N.H.; Tesema, A.A. Isolation and characterization of cellulase producing bacteria from forest, cow dung, Dashen brewery and agro-industrial waste. *PLoS ONE* **2024**, *19*, e0301607. [CrossRef]
50. Hussain, A.A.; Abdel-Salam, M.S.; Abo-Ghalia, H.H.; Hegazy, W.K.; Hafez, S.S. Optimization and molecular identification of novel cellulose degrading bacteria isolated from Egyptian environment. *J. Genet. Eng. Biotechnol.* **2017**, *15*, 77–85. [CrossRef]
51. Khatri, D.; Chhetri, S.B.B. Reducing Sugar, Total Phenolic Content, and Antioxidant Potential of Nepalese Plants. *BioMed Res. Int.* **2020**, *2020*, 7296859. [CrossRef] [PubMed]
52. Lanjekar, V.B.; Hivarkar, S.S.; Vasudevan, G.; Joshi, A.; Dhakephalkar, P.K.; Dagar, S.S. Actinomyces ruminis sp. nov., an obligately anaerobic bacterium isolated from the rumen of cattle. *Arch. Microbiol.* **2022**, *205*, 9. [CrossRef]
53. Stewart, R.D.; Auffret, M.D.; Warr, A.; Walker, A.W.; Roehe, R.; Watson, M. Compendium of 4941 rumen metagenome-assembled genomes for rumen microbiome biology and enzyme discovery. *Nat. Biotechnol.* **2019**, *37*, 953–961. [CrossRef] [PubMed]
54. Lagier, J.C.; Armougom, F.; Million, M.; Hugon, P.; Pagnier, I.; Robert, C.; Bittar, F.; Fournous, G.; Gimenez, G.; Maraninchi, M.; et al. Microbial culturomics: Paradigm shift in the human gut microbiome study. *Clin. Microbiol. Infect.* **2012**, *18*, 1185–1193. [CrossRef]
55. Kaminski, T.S.; Scheler, O.; Garstecki, P. Droplet microfluidics for microbiology: Techniques, applications and challenges. *Lab A Chip* **2016**, *16*, 2168–2187. [CrossRef] [PubMed]
56. Villa, M.M.; Bloom, R.J.; Silverman, J.D.; Durand, H.K.; Jiang, S.; Wu, A.; Dallow, E.P.; Huang, S.; You, L.; David, L.A. Interindividual Variation in Dietary Carbohydrate Metabolism by Gut Bacteria Revealed with Droplet Microfluidic Culture. *mSystems* **2020**, *5*. [CrossRef]
57. Zehavi, T.; Probst, M.; Mizrahi, I. Insights Into Culturomics of the Rumen Microbiome. *Front. Microbiol.* **2018**, *9*, 1999. [CrossRef]
58. Zhang, Z.; Li, F.; Ma, X.; Li, F.; Wang, Z. Effects of Barley Starch Level in Diet on Fermentation and Microflora in Rumen of Hu Sheep. *Animals* **2022**, *12*, 1941. [CrossRef]
59. Lee, P.; Lee, S.; Hong, S.; Chang, H. Isolation and characterization of a new succinic acid-producing bacterium, Mannheimia succiniciproducens MBEL55E, from bovine rumen. *Appl. Microbiol. Biotechnol.* **2002**, *58*, 663–668. [CrossRef]
60. Yeoman Carl, J.; Fields Christopher, J.; Lepercq, P.; Ruiz, P.; Forano, E.; White Bryan, A.; Mosoni, P. In Vivo Competitions between Fibrobacter succinogenes, Ruminococcus flavefaciens, and Ruminoccus albus in a Gnotobiotic Sheep Model Revealed by Multi-Omic Analyses. *mBio* **2021**, *12*. [CrossRef]
61. Dassa, B.; Borovok, I.; Ruimy-Israeli, V.; Lamed, R.; Flint, H.J.; Duncan, S.H.; Henrissat, B.; Coutinho, P.; Morrison, M.; Mosoni, P.; et al. Rumen Cellulosomics: Divergent Fiber-Degrading Strategies Revealed by Comparative Genome-Wide Analysis of Six Ruminococcal Strains. *PLoS ONE* **2014**, *9*, e99221. [CrossRef] [PubMed]
62. Boonsaen, P.; Poonko, S.; Kanjanapruetipong, J.; Phiriyangkul, P.; Sawanon, S. Isolation and partial characterization of Ruminococcus flavefaciens from the rumen of swamp buffalo. *Buffalo Bull.* **2019**, *38*, 311–325.
63. Odenyo, A.A.; Mackie, R.I.; Fahey, G.C., Jr.; White, B.A. Degradation of wheat straw and alkaline hydrogen peroxide-treated wheat straw by Ruminococcus albus 8 and Ruminococcus flavefaciens FD-1. *J. Anim. Sci.* **1991**, *69*, 819–826. [CrossRef]
64. van Erven, G.; Nayan, N.; Sonnenberg, A.S.M.; Hendriks, W.H.; Cone, J.W.; Kabel, M.A. Mechanistic insight in the selective delignification of wheat straw by three white-rot fungal species through quantitative 13C-IS py-GC–MS and whole cell wall HSQC NMR. *Biotechnol. Biofuels* **2018**, *11*, 262. [CrossRef] [PubMed]
65. Arora, D.S.; Chander, M.; Gill, P.K. Involvement of lignin peroxidase, manganese peroxidase and laccase in degradation and selective ligninolysis of wheat straw. *Int. Biodeterior. Biodegrad.* **2002**, *50*, 115–120. [CrossRef]
66. Tabka, M.G.; Herpoël-Gimbert, I.; Monod, F.; Asther, M.; Sigoillot, J.C. Enzymatic saccharification of wheat straw for bioethanol production by a combined cellulase xylanase and feruloyl esterase treatment. *Enzym. Microb. Technol.* **2006**, *39*, 897–902. [CrossRef]
67. Shrivastava, B.; Thakur, S.; Khasa, Y.P.; Gupte, A.; Puniya, A.K.; Kuhad, R.C. White-rot fungal conversion of wheat straw to energy rich cattle feed. *Biodegradation* **2011**, *22*, 823–831. [CrossRef]
68. van Erven, G.; Wang, J.; Sun, P.; de Waard, P.; van der Putten, J.; Frissen, G.E.; Gosselink, R.J.A.; Zinovyev, G.; Potthast, A.; van Berkel, W.J.H.; et al. Structural Motifs of Wheat Straw Lignin Differ in Susceptibility to Degradation by the White-Rot Fungus Ceriporiopsis subvermispora. *ACS Sustain. Chem. Eng.* **2019**, *7*, 20032–20042. [CrossRef]
69. Arantes, V.; Jellison, J.; Goodell, B. Peculiarities of brown-rot fungi and biochemical Fenton reaction with regard to their potential as a model for bioprocessing biomass. *Appl. Microbiol. Biotechnol.* **2012**, *94*, 323–338. [CrossRef]

70. Eastwood, D.C.; Floudas, D.; Binder, M.; Majcherczyk, A.; Schneider, P.; Aerts, A.; Asiegbu, F.O.; Baker, S.E.; Barry, K.; Bendiksby, M.; et al. The Plant Cell Wall–Decomposing Machinery Underlies the Functional Diversity of Forest Fungi. *Science* **2011**, *333*, 762–765. [CrossRef]
71. Gaskell, J.; Blanchette, R.A.; Stewart, P.E.; BonDurant, S.S.; Adams, M.; Sabat, G.; Kersten, P.; Cullen, D. Transcriptome and Secretome Analyses of the Wood Decay Fungus Wolfiporia cocos Support Alternative Mechanisms of Lignocellulose Conversion. *Appl. Environ. Microbiol.* **2016**, *82*, 3979–3987. [CrossRef] [PubMed]
72. Qi, J.; Zhang, X.; Zhou, Y.; Zhang, C.; Wen, J.; Deng, S.; Luo, B.; Fan, M.; Xia, Y. Selectively enzymatic conversion of wood constituents with white and brown rot fungi. *Ind. Crops Prod.* **2023**, *199*, 116703. [CrossRef]
73. Cömert, M.; Şayan, Y.; Özelçam, H.; Baykal, G.Y. Effects of Saccharomyces cerevisiae Supplementation and Anhydrous Ammonia Treatment of Wheat Straw on In-situ Degradability and, Rumen Fermentation and Growth Performance of Yearling Lambs. *Asian-Australas J. Anim. Sci.* **2015**, *28*, 639–646. [CrossRef]
74. Ben Saïd, S.; Jabri, J.; Amiri, S.; Aroua, M.; Najjar, A.; Khaldi, S.; Maalaoui, Z.; Kammoun, M.; Mahouachi, M. Effect of Saccharomyces cerevisiae Supplementation on Reproductive Performance and Ruminal Digestibility of Queue Fine de l'Ouest Adult Rams Fed a Wheat Straw-Based Diet. *Agriculture* **2022**, *12*, 1268. [CrossRef]
75. Ding, G.; Chang, Y.; Zhao, L.; Zhou, Z.; Ren, L.; Meng, Q. Effect of Saccharomyces cerevisiae on alfalfa nutrient degradation characteristics and rumen microbial populations of steers fed diets with different concentrate-to-forage ratios. *J. Anim. Sci. Biotechnol.* **2014**, *5*, 24. [CrossRef] [PubMed]
76. Bai, D.-P.; Lin, X.-Y.; Hu, Y.-Q.; Chen, Z.-Z.; Chen, L.; Huang, Y.-F.; Huang, X.-H.; Li, J. Metagenomics approach to identify lignocellulose-degrading enzymes in the gut microbiota of the Chinese bamboo rat cecum. *Electron. J. Biotechnol.* **2021**, *50*, 29–36. [CrossRef]
77. Chai, S.; Zhang, X.; Jia, Z.; Xu, X.; Zhang, Y.; Wang, S.; Feng, Z. Identification and characterization of a novel bifunctional cellulase/hemicellulase from a soil metagenomic library. *Appl. Microbiol. Biotechnol.* **2020**, *104*, 7563–7572. [CrossRef]
78. Wang, L.; Hatem, A.; Catalyurek, U.V.; Morrison, M.; Yu, Z. Metagenomic Insights into the Carbohydrate-Active Enzymes Carried by the Microorganisms Adhering to Solid Digesta in the Rumen of Cows. *PLoS ONE* **2013**, *8*, e78507. [CrossRef]
79. He, B.; Jin, S.; Cao, J.; Mi, L.; Wang, J. Metatranscriptomics of the Hu sheep rumen microbiome reveals novel cellulases. *Biotechnol. Biofuels* **2019**, *12*, 153. [CrossRef]
80. Yu, W.; Wu, Y.; Li, D. Oxidative cleavage of cellulose by fungi in the termite gut. *Int. J. Biol. Macromol.* **2025**, *284*, 138222. [CrossRef]
81. Xie, M.; An, F.; Wu, J.; Liu, Y.; Shi, H.; Wu, R. Meta-omics reveal microbial assortments and key enzymes in bean sauce mash, a traditional fermented soybean product. *J. Sci. Food Agric.* **2019**, *99*, 6522–6534. [CrossRef] [PubMed]
82. Lee, S.H.; Whon, T.W.; Roh, S.W.; Jeon, C.O. Unraveling microbial fermentation features in kimchi: From classical to meta-omics approaches. *Appl. Microbiol. Biotechnol.* **2020**, *104*, 7731–7744. [CrossRef] [PubMed]
83. Toyoda, A.; Iio, W.; Mitsumori, M.; Minato, H. Isolation and Identification of Cellulose-Binding Proteins from Sheep Rumen Contents. *Appl. Environ. Microbiol.* **2009**, *75*, 1667–1673. [CrossRef]
84. Liu, Q.; Gao, R.; Li, J.; Lin, L.; Zhao, J.; Sun, W.; Tian, C. Development of a genome-editing CRISPR/Cas9 system in thermophilic fungal Myceliophthora species and its application to hyper-cellulase production strain engineering. *Biotechnol. Biofuels* **2017**, *10*, 1. [CrossRef] [PubMed]
85. Liu, G.; Gong, H.; Tang, H.; Meng, Z.; Wang, Z.; Cui, W.; Zhang, K.; Chen, Y.; Yang, Y. Enhanced lignocellulose degradation in Bacillus subtilis RLI2019 through CRISPR/Cas9-mediated chromosomal integration of ternary cellulase genes. *Int. J. Biol. Macromol.* **2025**, *306*, 141727. [CrossRef]
86. Zhou, C.; Zhou, H.; Li, D.; Zhang, H.; Wang, H.; Lu, F. Optimized expression and enhanced production of alkaline protease by genetically modified Bacillus licheniformis 2709. *Microb. Cell Factories* **2020**, *19*, 45. [CrossRef]

animals

MDPI

Review

Anti-Nutritional Factors of Plant Protein Feeds for Ruminants and Methods for Their Elimination

Zhiyong Yan [1,2,3,†], Zixin Liu [1,3,†], Chuanshe Zhou [1,2,3,*] and Zhiliang Tan [1,2,3,4]

1 Hunan Provincial Key Laboratory of Animal Nutritional Physiology and Metabolic Process, National Engineering Laboratory for Pollution Control and Waste Utilization in Livestock and Poultry Production, Institute of Subtropical Agriculture, Chinese Academy of Sciences, Changsha 410125, China; yanzhiyong23@mails.ucas.ac.cn (Z.Y.); lzx@isa.ac.cn (Z.L.); zltan@isa.ac.cn (Z.T.)

2 College of Advanced Agricultural Sciences, University of Chinese Academy of Sciences, Beijing 100049, China

3 State Key Laboratory of Forage Breeding-by-Design and Utilization, National Engineering Laboratory for Pollution Control and Waste Utilization in Livestock and Poultry Production, and Hunan Provincial Key Laboratory of Animal Nutritional Physiology and Metabolic Process, Institute of Subtropical Agriculture, Chinese Academy of Sciences, Changsha 410125, China

4 Yuelushan Laboratory, Changsha 410125, China

* Correspondence: zcs@isa.ac.cn

† These authors contributed equally to this work.

Simple Summary: Ruminants are closely related to human life because they are able to convert plants into nutrients available to humans, including meat and milk. In modern agriculture, ruminant feeds include protein feeds, energy feeds, roughages, and additive feeds, where protein feeds can be categorised into plant proteins, animal proteins, microbial proteins, and non-protein nitrogen, according to the feed source. Only more plant protein feeds such as soybean meal can maintain a higher level of ruminant feeding to meet people's demand for meat and milk; however, some anti-nutritional factors such as gossypol and tannins present in the plants themselves can affect the utilisation efficiency of ruminant feeds and even the health of ruminants, and it has become an imperative task to improve and solve this issue. Different plants contain different kinds of anti-nutritional factors, and different anti-nutritional factors have different physicochemical properties, so the corresponding treatment methods are also different. In order to minimise the impact of anti-nutritional factors in plants on ruminant production, we have to find out the distribution of anti-nutritional factors in different plants, the physicochemical properties of different anti-nutritional factors, and the most reasonable methods to eliminate them.

Abstract: In recent years, the rapid development of the ruminant feeding industry and the limited availability and rising prices of traditional protein feed ingredients have renewed the focus on protein feeds in ruminant diets. Plant protein feeds are a core component of protein feeds for ruminants; however, the utilisation of both conventional and non-conventional plant protein feeds is limited by the presence of anti-nutritional factors (ANFs). In order to maximise the use of plant protein feeds and to promote their application in ruminant production, it is important to have a comprehensive understanding of the types and nature of their ANFs, their anti-nutritional mechanisms, and current effective methods of eliminating ANFs. Therefore, the types, anti-nutritional mechanisms, and elimination methods of ANFs in major plant protein feeds for ruminants are initially summarised in this review, which provides a reference for anti-nutritional factor elimination and the production of full-price compound feeds for ruminants.

Keywords: plant protein feeds; anti-nutritional factors; ruminants; microbial degradation

1. Introduction

In recent years, the price of feed raw materials has increased, the feed cost of farming has increased by 70~80%, and the increase in meat consumption requires more protein feed to maintain the level of animal production [1]. The overall shortage of feed raw materials and the expensive price of high-quality protein feed resources have become the main problems limiting the efficient farming of ruminants, and improving the digestive and absorptive efficiency of the feeds in ruminants is an effective way to alleviate the above problems. The feed formulation for ruminants is a complex and delicate process, aiming to fully satisfy their nutritional needs for growth, development, health maintenance, and performance enhancement. This integrated system consists of the following four main feed categories: protein feeds, energy feeds, roughages, and feed additives, each of which carries specific nutritional and physiological functions that together support a healthy feeding system for ruminants. Protein feeds are the core component of ruminant feeds. Protein feeds can be divided into plant protein, animal protein, microbial protein, and non-protein nitrogen according to the feed source. Plant protein feed sources mainly include all kinds of oilseed crops after the oil extraction of the remaining cake meal, including soybean meal (SBM), rapeseed meal (RSM), cottonseed meal (CSM), peanut meal, sunflower kernel meal, flax kernel meal, and so on. RSM, CSM, and other types of miscellaneous meal follows the 'SBM reduction in feed' trend. The application of the proportion of plant protein feed sources increased year by year, but it contains a variety of anti-nutritional components ANFs to limit the various types of miscellaneous meal as well as the application of SBM [2]. When livestock and poultry consume excessive amounts of ANFs, it not only affects the digestibility and absorption of nutrients, but it may even have adverse effects on growth and health [3,4]. For example, free gossypol, glucosinolates, trypsin inhibiting factor, antigenic proteins, and phytohemagglutinins present in cottonseed will inhibit the absorption of amino acids, and RSM contains a large amount of glucosinolates, tannins, phytic acid, erucic acid, and other ANFs [5,6]. At the same time, the nutrient composition of various types of miscellaneous meals is also defective, for instance, the lysine and threonine content of RSM is lower than that of SBM, and the digestibility in ruminants is significantly lower than that of SBM [7]. In order to improve the utilisation efficiency of plant protein feeds and their processing by-products, the inactivation of ANFs in them is a critical step. Currently, a number of different methods have been investigated to eliminate or inactivate these ANFs, mainly covering the following three techniques: physicochemical treatment, enzymatic treatment, and microbial decomposition [8]. This paper comprehensively reviews the classification, anti-nutritional mechanisms, and mitigation strategies of ANFs in plant-based protein feeds for ruminants. By synthesising current knowledge, this work aims to provide actionable insights for optimising feed formulations, enhancing nutrient utilisation, and advancing the development of balanced, high-performance compound feeds. Ultimately, this review serves as a foundational resource to guide future research and industry practices in sustainable ruminant nutrition.

2. Overview of Plant Protein Feeds and ANFs

Plant protein feed sources mainly include the cake meal remaining after the oil extraction from various oilseed crops, including SBM, RSM, CSM, and so on. They have great potential as a protein source for ruminant protein feeds due to their richness in crude protein, amino acids, and good availability [9]. These plant protein feeds contain a variety of ANFs, such as protease inhibitors, antigenic proteins, and gossypol, which limit the efficiency of protein and carbohydrate utilisation in plant protein feeds and adversely affect the growth and reproduction of animals. The content and physicochemical properties of different ANFs in different feeds are varied, so the corresponding elimination methods are also different [10].

2.1. ANFs in SBM

SBM is the most commonly used plant-based protein feed in various animal feeds because of its high yield, high crude protein content, good quality, reasonable composition of essential amino acids, and high digestibility [3]. However, soybean also contains ANFs such as protease inhibitors, soybean antigens, soybean agglutinin, etc., which reduce its nutritional value.

2.1.1. Soybean Antigenic Protein

The main ANFs in soybeans are antigenic proteins, which are immunoreactive and thermostable, may cause allergic reactions in animals, and may affect their productive performance [11]. Soybean antigenic proteins can be classified into four categories according to the centrifugal sedimentation coefficient as follows: 2 S, 7 S, 11 S, and 15 S, with the highest proportion of soybean antigenic proteins being 11 S at approximately 40% [12]. The main components of soybeans that trigger the immune response of animals are β-accompanied by soybean globulin (7 S) and soybean globulin (11 S), with the 7 S protein having the greater immunoreactivity, and 11 S proteins consisting of six subunits [13]. Each of these is connected to a disulfide bond, subunits, connecting basic and acidic peptide chains via disulfide bonds [11].

2.1.2. Protease Inhibitors

Protease inhibitors are capable of hindering the function of a variety of proteases, such as trypsin, pepsin, and coagulation factors, etc. There are numerous types of protease inhibitors in nature, among which, the main ones contained in SBMs are trypsin inhibitors (TIs), classified into Bowman-Birk inhibitors (BBIs) and Kunitz inhibitors (KTIs) [14]. The KTI has a variable structure containing four cysteine residues, which form two disulfide bonds, a structure that makes the KTI easy to destroy when heated. The BBI contains a higher number of cysteines and seven disulfide bonds, and it has multiple reactive groups that can inhibit the activity of multiple proteases simultaneously. Although the content of TIs in soybean is only 2% of the total mass (about 0.6% for BBI and 1.4% for KTI), they account for 40% of the total anti-nutritional effect of all ANFs in soybean [4].

2.1.3. Soybean Agglutinin

Soybean agglutinin (SBA), as the name indicates, is a class of ANFs that promotes erythrocyte aggregation and is a major anti-nutrient in soybean, which is heat-insensitive, stable, and undamaged. In SBM, the concentration of SBA is typically 3% [15]. SBA impairs animal growth by interfering with the digestion and absorption of nutrients in the gastrointestinal tract. In addition, one of the most important characteristics of SBA considered to be ANFs is the remarkably high resistance and stability to protein hydrolysis over a large physiological pH range [16]. When SBA is ingested by livestock, SBA specifically binds to glycoproteins on the surface of small intestinal epithelial cells, thereby adversely affecting the animal's intestinal epithelial cells and immune function. Some studies based on intestinal porcine epithelial cell lines (IPEC-J2) have shown that SBA exerts its toxic effects on the intestine through a variety of apoptosis-related pathways [17]. Other studies have shown that SBA have an effect on growth performance in young ruminants, but not in adult ruminants [18]. SBA may be degraded by rumen microorganisms.

2.1.4. Tannins

Tannins are water-soluble natural polyphenolic compounds, mainly found in the shell of soybeans, which can be divided into hydrolysed tannins (HTs) and condensed tannins (CTs). CTs cannot be hydrolysed and can react with amylase, trypsin, and their substrates, thus decreasing the utilisation rate of carbohydrates and proteins, and affecting the intake of the animal [19]. Tannins not only reduce the digestibility of feed proteins, but also affect the intestinal micro-ecosystem. When tannins are increased in feeds, they significantly reduce total gas production and the concentration of ammonia and volatile fatty acids in the intestinal tract, and they increase the number of viable bacteria in the enterococci and coliform flora [20].

2.1.5. Phytic Acid

Phytic acid (PA), chemically known as inositol hexakisphosphate, exists mainly in the form of phosphorus storage from seeds of cereal and oilseed crops, and its molecular structure contains an inositol ring and six phosphoric acid groups. PA is present not only in legume feeds, but also in rapeseed and its cake meal [21]. The PA molecule consists of six phosphate groups, which can interact with metal ions and proteins to produce insoluble complexes, which are not easily digested and absorbed by animals, reducing the nutritional value of the feed and affecting its palatability, which has a negative impact on the healthy growth of animals [22].

2.1.6. Non-Starch Polysaccharides

SBM contains many non-starch polysaccharides (NSPs), of which β-glucan, arabinoxylan, and cellulose are the main components, usually accounting for less than 10% of its weight [23]. The high proportion of ether bonds in lignin makes it more difficult to degrade than cellulose and hemicellulose, and it requires 150–400 °C to break [24]. Degradation of NSPs in ruminants can occur in the rumen, small intestine, and large intestine, but the rumen is the most important degradation site for feed NSPs, most of which are rapidly degraded mainly by rumen microorganisms. It has been shown that the ingestion of NSPs can have a systemic viscosity effect, which can inhibit glucose absorption or form a physical barrier to inhibit starch digestion by amylase and reduce its digestion rate [25–27].

2.1.7. ANFs in Miscellaneous Meals

Miscellaneous meals contain high levels of protein and a variety of vitamins and minerals, but their amino acid ratios are unbalanced, and their use in ruminant feeds is limited by the presence of free lintol, thioglucosides, PA, and other ANFs in miscellaneous meals [24].

2.1.8. Gossypol

Gossypol is a phenolic substance that is particularly abundant in the seeds of the cotton plant, with the molecular formula $C_{50}H_{50}O_8$. In CSM, gossypol exists in both free and bound forms, and the total amount of gossypol in both forms is referred to as total gossypol [28,29]. The World Health Organisation (WHO) in 2018 established a maximum threshold of 450 mg/kg for the content of free gossypol (FG) in CSM protein products [30]. Gossypol damages the gastrointestinal mucosa of animals and the reproductive capacity of males, and tolerance to gossypol varies in different animals, with pigs having a low tolerance and broilers and laying hens having a relatively high tolerance [29]. Ruminants have a better tolerance to gossypol due to their special digestive system, and it was found that CSM can replace SBM as the main protein source for beef cattle and goats, without affecting their growth and slaughter performance [31]. However, gossypol has toxic effects on the reproductive system of ruminants. Tiago et al. collected testicular samples from young male goats after feeding them with gossypol for 95 days, analysed them, and found that there was a significant decrease in the concentration of testosterone after feeding gossypol, as well as higher total sperm defects, poor sperm motility, and a high rate of lesions [32]. This indicates that the intake of gossypol in ruminants used for breeding is unsafe.

2.1.9. Glucosinolates

There are many glucosinolates in RSM that are decomposed by mustard enzymes to produce oxazolidinethione (OZT), isothiocyanate (ITC), and nitrile, with OZT being the main toxin in RSM, and it is more toxic to the thyroid gland. OZT is the main toxin in RSM and has a strong toxic effect on the thyroid gland [33]. ITC can damage the mucous membranes of the gastrointestinal tract of animals or cause substantial damage to the lungs and kidneys due to its own irritant properties [34]. Nitrile is the most toxic of the glucosinolates decomposition products, and prolonged or large intake can damage the liver and kidneys of animals [35]. Tolerance to glucosinolates varies among animals, being higher in ruminants and lower in pigs and poultry [36,37]. The treatment of CSM and RSM by physical, chemical, and biological methods can reduce the content of ANFs and improve the digestion and absorption of nutrients in animals [38]. The anti-nutritional mechanisms of ANFs in plant protein feeds are shown in Table 1.

Table 1. Anti-nutritional mechanisms of ANFs in plant protein feeds.

ANFs	Anti-Nutritional Mechanisms	References
Soybean antigenic protein	It can cross the epithelium of the small intestine and enter the internal circulatory system, activating the immune response and promoting the production of specific antibodies by B-cells, leading to intestinal allergy and disruption of the intestinal barrier, which can lead to diarrhoea and a decline in production performance.	[12,39–42]
Protease inhibitors	It binds to the active site of trypsin to form a stable inhibitory complex, which renders trypsin and pancreatic rennet ineffective, thereby affecting the growth rate of the animal and possibly leading to abnormal pancreatic function.	[43,44]
SBA	SBA disrupts the intestinal flora environment and negatively affects the immune system by inhibiting the intestinal production of immunoglobulin A (IgA). SBA also acts by inducing apoptosis through the FAK signalling pathway.	[17,45,46]
Tannins	High levels of tannins react with salivary mucins to bring about astringency, which in turn reduces their intake and decreases the efficiency of nutrient transport by the intestinal mucosa. Tannins also affect the metabolism of indole compounds and androstenone in animals, which in turn affects the development of the secondary gonads.	[47–49]
Gossypol	FG inhibits cardiovascular diastole and contraction, hinders neurotransmitter transmission, complexes with amino acids, and reduces lysine utilisation efficiency. FG causes loss of electrical rhythm in animals, leading to heart failure. FG affects the function of various enzymes, alters the nature of cell membranes, forms a strong bond with metal ions, and interferes with the bioavailability of mineral elements.	[6,28,50,51]
NSPs	They have a complex structure that requires the simultaneous presence of multiple digestive enzymes for degradation, so they are less fermentable in the digestive tract. Insoluble cellulose consists of glucose linked by β-1,4-glycosidic bonds, which cannot interact with water after the formation of hydrogen bonds and is therefore degraded at a slower rate.	[25,26,52]

Table 1. *Cont.*

ANFs	Anti-Nutritional Mechanisms	References
PA	PA has the ability to reduce the bioavailability of some essential minerals by chelating with divalent or multivalent metal ions (e.g., zinc, calcium, copper) to form insoluble PA salts. PA also directly affects the activity of digestive enzymes, such as proteases and amylases, which can inhibit protein breakdown and absorption during rumen fermentation in ruminants.	[53–56]
Glucosinolates	Glucosinolates hydrolysed to produce thiocyanate, isothiocyanate, and oxazolidinethione, which affects thyroid development and hormone secretion in animals.	[35,57]

3. Advances in ANF Elimination Methods

Currently, there are many techniques to eliminate ANFs in plants, such as high temperature and pressure, fermentation, etc., but they can be broadly categorised as physical, chemical, and biological methods [56]. Physical methods completely deactivate ANFs through high temperature, high pressure, radiation, etc. The chemical method involves adding chemical reagents (e.g., $CuSO_4$ and H_2O_2, etc.) to feeds containing ANFs under appropriate conditions to inactivate ANFs through chemical reactions. Biological methods, on the other hand, include enzymatic degradation by enzymes and microbial fermentation to inactivate ANFs or degrade them into other small molecules.

3.1. Physical Methods

For heat-stable ANFs, such as PA and tannins, physical methods such as peeling and crushing can be used to reduce their content. For example, soybean can remove most of the tannins and reduce PA content by removing the outer hulls [58]. For thermally unstable ANFs, such as agglutinins and protease inhibitors, a common method is heating [59].

Heating methods include dry heat treatments (e.g., baking, microwave heating, and infrared irradiation) and moist heat treatments (e.g., boiling, autoclaving, and extrusion heating), which are only suitable for thermally unstable ANFs. They are not effective for thermally stable ANFs, such as PA, cyanogenic compounds, and oligosaccharides. RSM heated at 100 °C for 5 min or 2 h increased the removal of glucosinolates from 24% to 95% but protein digestibility decreased from 79% to 71% [60]. Duodu et al. used soaking and autoclaving techniques, leading to a reduction in FG levels by 34.48% and 27.59%, respectively, as well as a significant reduction in PA [61]. Water immersion also removed soluble NSPs [62]. In addition to this, studies have also reported the elimination of NSPs using electron beam (EB) irradiation methods [63]. The method of electron radiation also effectively eliminates FG and total gossypol in CSM. Bahraini et al. compared the effects of EB and gamma ray (GR) irradiation treatments at doses of 10, 20, and 30 kGy on the chemical composition, protein quality, and protein digestibility of CSM, significantly reducing FG and total gossypol content in CSM in a dose-dependent manner compared with unirradiated CSM. In addition, ER irradiation induced a greater decrease in FG and total gossypol content than GR irradiation [64].

The processing of feeds also affects the degradation or deposition of ANFs over time. The first step in feed production is comminution, which also destroys the fibrous structure of the raw material and increases the solubility of NSPs, making them available to digestive enzymes. Extrusion is a high-temperature, short-duration process that combines several processes, including heat and mass transfer, mixing, shearing, particle size reduction, melting, texturizing, caramelising, and moulding. Extrusion is also a very effective method to inactivate amylase inhibitors, trypsin, pancreatic rennet, and SBA activity without altering the protein levels in the food [65]. In general, the moisture content and composition of raw materials, barrel temperature, and extruder feed rate are the most important factors affecting the extrusion process in terms of reducing the amount of ANFs [66]. In another study, the effectiveness of extrusion processing was compared with conventional non-thermal processing methods such as dehulling, soaking (in double deionised water at 30 °C for 12 h), and germination (at 25 °C for 72 h) in an attempt to reduce ANFs in broad beans and kidney beans. The results showed that germination for 72 h resulted in a significant reduction in phytates in broad beans (60.8%) and kidney beans (30.2%), whereas extrusion alone resulted in a reduction in phytates by about 26.7% and 21.4% in broad beans and kidney beans, respectively [67].

3.2. Chemical Methods

Chemical methods refer to the addition of acids, bases, alcohols, ammonia, and special substances to react with ANFs or purify ANFs by extraction. Chemical methods can deactivate ANFs by decomposing them, neutralising them, dissolving them, binding them, destroying the structure of the ANFs, or breaking the bonds that maintain the structure. For example, the treatment of SBM with ascorbic acid, $CuSO_4$, $FeSO_4$, and urea can inactivate TIs, and the addition of $FeSO_4$ and $CuSO_4$ to RSM results in a glucosinolates detoxification rate of up to 85.2%, which is less costly, but also suffers from poor palatability, severe nutrient loss, and some contamination [68].

Barraza et al. found that the addition of 0.5%, 1%, and 2% calcium hydroxide to CSM led to a reduction in FG by 21.25%, 28.15%, and 40.52%, respectively, but the use of calcium hydroxide often reduces the bioactivity and detoxification efficiency of the vitamins in the feed [69]. In addition, supercritical CO_2 extraction and solvent extraction were more effective in removing FG. Bhattacharjee et al. used supercritical fluid extraction with CO_2 to extract oil from CSM and found that when the extraction parameters were set at a pressure higher than 550 bar, a temperature range of 70–80 °C, and an extraction time of 2–3 h, it was possible to maximise the oil yield while minimising FG extraction [70]. Pelitire et al. used acetone and ethanol as solvents to remove FG from CSM and found that both solvents were effective in reducing the total FG level in CSM to between 5% and 10% of its initial value, with the rate of FG removal being much faster via ethanol extraction than via acetone extraction [71]. It is suggested that the process of supercritical CO_2 extraction and solvent extraction can be used to produce CSM with low FG.

The chemical method is convenient in operation, has a good inactivation effect, and saves equipment and energy, but the chemical reagents may remain in the raw materials and need to be cleaned and dried after treatment, which not only increases the production cost, but it may also cause pollution to the environment. This limits the use of chemical methods in feed processes.

3.3. Biological Methods

3.3.1. Enzymatic Method

The enzymatic method utilises enzyme preparations such as protease, NSPs enzyme, phytase, and tannase to degrade ANFs and thus reduce their content. Since the implementation of China's policy of banning antibiotics in feed in 2020, enzyme preparation has been more and more widely used in the feed industry, which not only serves to eliminate ANFs in feed ingredients but also solves the problem of insufficient endogenous enzymes in young animals, promotes the secretion of endogenous enzymes, and improves the utilisation rate of proteins. Phytase is a critical feed additive for monogastric animals, as their gastrointestinal tracts naturally lack this enzyme. To enhance phytate hydrolysis in diets, exogenous phytase must be supplemented—primarily derived from fungal and bacterial strains [72]. Up to dozens more enzymes are known to catalyse PA cleavage, including phytase secreted by grains, intestinal mucosa, and gastrointestinal microbial communities and a large number of commercial preparations of phytase with modified properties, with the role of each phytase type and its maximum activity depending on the pH of the enzymatic environment [73]. On this basis, phytase is classified as either acidic or alkaline. Acidic phytase shows broad substrate specificity for metal-free phytates, while alkaline phytase is specific for metal-bound PA. Phytase promotes balanced animal mineral nutrition and significantly reduces phosphorus excretion from animal faeces with the addition of appropriate amounts of phytase to feeds, thus mitigating the ecological problem of phosphorus pollution [74].

3.3.2. Microbial Degradation

Microbial fermented feed is a type of feed produced by the metabolism of microorganisms, which has the characteristics and effects of reducing the content of ANFs in raw materials, producing beneficial metabolites, and improving the digestion and absorption of feed nutrients by animals [75,76]. Fermentation is a dynamic process that converts complex substrates into simple compounds, and its products are mainly influenced by the microbial species, fermentation substrate characteristics, and fermentation parameters [77]. Probiotic species commonly used in fermented feeds include yeasts, bacilli, lactobacilli, and moulds [78]. Zheng et al. fermented SBM using *Bacillus sphaericus* and found that *Bacillus sphaericus* was able to degrade ANFs as well as change the microstructure of fermented SBM proteins to improve the nutritional quality of fermented SBM [79]. The fermentation of whole feeds is a more convenient way compared to the fermentation of single protein feed ingredients, but this type of fermentation results in a higher loss of amino acids such as lysine [80]. The fermentation process is catalysed by enzymes, and temperature is an important factor to ensure enzyme activity. Appropriate fermentation temperatures can help shorten the reaction stabilisation time, and Dujardin et al. showed that increasing the temperature of liquid fermented feeds from 15 °C to 30 °C reduced the time taken for the pH of the feeds to drop to 4.0 by six-fold [81]. As understood from the perspective of enzyme kinetics, increasing temperature accelerates microbial growth and metabolism [82]. In conclusion, microbial fermentation is a complex and dynamic fermentation process, and the fermentation end products are regulated by the interactions of its influencing factors. In order to obtain the desired fermentation products, the precise optimisation of fermentation conditions is imperative. The process of bacterial growth can decompose some of the ANFs, which significantly improves the efficiency of feed utilisation [83]. Protozoa and bacteria in the rumen of ruminants will compete in the decomposition of starch granules, slowing down the process of rumen fermentation and thus reducing problems such as acidosis [84].

The degradation of ANFs by microbial fermentation seems to be a promising and environmentally friendly method compared to physicochemical methods. The key to the degradation of ANFs by microbial degradation is the selection of bacterial strains. Many strains of bacteria (including Aspergillus, yeasts, and Bacillus) have been used in the microbial fermentation of plant protein feeds, since the fermentation process can secrete a variety of enzymes, including proteases, cellulases, and growth factors [85]. Improvement varies between strains, with some strains only improving roughage nutrient content and positively affecting the reduction in ANFs. Some strains affect only positively the improvement of enzyme activity in the gut, while other microorganisms can improve both [86]. For example, Bacillus sp. is able to degrade ANFs [79]. The effect of mixed-strain fermentation is generally better than that of single-strain fermentation, so mixed-strain fermentation is mostly used in practical production applications [75]. The microbial fermentation of plant protein feeds for ruminant production has also made progress with the widespread use of fermented feed technology. Wang et al. found that fermented

SBM improved the rumen microecology of lactating dairy cows by increasing the number of copies of filamentous *Bacillus succinicus*, *Leptospira ruminantium*, and Prevotella [87]. Kim et al. reported that fermented SBM also reduced diarrhoea in calves [88]. Microbial fermentation not only improves the nutritional value and utilisation of feeds, but also improves the intestinal microbiota of animals, which plays a growth-promoting role [89].

Microbial fermentation can also destroy ANFs in miscellaneous meal. Khalaf et al. found that microbial fermentation significantly reduced FG in CSM by studying the effect of microorganisms screened during the solid-state fermentation of CSM. However, there were differences between different strains of bacteria, among which the optimal conditions for the reduction in FG content by Pseudohyphae tropicalis yeast fermentation were incubation for 48 h, incubation temperature of 30 °C, 55% moisture content of solid substrate, and pH value around 5.2. Under these optimised conditions, the crude protein and amino acid content of fermentation substrate was significantly increased [90]. Olukomaiya et al. used *Aspergillus soya* and *Aspergillus figlii* in a complex to ferment RSM and reduced the PA content from 27.06 mg/g to 22.13 mg/g [91]. Bacillus subtilis was able to produce extracellular enzymes, thereby promoting the degradation of acid detergent fibres (ADF) [86]. Lactobacilli were able to reduce the activity of trypsin inhibiting factor and reduce the content of ANFs in feeds [92]. After microbial fermentation, the content of ANFs decreased significantly [93,94]. Studies have shown that fermentation can remove 85% of ANFs and toxic substances from CSM and RSM [95]. However, the mechanism of the microbial fermentation of feeds for the removal of ANFs and the degradation of harmful substances in raw materials is not fully understood, and it may be related to microbial metabolism. Mi et al. showed that the addition of 19.04% fermented RSM to the diet of black goats significantly increased the abundance of propionate, butyrate, and volatile fatty acids (VFA) in the rumen, and it was also able to increase the number of beneficial bacteria in black goat rumen microbiota [96].

It is worth noting that microbial fermentation can not only reduce ANFs in miscellaneous meals, but it can also reduce costs and increase efficiency [97]. In addition, the process of feed fermentation also produces aromatic substances that enhance the flavour of feed ingredients to promote animal feeding, so it is more advantageous to use microbial degradation to treat ANFs. However, it is necessary to assess the safety of certain metabolites produced during fermentation to prevent them from posing any toxicity risks. The application of microbial degradation on different ANFs is shown in Table 2.

Table 2. Application of microbial degradation to different ANFs.

ANFs	Microbiological Species	Experimental Methods	Effects	References
Soybean antigenic protein	*Bacillus licheniformis*	SBM fermented for 24 h	Soybean antigenic protein reduced by 25.5%	[98]
Soybean antigenic protein	*Bacillus cereus*	SBM fermented for 24 h	Soybean antigenic protein reduced by 68.14%	[99]
Protease inhibitors	*Planctomyces militaris*	De-oiled rice bran fermented for 3 d	TIs reduced by 24.8%	[100]
Protease inhibitors	*Bacillus Siamensis jl8*	SBM fermented for 24 h	TIs reduced by 95.01%	[79]
PA	*Bacillus subtilis*	Wheat flour fermentation for 24 h	PA content reduced by 55%	[101]

Table 2. *Cont.*

ANFs	Microbiological Species	Experimental Methods	Effects	References
Glucosinolates	*Lactobacillus acidophilus*	RSM fermented for 25 d	Glucosinolates levels decreased by 67.81%	[102]
Tannins	*Saccharomyces cerevisiae*	SBM incubated at 28 °C for 48 h	Tannin degradation rate of 37.9%	[103]
NSPs	*Lactobacillus plantarum*	SBM fermented for 48 h	NSPs decreased by 39.15%	[104]
NSPs	*Bacillus licheniformis B4*	SBM fermented for 48 h	NSPs decreased by 30.20%	[98]
Gossypol	*Bacillus coagulans*	CSM fermented for 14 d	The detoxification efficiency of FG was 93.46%	[97]
Gossypol	*Lactobacillus agalactiae*	CSM anaerobically fermented for 5 d	The detoxification efficiency of FG was 80%	[105]

4. Prospects

In the future, the chemical structure, biological activity, and mechanism of action of ANFs in feeds need to be deeply investigated to explore more efficient inactivation strategies. The currently available elimination methods have their obvious advantages and disadvantages. High-temperature treatments are particularly effective for unstable ANFs such as antigenic proteins and protease inhibitors, and dehulling also facilitates the removal of ANFs, such as tannins, which are mainly found in seed shells. While chemical methods have limited their use due to their toxicity and possible environmental stress, enzymes are more effective in removing stable ANFs and are now widely used in the feed industry. Microbial degradation is more effective than enzymatic degradation, but this depends on the type of microorganism, the amount added, and the fermentation conditions. In addition, certain emerging technologies such as gamma ray and irradiation have multiple benefits, such as the elimination of microbial and fungal contamination of SBM and miscellaneous meal while retaining protein and amino acid integrity, reduced energy consumption to reduce chemical reactions, etc. These should be further investigated in more detail in the future. At present, the specific mechanism of action of strains in the fermentation process has not been fully studied, and the effect of their application in production varies. Therefore, further research on the specific mechanism of action of different strains in the fermentation process, as well as the interaction effects between different strains, will help develop microbial fermented feeds that are more meaningful for ruminant nutrition. Nevertheless, the mechanism of microbial fermentation feed to reduce ANFs and toxic and harmful substances has not been clarified yet. Moreover, the preservation problem of strains used for fermentation has not been effectively solved, the ratio of fermentation of mixed bacteria needs to be studied in depth, and the speed of fermentation is difficult to be controlled. At the same time, the secondary fermentation of feeds will also bring about a certain amount of economic losses [106]. In the future, during the development and utilisation of microbial fermented feed, it is necessary to establish a special microbial fermented feed research database, improve the quality evaluation system, thoroughly study the action mechanism and advantages of microbial fermented feed, and promote the sustainable development of microbial fermented feed. Microbial fermented plant protein feeds have significantly improved the production performance of ruminants, but there is no significant improvement in the quality of ruminant products. Therefore, it will be of great significance to carry out the research work related to bacterial strain research, animal immunity, intestinal flora, nutrient metabolism, etc., in order to gain a deeper understanding of the mechanism of action of microbial fermented plant protein feeds, and to develop different strains of bacterial strains, different roughage, and different animal species fermented feeds according to the local conditions. Various types of miscellaneous meal due to the price advantage in the proportion of feed gradually increased. In order to solve the issue of the accumulation of gossypol, glucosinolates, tannins, and other ANFs, as well as other problems, biological fermentation and bacteria and enzyme synergistic treatment have been widely introduced into the production process, using various types of enzyme specificity and the high efficiency of ANFs for decomposition and transformation [2]. Biological methods have obvious advantages, as they do not need to use difficult to separate solvents, can significantly reduce costs, and can rely on the physiological activities of the bacterial colony to decompose various types of ANFs. Moreover, the bacterial colony can increase the protein content of the product after moderate reproduction to improve the composition of the product's nutrient composition. However, the degradation rate of ANFs in the biological method process based on strain compounding is lower than that of the chemical method, and the synergistic treatment of bacteria and enzymes can improve the decomposition rate. However, there is no cost advantage compared with the chemical method, so the method urgently needs corresponding high enzyme-producing bacterial strains. The main ANFs of plant protein feeds and the common elimination methods are shown in Figure 1.

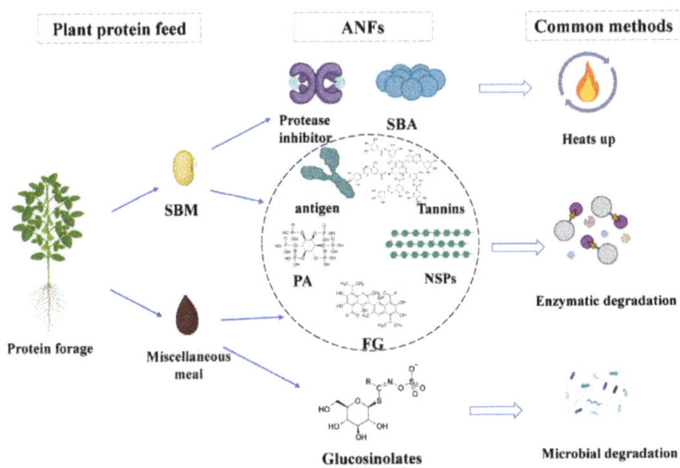

Figure 1. ANFs in plant protein feeds and common elimination methods (figure created in BioRender, https://biorender.com, accessed on 15 December 2024).

5. Conclusions

The types, anti-nutritional mechanisms and elimination methods of ANFs in major plant protein feeds for ruminants were initially summarised in this review, which provides a reference for antinutritional factor elimination and production of full-price compound feeds for ruminants. Currently, the variable quality of fermentation products from protein feed resources, the unclear association between fermentation effect and microbial strains, and the standardisation of the fermentation process remain the main challenges for the efficient utilisation of protein feed resources for ruminants. With the continuous research on fermentation technology, more effective methods of eliminating ANFs from plant protein feeds will be developed, which is important for alleviating the shortage of high-quality protein resources and enhancing the efficiency of the use of plant protein raw materials for ruminants.

Author Contributions: Conceptualization, Z.Y., Z.L., C.Z. and Z.T.; methodology, Z.Y.; software, Z.Y.; validation, Z.Y., Z.L., C.Z. and Z.T.; formal analysis, Z.Y.; investigation, Z.Y.; resources, Z.Y.; data curation, Z.L.; writing—original draft preparation, Z.Y.; writing—review and editing, Z.Y.; visualization, Z.Y.; supervision, C.Z.; project administration, C.Z.; funding acquisition, C.Z. All authors have read and agreed to the published version of the manuscript.

Funding: This research was jointly supported by the Projects of International Cooperation and Exchanges NSFC (32261143467), the Projects of Hunan Province Science and Technology Innovation Program (2023WK2003), National Key Research and Development Program (2022YFD1300805), and the Opening Fund of the Key Laboratory of South Yuelu Mountain Biological Resources Protection and Utilization (2022HSKFJJ02).

Institutional Review Board Statement: Not applicable.

Informed Consent Statement: Not applicable.

Data Availability Statement: Not applicable.

Conflicts of Interest: The authors declare no conflicts of interest.

Abbreviations

The following abbreviations are used in this manuscript:

ANFs	Anti-Nutritional Factors
SBM	Soybean Meal
RSM	Rapeseed Meal
CSM	Cottonseed Meal
TIs	Trypsin Inhibitors
BBI	Bowman-Birk Inhibitor
KTI	Kunitz Inhibitor
SBA	Soybean Agglutinin
HTs	Hydrolysed Tannins
CTs	Condensed Tannins
PA	Phytic Acid
NSPs	Non-Starch Polysaccharides
FG	Free Gossypol
OZT	Oxazolidinethione
ITC	Isothiocyanate
EB	Electron Beam
GR	Gamma Ray
ADF	Acid Detergent Fibres

References

1. Kim, T.-I.; Mayakrishnan, V.; Lim, D.-H.; Yeon, J.-H.; Baek, K.-S. Effect of Fermented Total Mixed Rations on the Growth Performance, Carcass and Meat Quality Characteristics of Hanwoo Steers. *Anim. Sci. J.* **2018**, *89*, 606–615. [CrossRef] [PubMed]
2. Górka, P.; Penner, G.B. Rapeseed and Canola Meal as Protein Sources in Starter Diets for Calves: Current Knowledge and Directions of Future Studies. *Ank. Üniversitesi Vet. Fakültesi Derg.* **2020**, *67*, 313–321. [CrossRef]
3. Di, D.; He, S.; Zhang, R.; Gao, K.; Qiu, M.; Li, X.; Sun, H.; Xue, S.; Shi, J. Exploring the Dual Role of Anti-Nutritional Factors in Soybeans: A Comprehensive Analysis of Health Risks and Benefits. *Crit. Rev. Food Sci. Nutr.* **2024**, 1–18. [CrossRef] [PubMed]
4. Bondi, A.; Alumot, E. Anti-Nutritive Factors in Animal Feedstuffs and Their Effects on Livestock. *Prog. Food Nutr. Sci.* **1987**, *11*, 115–151.
5. Yusuf, H.A.; Piao, M.; Ma, T.; Huo, R.; Tu, Y. Enhancing the Quality of Total Mixed Ration Containing Cottonseed or Rapeseed Meal by Optimization of Fermentation Conditions. *Fermentation* **2021**, *7*, 234. [CrossRef]

6. Santosh, S.; Raghavendra, K.P.; Velmourougane, K.; Mageshwaran, V.; Blaise, D.; Waghmare, V.N. Microbial Detoxification of Gossypol in Cotton Seed Meal by Solid Substrate Fermentation. *Int. J. Curr. Microbiol. App. Sci.* **2020**, *9*, 1654–1663. [CrossRef]
7. Heendeniya, R.G.; Christensen, D.A.; Maenz, D.D.; McKinnon, J.J.; Yu, P. Protein Fractionation Byproduct from Canola Meal for Dairy Cattle. *J. Dairy Sci.* **2012**, *95*, 4488–4500. [CrossRef]
8. Enneking, D.; Wink, M. Towards the Elimination of Anti-Nutritional Factors in Grain Legumes. In *Linking Research and Marketing Opportunities for Pulses in the 21st Century: Proceedings of the Third International Food Legumes Research Conference*; Knight, R., Ed.; Springer: Dordrecht, The Netherlands, 2000; pp. 671–683. ISBN 978-94-011-4385-1.
9. Tao, A.; Wang, J.; Luo, B.; Liu, B.; Wang, Z.; Chen, X.; Zou, T.; Chen, J.; You, J. Research Progress on Cottonseed Meal as a Protein Source in Pig Nutrition: An Updated Review. *Anim. Nutr.* **2024**, *18*, 220–233. [CrossRef]
10. Yin, X.; Chen, M.; Yang, C.; Duan, C.; Ji, S.; Yan, H.; Liu, Y.; Zhang, Y. Effects of Replacing Soybean Meal with Cottonseed Meal, Peanut Meal, Rapeseed Meal, or Distillers' Dried Grains with Solubles on the Growth Performance, Nutrient Digestibility, Serum Parameters, and Rumen Fermentation in Growing Lambs. *Vet. Sci.* **2024**, *11*, 322. [CrossRef]
11. Wang, L.; Liu, Y.; Zhang, D.; Ding, H.; Feng, S.; Zhao, C.; Wu, J.; Wang, X. Soybean Antigen Protein-Induced Intestinal Barrier Damage by Trigging Endoplasmic Reticulum Stress and Disordering Gut Microbiota in Weaned Piglets. *Molecules* **2023**, *28*, 6500. [CrossRef]
12. Ogawa, T.; Bando, N.; Tsuji, H.; Nishikawa, K.; Kitamura, K. α-Subunit of β-Conglycinin, an Allergenic Protein Recognized by IgE Antibodies of Soybean-Sensitive Patients with Atopic Dermatitis. *Biosci. Biotechnol. Biochem.* **1995**, *59*, 831–833. [CrossRef] [PubMed]
13. Zheng, S.; Qin, G.; Tian, H.; Zhang, F. Three-Dimensional Structure of Gly m 5 (β-Conglycinin) Plays an Important Role in Its Stability and Overall Allergenicity. *Food Chem.* **2017**, *234*, 381–388. [CrossRef] [PubMed]
14. Arsov, A.; Tsigoriyna, L.; Batovska, D.; Armenova, N.; Mu, W.; Zhang, W.; Petrov, K.; Petrova, P. Bacterial Degradation of Antinutrients in Foods: The Genomic Insight. *Foods* **2024**, *13*, 2408. [CrossRef] [PubMed]
15. Lin, P.; Ye, X.; Ng, T. Purification of Melibiose-Binding Lectins from Two Cultivars of Chinese Black Soybeans. *Acta Biochim. Biophys. Sin.* **2008**, *40*, 1029–1038. [CrossRef]
16. Lam, S.K.; Ng, T.B. Lectins: Production and Practical Applications. *Appl. Microbiol. Biotechnol.* **2011**, *89*, 45–55. [CrossRef]
17. Tianjiao, E.; Xu, C.; Fan, X.; Liu, J.; Zhao, J.; Bao, N.; Zhao, Y.; Farouk, M.H.; Ji, Y.; Wu, Z.; et al. Soybean Agglutinin Induced Apoptotic Effects by Down-Regulating ANXA2 Through FAK Pathway in IPEC-J2 Cells. *J. Anim. Physiol. Anim. Nutr.* **2024**, *109*, 350–361. [CrossRef]
18. Lalles, J.P.; Tukur, H.M.; Toullec, R.; Miller, B.G. Analytical Criteria for Predicting Apparent Digestibility of Soybean Protein in Preruminant Calves. *J. Dairy Sci.* **1996**, *79*, 475–482. [CrossRef]
19. Besharati, M.; Maggiolino, A.; Palangi, V.; Kaya, A.; Jabbar, M.; Eseceli, H.; De Palo, P.; Lorenzo, J.M. Tannin in Ruminant Nutrition: Review. *Molecules* **2022**, *27*, 8273. [CrossRef]
20. Biagia, G.; Cipollini, I.; Paulicks, B.R.; Roth, F.X. Effect of Tannins on Growth Performance and Intestinal Ecosystem in Weaned Piglets. *Arch. Anim. Nutr.* **2010**, *64*, 121–135. [CrossRef]
21. Khajali, F.; Slominski, B.A. Factors That Affect the Nutritive Value of Canola Meal for Poultry. *Poult. Sci.* **2012**, *91*, 2564–2575. [CrossRef]
22. Sun, X.; Tiffany, D.G.; Urriola, P.E.; Shurson, G.G.; Hu, B. Nutrition Upgrading of Corn-Ethanol Co-Product by Fungal Fermentation: Amino Acids Enrichment and Anti-Nutritional Factors Degradation. *Food Bioprod. Process.* **2021**, *130*, 1–13. [CrossRef]
23. Kiszonas, A.M.; Fuerst, E.P.; Morris, C.F. Wheat Arabinoxylan Structure Provides Insight into Function. *Cereal Chem.* **2013**, *90*, 387–395. [CrossRef]
24. Chio, C.; Sain, M.; Qin, W. Lignin Utilization: A Review of Lignin Depolymerization from Various Aspects. *Renew. Sustain. Energy Rev.* **2019**, *107*, 232–249. [CrossRef]
25. Wee, M.S.M.; Henry, C.J. Reducing the Glycemic Impact of Carbohydrates on Foods and Meals: Strategies for the Food Industry and Consumers with Special Focus on Asia. *Compr. Rev. Food Sci. Food Saf.* **2020**, *19*, 670–702. [CrossRef] [PubMed]
26. Marcobal, A.M.; McConnell, B.R.; Drexler, R.A.; Ng, K.M.; Maldonado-Gomez, M.X.; Conner, A.M.S.; Vierra, C.G.; Krishnakumar, N.; Gerber, H.M.; Garcia, J.K.A.; et al. Highly Soluble β-Glucan Fiber Modulates Mechanisms of Blood Glucose Regulation and Intestinal Permeability. *Nutrients* **2024**, *16*, 2240. [CrossRef]
27. Recharla, N.; Kim, D.; Ramani, S.; Song, M.; Park, J.; Balasubramanian, B.; Puligundla, P.; Park, S. Dietary Multi-Enzyme Complex Improves In Vitro Nutrient Digestibility and Hind Gut Microbial Fermentation of Pigs. *PLoS ONE* **2019**, *14*, e0217459. [CrossRef]
28. Basini, G.; Bussolati, S.; Baioni, L.; Grasselli, F. Gossypol, a Polyphenolic Aldehyde from Cotton Plant, Interferes with Swine Granulosa Cell Function. *Domest. Anim. Endocrinol.* **2009**, *37*, 30–36. [CrossRef]
29. Liu, Y.; Wang, L.; Zhao, L.; Zhang, Y. Structure, Properties of Gossypol and Its Derivatives—From Physiological Activities to Drug Discovery and Drug Design. *Nat. Prod. Rep.* **2022**, *39*, 1282–1304. [CrossRef]
30. Ma, M.; Ren, Y.; Xie, W.; Zhou, D.; Tang, S.; Kuang, M.; Wang, Y.; Du, S. Physicochemical and Functional Properties of Protein Isolate Obtained from Cottonseed Meal. *Food Chem.* **2018**, *240*, 856–862. [CrossRef]

31. Noftsger, S.M.; Hopkins, B.A.; Diaz, D.E.; Brownie, C.; Whitlow, L.W. Effect of Whole and Expanded-Expelled Cottonseed on Milk Yield and Blood Gossypol. *J. Dairy Sci.* **2000**, *83*, 2539–2547. [CrossRef]
32. Paim, T.; Viana, P.; Brandão, E.; Amador, S.; Barbosa, T.; Cardoso, C.; Lucci, C.; DeSouza, J.; Mcmanus, C.; Abdalla, A.; et al Impact of Feeding Cottonseed Coproducts on Reproductive System of Male Sheep during Peripubertal Period. *Sci. Agric.* **2016**, *73*, 489–497. [CrossRef]
33. Yadav, S.; Teng, P.-Y.; Choi, J.; Singh, A.K.; Vaddu, S.; Thippareddi, H.; Kim, W.K. Influence of Rapeseed, Canola Meal and Glucosinolate Metabolite (AITC) as Potential Antimicrobials: Effects on Growth Performance, and Gut Health in Salmonella Typhimurium Challenged Broiler Chickens. *Poult. Sci.* **2022**, *101*, 101551. [CrossRef] [PubMed]
34. Wlazło, Ł.; Kowalska, D.; Bielański, P.; Chmielowiec-Korzeniowska, A.; Ossowski, M.; Łukaszewicz, M.; Czech, A.; Nowakowicz-Dębek, B. Effect of Fermented Rapeseed Meal on the Gastrointestinal Microbiota and Immune Status of Rabbit (*Oryctolagus cuniculus*). *Animals* **2021**, *11*, 716. [CrossRef] [PubMed]
35. Lee, J.W.; Kim, I.H.; Woyengo, T.A. Toxicity of Canola-Derived Glucosinolate Degradation Products in Pigs—A Review. *Animals* **2020**, *10*, 2337. [CrossRef]
36. Hamadi, S.; Salari, S.; Aghaei, A.; Ghorbani, M.R. Changes in Performance and Apparent Ileal Digestibility of Broiler Chickens Fed Diets Containing Electron-Irradiated Full-Fat Canola Seed. *Radiat. Phys. Chem.* **2023**, *210*, 111046. [CrossRef]
37. Hajiazizi, F.; Sadeghi, A.; Ibrahim, S. *Camelina sativa* (L. Crantz) Products; an Alternative Feed Ingredient for Poultry Diets with Its Nutritional and Physiological Consequences. *Trop. Anim. Health Prod.* **2024**, *56*, 59. [CrossRef]
38. Grela, E.R.; Czech, A.; Kiesz, M.; Wlazło, Ł.; Nowakowicz-Dębek, B. A Fermented Rapeseed Meal Additive: Effects on Production Performance, Nutrient Digestibility, Colostrum Immunoglobulin Content and Microbial Flora in Sows. *Anim. Nutr.* **2019**, *5*, 373–379. [CrossRef]
39. Kelly, D.; O'Brien, J.J.; McCracken, K.J. Effect of Creep Feeding on the Incidence, Duration and Severity of Post-Weaning Diarrhoea in Pigs. *Res. Vet. Sci.* **1990**, *49*, 223–228. [CrossRef]
40. Hampson, D.J.; Smith, W.C. Influence of Creep Feeding and Dietary Intake after Weaning on Malabsorption and Occurrence of Diarrhoea in the Newly Weaned Pig. *Res. Vet. Sci.* **1986**, *41*, 63–69. [CrossRef]
41. Peng, C.; Cao, C.; He, M.; Shu, Y.; Tang, X.; Wang, Y.; Zhang, Y.; Xia, X.; Li, Y.; Wu, J. Soybean Glycinin- and β-Conglycinin-Induced Intestinal Damage in Piglets via the P38/JNK/NF-κB Signaling Pathway. *J. Agric. Food Chem.* **2018**, *66*, 9534–9541. [CrossRef]
42. Deng, H.; Ye, T.; Deng, Y.; Cui, Y.; Guo, H.; Deng, J. miRNA Expression Analysis of IPEC-J2 Cells Damaged by Soybean 7S Globulin Reveals Ssc-miR-221-5p as the Factor Alleviating Cell Damage. *J. Agric. Food Chem.* **2024**, *72*, 11694–11705. [CrossRef] [PubMed]
43. Gillman, J.D.; Kim, W.-S.; Krishnan, H.B. Identification of a New Soybean Kunitz Trypsin Inhibitor Mutation and Its Effect on Bowman-Birk Protease Inhibitor Content in Soybean Seed. *J. Agric. Food Chem.* **2015**, *63*, 1352–1359. [CrossRef] [PubMed]
44. Moreau, T.; Recoules, E.; De Pauw, M.; Labas, V.; Réhault-Godbert, S. Evidence That the Bowman-Birk Inhibitor from Pisum Sativum Affects Intestinal Proteolytic Activities in Chickens. *Poult. Sci.* **2024**, *103*, 103182. [CrossRef] [PubMed]
45. Kabir, S.R.; Rahman, M.M.; Amin, R.; Karim, M.R.; Mahmud, Z.H.; Hossain, M.T. Solanum Tuberosum Lectin Inhibits Ehrlich Ascites Carcinoma Cells Growth by Inducing Apoptosis and G2/M Cell Cycle Arrest. *Tumour Biol.* **2016**, *37*, 8437–8444. [CrossRef]
46. Qiao, W.-L.; Hu, H.-Y.; Shi, B.-W.; Zang, L.-J.; Jin, W.; Lin, Q. Lentivirus-Mediated Knockdown of TSP50 Suppresses the Growth of Non-Small Cell Lung Cancer Cells via G0/G1 Phase Arrest. *Oncol. Rep.* **2016**, *35*, 3409–3418. [CrossRef]
47. Xu, H.; Fu, J.; Luo, Y.; Li, P.; Song, B.; Lv, Z.; Guo, Y. Effects of Tannic Acid on the Immunity and Intestinal Health of Broiler Chickens with Necrotic Enteritis Infection. *J. Anim. Sci. Biotechnol.* **2023**, *14*, 72. [CrossRef]
48. Soares, S.; Brandão, E.; Guerreiro, C.; Soares, S.; Mateus, N.; De Freitas, V. Tannins in Food: Insights into the Molecular Perception of Astringency and Bitter Taste. *Molecules* **2020**, *25*, 2590. [CrossRef]
49. Bee, G.; Silacci, P.; Ampuero-Kragten, S.; Čandek-Potokar, M.; Wealleans, A.L.; Litten-Brown, J.; Salminen, J.-P.; Mueller-Harvey, I. Hydrolysable Tannin-Based Diet Rich in Gallotannins Has a Minimal Impact on Pig Performance but Significantly Reduces Salivary and Bulbourethral Gland Size. *Animal* **2017**, *11*, 1617–1625. [CrossRef]
50. Li, J.; Gao, T.; Hao, Z.; Guo, X.; Zhu, B. Anaerobic Solid-State Fermentation with Bacillus Subtilis for Digesting Free Gossypol and Improving Nutritional Quality in Cottonseed Meal. *Front. Nutr.* **2022**, *9*, 1017637. [CrossRef]
51. Rathore, K.; Pandeya, D.; Campbell, L.; Wedegaertner, T.; Puckhaber, L.; Stipanovic, R.; Thenell, S.; Hague, S.; Hake, K. Ultra-Low Gossypol Cottonseed: Selective Gene Silencing Opens Up a Vast Resource of Plant-Based Protein to Improve Human Nutrition *Crit. Rev. Plant Sci.* **2020**, *39*, 2539–2547. [CrossRef]
52. Chen, Z.; Li, S.; Fu, Y.; Li, C.; Chen, D.; Chen, H. Arabinoxylan Structural Characteristics, Interaction with Gut Microbiota and Potential Health Functions. *J. Funct. Foods* **2019**, *54*, 536–551. [CrossRef]
53. Ravindran, V.; Cabahug, S.; Ravindra, G.; Selle, P.H.; Bryden, W.L. Response of Broiler Chickens to Microbial Phytase Supplementation as Influenced by Dietary Phytic Acid and Non-Phytate Phosphorous Levels. II. Effects on Apparent Metabolisable Energy, Nutrient Digestibility and Nutrient Retention. *Br. Poult. Sci.* **2000**, *41*, 193–200. [CrossRef] [PubMed]

54. Babatunde, O.O.; Bello, A.; Dersjant-Li, Y.; Adeola, O. Evaluation of the Responses of Broiler Chickens to Varying Concentrations of Phytate Phosphorus and Phytase. II. Grower Phase (Day 12–23 Post Hatching). *Poult. Sci.* **2022**, *101*, 101616. [CrossRef] [PubMed]
55. Darambazar, E.; Damiran, D.; Beaulieu, D. Effect of Hydrothermal, Phytase, or Organic Acid Pretreatments of Canola Meal on Phytate Level of the Meal. *Sustain. Agric. Res.* **2019**, *8*, 35. [CrossRef]
56. Samtiya, M.; Aluko, R.E.; Dhewa, T. Plant Food Anti-Nutritional Factors and Their Reduction Strategies: An Overview. *Food Prod. Process Nutr.* **2020**, *2*, 6. [CrossRef]
57. Xie, C.; Li, W.; Gao, R.; Yan, L.; Wang, P.; Gu, Z.; Yang, R. Determination of Glucosinolates in Rapeseed Meal and Their Degradation by Myrosinase from Rapeseed Sprouts. *Food Chem.* **2022**, *382*, 132316. [CrossRef]
58. Korompokis, K.; De Brier, N.; Delcour, J.A. Differences in Endosperm Cell Wall Integrity in Wheat (*Triticum aestivum* L.) Milling Fractions Impact on the Way Starch Responds to Gelatinization and Pasting Treatments and Its Subsequent Enzymatic in Vitro Digestibility. *Food Funct.* **2019**, *10*, 4674–4684. [CrossRef]
59. Andrade, J.C.; Mandarino, J.M.G.; Kurozawa, L.E.; Ida, E.I. The Effect of Thermal Treatment of Whole Soybean Flour on the Conversion of Isoflavones and Inactivation of Trypsin Inhibitors. *Food Chem.* **2016**, *194*, 1095–1101. [CrossRef]
60. Jensen, S.K.; Liu, Y.-G.; Eggum, B.O. The Effect of Heat Treatment on Glucosinolates and Nutritional Value of Rapeseed Meal in Rats. *Anim. Feed Sci. Technol.* **1995**, *53*, 17–28. [CrossRef]
61. Duodu, C.P.; Adjei-Boateng, D.; Edziyie, R.E.; Agbo, N.W.; Owusu-Boateng, G.; Larsen, B.K.; Skov, P.V. Processing Techniques of Selected Oilseed By-Products of Potential Use in Animal Feed: Effects on Proximate Nutrient Composition, Amino Acid Profile and Antinutrients. *Anim. Nutr.* **2018**, *4*, 442–451. [CrossRef]
62. Acosta, J.A.; Petry, A.L.; Gould, S.A.; Jones, C.K.; Stark, C.R.; Fahrenholz, A.; Patience, J.F. Effects of Grinding Method and Particle Size of Wheat Grain on Energy and Nutrient Digestibility in Growing and Finishing Pigs. *Transl. Anim. Sci.* **2020**, *4*, txaa062. [CrossRef] [PubMed]
63. Bornaei, L.; Salari, S.; Erfani Majd, N. Effect of Electron Beam Irradiated Barley Grains on Growth Performance, Blood Parameters, Nutrient Digestibility, Microbial Population, and Intestinal Histomorphometry in Broiler Chickens. *J. Appl. Anim. Res.* **2022**, *50*, 408–419. [CrossRef]
64. Bahraini, Z.; Salari, S.; Sari, M.; Fayazi, J.; Behgar, M. Effect of Radiation on Chemical Composition and Protein Quality of Cottonseed Meal. *Anim. Sci. J.* **2017**, *88*, 1425–1435. [CrossRef] [PubMed]
65. Soetan, K.O.; Oyewole, O.E. The Need for Adequate Processing to Reduce the Anti-Nutritional Factors in Plants Used as Human Foods and Animal Feeds: A Review. *Afr. J. Food Sci.* **2009**, *3*, 223–232.
66. Nikmaram, N.; Leong, S.Y.; Koubaa, M.; Zhu, Z.; Barba, F.J.; Greiner, R.; Oey, I.; Roohinejad, S. Effect of Extrusion on the Anti-Nutritional Factors of Food Products: An Overview. *Food Control* **2017**, *79*, 62–73. [CrossRef]
67. Alonso, R.; Aguirre, A.; Marzo, F. Effects of Extrusion and Traditional Processing Methods on Antinutrients and in Vitro Digestibility of Protein and Starch in Faba and Kidney Beans. *Food Chem.* **2000**, *68*, 159–165. [CrossRef]
68. Rodriguez, D.A.; Lee, S.A.; Jones, C.K.; Htoo, J.K.; Stein, H.H. Digestibility of Amino Acids, Fiber, and Energy by Growing Pigs, and Concentrations of Digestible and Metabolizable Energy in Yellow Dent Corn, Hard Red Winter Wheat, and Sorghum May Be Influenced by Extrusion. *Anim. Feed Sci. Technol.* **2020**, *268*, 114602. [CrossRef]
69. Barraza, M.L.; Coppock, C.E.; Brooks, K.N.; Wilks, D.L.; Saunders, R.G.; Latimer, G.W. Iron Sulfate and Feed Pelleting to Detoxify Free Gossypol in Cottonseed Diets for Dairy Cattle1. *J. Dairy Sci.* **1991**, *74*, 3457–3467. [CrossRef]
70. Bhattacharjee, P.; Singhal, R.S.; Tiwari, S.R. Supercritical Carbon Dioxide Extraction of Cottonseed Oil. *J. Food Eng.* **2007**, *79*, 892–898. [CrossRef]
71. Pelitire, S.M.; Dowd, M.K.; Cheng, H.N. Acidic Solvent Extraction of Gossypol from Cottonseed Meal. *Anim. Feed Sci. Technol.* **2014**, *195*, 120–128. [CrossRef]
72. Dersjant-Li, Y.; Awati, A.; Schulze, H.; Partridge, G. Phytase in Non-Ruminant Animal Nutrition: A Critical Review on Phytase Activities in the Gastrointestinal Tract and Influencing Factors. *J. Sci. Food Agric.* **2015**, *95*, 878–896. [CrossRef] [PubMed]
73. Kryukov, V.S.; Glebova, I.V.; Zinoviev, S.V. Reevaluation of Phytase Action Mechanism in Animal Nutrition. *Biochemistry* **2021**, *86*, S152–S165. [CrossRef]
74. Jorquera, M.; Martínez, O.; Maruyama, F.; Marschner, P.; de la Luz Mora, M. Current and Future Biotechnological Applications of Bacterial Phytases and Phytase-Producing Bacteria. *Microbes Environ.* **2008**, *23*, 182–191. [CrossRef] [PubMed]
75. Olukomaiya, O.; Fernando, C.; Mereddy, R.; Li, X.; Sultanbawa, Y. Solid-State Fermented Plant Protein Sources in the Diets of Broiler Chickens: A Review. *Anim. Nutr.* **2019**, *5*, 319–330. [CrossRef] [PubMed]
76. Missotten, J.A.; Michiels, J.; Degroote, J.; De Smet, S. Fermented Liquid Feed for Pigs: An Ancient Technique for the Future. *J. Anim. Sci. Biotechnol.* **2015**, *6*, 4. [CrossRef]
77. Canibe, N.; Jensen, B.B. Fermented Liquid Feed—Microbial and Nutritional Aspects and Impact on Enteric Diseases in Pigs. *Anim. Feed Sci. Technol.* **2012**, *173*, 17–40. [CrossRef]

78. Yang, L.; Zeng, X.; Qiao, S. Advances in Research on Solid-State Fermented Feed and Its Utilization: The Pioneer of Private Customization for Intestinal Microorganisms. *Anim. Nutr.* **2021**, *7*, 905–916. [CrossRef]
79. Zheng, L.; Li, D.; Li, Z.-L.; Kang, L.-N.; Jiang, Y.-Y.; Liu, X.-Y.; Chi, Y.-P.; Li, Y.-Q.; Wang, J.-H. Effects of Bacillus Fermentation on the Protein Microstructure and Anti-Nutritional Factors of Soybean Meal. *Lett. Appl. Microbiol.* **2017**, *65*, 520–526. [CrossRef]
80. O'Meara, F.M.; Gardiner, G.E.; O'Doherty, J.V.; Clarke, D.; Cummins, W.; Lawlor, P.G. Effect of Wet/Dry, Fresh Liquid, Fermented Whole Diet Liquid, and Fermented Cereal Liquid Feeding on Feed Microbial Quality and Growth in Grow-Finisher Pigs. *J. Anim. Sci.* **2020**, *98*, skaa166. [CrossRef]
81. Dujardin, M.; Elain, A.; Lendormi, T.; Le Fellic, M.; Le Treut, Y.; Sire, O. Keeping under Control a Liquid Feed Fermentation Process for Pigs: A Reality Scale Pilot Based Study. *Anim. Feed Sci. Technol.* **2014**, *194*, 81–88. [CrossRef]
82. Pandey, A. Solid-State Fermentation. *Biochem. Eng. J.* **2003**, *13*, 81–84. [CrossRef]
83. Verni, M.; Rizzello, C.G.; Coda, R. Fermentation Biotechnology Applied to Cereal Industry By-Products: Nutritional and Functional Insights. *Front. Nutr.* **2019**, *6*, 42. [CrossRef] [PubMed]
84. Huws, S.A.; Creevey, C.J.; Oyama, L.B.; Mizrahi, I.; Denman, S.E.; Popova, M.; Muñoz-Tamayo, R.; Forano, E.; Waters, S.M.; Hess, M.; et al. Addressing Global Ruminant Agricultural Challenges Through Understanding the Rumen Microbiome: Past, Present, and Future. *Front. Microbiol.* **2018**, *9*, 2161. [CrossRef] [PubMed]
85. Garrido-Galand, S.; Asensio-Grau, A.; Calvo-Lerma, J.; Heredia, A.; Andrés, A. The Potential of Fermentation on Nutritional and Technological Improvement of Cereal and Legume Flours: A Review. *Food Res. Int.* **2021**, *145*, 110398. [CrossRef]
86. Chi, C.-H.; Cho, S.-J. Improvement of Bioactivity of Soybean Meal by Solid-State Fermentation with *Bacillus Amyloliquefaciens* versus *Lactobacillus* Spp. and *Saccharomyces Cerevisiae*. *LWT Food Sci. Technol.* **2016**, *68*, 619–625. [CrossRef]
87. Wang, Z.; Yu, Y.; Li, X.; Xiao, H.; Zhang, P.; Shen, W.; Wan, F.; He, J.; Tang, S.; Tan, Z.; et al. Fermented Soybean Meal Replacement in the Diet of Lactating Holstein Dairy Cows: Modulated Rumen Fermentation and Ruminal Microflora. *Front. Microbiol.* **2021**, *12*, 625857. [CrossRef]
88. Kim, M.H.; Yun, C.H.; Kim, H.S.; Kim, J.H.; Kang, S.J.; Lee, C.H.; Ko, J.Y.; Ha, J.K. Effects of Fermented Soybean Meal on Growth Performance, Diarrheal Incidence and Immune-Response of Neonatal Calves. *Anim. Sci. J.* **2010**, *81*, 475–481. [CrossRef]
89. Shi, C.; Zhang, Y.; Yin, Y.; Wang, C.; Lu, Z.; Wang, F.; Feng, J.; Wang, Y. Amino Acid and Phosphorus Digestibility of Fermented Corn-Soybean Meal Mixed Feed with Bacillus Subtilis and Enterococcus Faecium Fed to Pigs. *J. Anim. Sci.* **2017**, *95*, 3996–4004. [CrossRef]
90. Khalaf, M.A.; Meleigy, S.A. Reduction of Free Gossypol Levels in Cottonseed Meal by Microbial Treatment. *Int. J. Agric. Biol.* **2008**, *10*, 1560–8530.
91. Olukomaiya, O.O.; Fernando, W.C.; Mereddy, R.; Li, X.; Sultanbawa, Y. Solid-State Fermentation of Canola Meal with *Aspergillus Sojae*, *Aspergillus Ficuum* and Their Co-Cultures: Effects on Physicochemical, Microbiological and Functional Properties. *LWT* **2020**, *127*, 109362. [CrossRef]
92. Dzialo, M.C.; Park, R.; Steensels, J.; Lievens, B.; Verstrepen, K.J. Physiology, Ecology and Industrial Applications of Aroma Formation in Yeast. *FEMS Microbiol. Rev.* **2017**, *41*, S95–S128. [CrossRef] [PubMed]
93. Mukherjee, R.; Chakraborty, R.; Dutta, A. Role of Fermentation in Improving Nutritional Quality of Soybean Meal—A Review. *Asian-Australas. J. Anim. Sci.* **2016**, *29*, 1523–1529. [CrossRef] [PubMed]
94. Murekatete, N.; Hua, Y.; Kong, X.; Zhang, C. Effects of Fermentation on Nutritional and Functional Properties of Soybean, Maize, and Germinated Sorghum Composite Flour. *Int. J. Food Eng.* **2012**, *8*, 1–15. [CrossRef]
95. Deng, W.; Dong, X.F.; Tong, J.M.; Zhang, Q. The Probiotic Bacillus Licheniformis Ameliorates Heat Stress-Induced Impairment of Egg Production, Gut Morphology, and Intestinal Mucosal Immunity in Laying Hens. *Poult. Sci.* **2012**, *91*, 575–582. [CrossRef]
96. Mi, H.; Ren, A.; Zhu, J.; Ran, T.; Shen, W.; Zhou, C.; Zhang, B.; Tan, Z. Effects of Different Protein Sources on Nutrient Disappearance, Rumen Fermentation Parameters and Microbiota in Dual-Flow Continuous Culture System. *AMB Express* **2022**, *12*, 15. [CrossRef]
97. Zhang, Z.; Yang, D.; Liu, L.; Chang, Z.; Peng, N. Effective Gossypol Removal from Cottonseed Meal through Optimized Solid-State Fermentation by Bacillus Coagulans. *Microb. Cell Factories* **2022**, *21*, 252. [CrossRef]
98. Qi, N.; Zhan, X.; Milmine, J.; Sahar, M.; Chang, K.-H.; Li, J. Isolation and Characterization of a Novel Hydrolase-Producing Probiotic Bacillus Licheniformis and Its Application in the Fermentation of Soybean Meal. *Front. Nutr.* **2023**, *10*, 1123422. [CrossRef]
99. Chen, L.; Zhao, Z.; Yu, W.; Zheng, L.; Li, L.; Gu, W.; Xu, H.; Wei, B.; Yan, X. Nutritional Quality Improvement of Soybean Meal by Bacillus Velezensis and Lactobacillus Plantarum during Two-Stage Solid-State Fermentation. *AMB Express* **2021**, *11*, 23. [CrossRef]
100. Ranjan, A.; Sahu, N.P.; Deo, A.D.; Kumar, S. Solid State Fermentation of De-Oiled Rice Bran: Effect on in Vitro Protein Digestibility, Fatty Acid Profile and Anti-Nutritional Factors. *Food Res. Int.* **2019**, *119*, 1–5. [CrossRef]
101. Jain, J.; Kumar, A.; Singh, D.; Singh, B. Purification and Kinetics of a Protease-Resistant, Neutral, and Thermostable Phytase from Bacillus Subtilis Subsp. Subtilis JJBS250 Ameliorating Food Nutrition. *Prep. Biochem. Biotechnol.* **2018**, *48*, 718–724. [CrossRef]

102. Ashayerizadeh, A.; Dastar, B.; Shams Shargh, M.; Sadeghi Mahoonak, A.; Zerehdaran, S. Fermented Rapeseed Meal Is Effective in Controlling Salmonella Enterica Serovar Typhimurium Infection and Improving Growth Performance in Broiler Chicks. *Vet. Microbiol.* **2017**, *201*, 93–102. [CrossRef] [PubMed]
103. Puyo, M.; Simonin, S.; Klein, G.; David-Vaizant, V.; Quijada-Morín, N.; Alexandre, H.; Tourdot-Maréchal, R. Use of Oenological Tannins to Protect the Colour of Rosé Wine in a Bioprotection Strategy with Metschnikowia Pulcherrima. *Foods* **2023**, *12*, 735. [CrossRef] [PubMed]
104. Refstie, S.; Sahlström, S.; Bråthen, E.; Baeverfjord, G.; Krogedal, P. Lactic Acid Fermentation Eliminates Indigestible Carbohydrates and Antinutritional Factors in Soybean Meal for Atlantic Salmon (*Salmo Salar*). *Aquaculture* **2005**, *246*, 331–345. [CrossRef]
105. Wang, W.-K.; Li, W.-J.; Wu, Q.-C.; Wang, Y.-L.; Li, S.-L.; Yang, H.-J. Isolation and Identification of a Rumen Lactobacillus Bacteria and Its Degradation Potential of Gossypol in Cottonseed Meal during Solid-State Fermentation. *Microorganisms* **2021**, *9*, 2200. [CrossRef]
106. Ezekiel, C.N.; Ayeni, K.I.; Ezeokoli, O.T.; Sulyok, M.; van Wyk, D.A.B.; Oyedele, O.A.; Akinyemi, O.M.; Chibuzor-Onyema, I.E.; Adeleke, R.A.; Nwangburuka, C.C.; et al. High-Throughput Sequence Analyses of Bacterial Communities and Multi-Mycotoxin Profiling During Processing of Different Formulations of Kunu, a Traditional Fermented Beverage. *Front. Microbiol.* **2018**, *9*, 3282. [CrossRef]